The Production of Space

1

Plan of the Present Work

I

Not so many years ago, the word 'space' had a strictly geometrical meaning: the idea it evoked was simply that of an empty area. In scholarly use it was generally accompanied by some such epithet as 'Euclidean', 'isotropic', or 'infinite', and the general feeling was that the concept of space was ultimately a mathematical one. To speak of 'social space', therefore, would have sounded strange.

Not that the long development of the concept of space had been forgotten, but it must be remembered that the history of philosophy also testified to the gradual emancipation of the sciences – and especially of mathematics – from their shared roots in traditional metaphysics. The thinking of Descartes was viewed as the decisive point in the working-out of the concept of space, and the key to its mature form. According to most historians of Western thought, Descartes had brought to an end the Aristotelian tradition which held that space and time were among those *categories* which facilitated the naming and classing of the evidence of the senses. The status of such categories had hitherto remained unclear, for they could be looked upon either as simple empirical tools for ordering sense data or, alternatively, as generalities in some way superior to the evidence supplied by the body's sensory organs. With the advent of Cartesian logic, however, space had entered the realm of the absolute. As Object opposed to Subject, as *res extensa* opposed to, and present to, *res cogitans*, space came to dominate, by containing them, all senses and all bodies. Was space therefore a divine attribute? Or was it an order immanent to the totality of what existed? Such were the terms in which the problem was couched for those philosophers who came in Descartes's wake – for Spinoza, for Leibniz, for the Newtonians. Then Kant revived, and revised, the old notion of the category. Kantian space, albeit relative, albeit a tool of knowledge, a means of classifying phenomena, was yet quite clearly separated (along with time) from the empirical sphere: it belonged to the *a priori* realm of consciousness (i.e. of the 'subject'), and partook of that realm's internal, ideal – and hence transcendental and essentially ungraspable – structure.

These protracted debates marked the shift from the philosophy to the science of space. It would be mistaken to pronounce them outdated, however, for they have an import beyond that of moments or stages in the evolution of the Western Logos. So far from being confined within the abstractness with which that Logos in its decline endowed so-called pure philosophy, they raise precise and concrete issues, among them the questions of symmetry versus asymmetry, of symmetrical objects, and of the *objective* effects of reflections and mirrors. These are all questions to which I shall be returning because of their implications for the analysis of social space.

II

Mathematicians, in the modern sense of the word, emerged as the proprietors of a science (and of a claim to scientific status) quite clearly detached from philosophy – a science which considered itself both necessary and self-sufficient. Thus mathematicians appropriated space, and time, and made them part of their domain, yet they did so in a rather paradoxical way. They invented spaces – an 'indefinity', so to speak, of spaces: non-Euclidean spaces, curved spaces, x-dimensional spaces (even spaces with an infinity of dimensions), spaces of configuration, abstract spaces, spaces defined by deformation or transformation, by a topology, and so on. At once highly general and highly specialized, the language of mathematics set out to discriminate between and classify all these innumerable spaces as precisely as possible. (Apparently the *set* of spaces, or 'space of spaces', did not lend itself very readily to conceptualization.) But the relationship between mathematics and reality – physical or social reality – was not obvious, and indeed a deep rift had developed between these two realms. Those mathematicians who had opened up this 'problematic' subsequently abandoned it to the philosophers, who were only too happy to seize upon it as a means of making up a little of the ground they had lost. In this way space became – or, rather, once more became – the very thing which an earlier philosophical tradition, namely Platonism, had proposed in opposition to the doctrine of categories: it became what Leonardo da Vinci had called a 'mental thing'. The proliferation of mathematical theories (topologies) thus aggravated the old 'problem of knowledge': how were transitions to be made from mathematical spaces (i.e. from the mental capacities of the human species, from logic) to nature in the first place, to practice in the second, and thence to the theory of social life – which also presumably must unfold in space?

III

From the tradition of thought just described – that is, from a philosophy of space revised and corrected by mathematics – the modern field of inquiry known as epistemology has inherited and adopted the notion that the status of space is that of a 'mental thing' or 'mental place'. At the same time, set theory, as the supposed logic of that place, has exercised a fascination not only upon philosophers but also upon writers and linguists. The result has been a broad proliferation of 'sets' (*ensembles*), some practical,[1] some historical,[2] but all inevitably accompanied by their appropriate 'logic'. None of these sets, or their 'logics', have anything in common with Cartesian philosophy.

No limits at all have been set on the generalization of the concept of *mental space*: no clear account of it is ever given and, depending on the author one happens to be reading, it may connote logical coherence, practical consistency, self-regulation and the relations of the parts to the whole, the engendering of like by like in a set of places, the logic of container *versus* contents, and so on. We are forever hearing about the space of this and/or the space of that: about literary space,[3] ideological spaces, the space of the dream, psychoanalytic topologies, and so on and so forth. Conspicuous by its absence from supposedly fundamental epistemological studies is not only the idea of 'man' but also that of space – the fact that 'space' is mentioned on every page notwithstanding.[4] Thus Michel Foucault can calmly assert that 'knowledge [*savoir*] is also the

---

[1] See J.-P. Sartre, *Critique de la raison dialectique*, I: *Théorie des ensembles pratiques* (Paris: Gallimard, 1960).

[2] See Michel Clouscard, *L'être et le code: procès de production d'un ensemble précapitaliste* (The Hague: Mouton, 1972).

[3] See Maurice Blanchot, *L'espace littéraire* (Paris: Gallimard, 1955).

[4] This is the least of the faults of an anthology entitled *Panorama des sciences humaines* (Paris: Gallimard, 1973).

space in which the subject may take up a position and speak of the objects with which he deals in his discourse'.[5] Foucault never explains what space it is that he is referring to, nor how it bridges the gap between the theoretical (epistemological) realm and the practical one, between mental and social, between the space of the philosophers and the space of people who deal with material things. The scientific attitude, understood as the application of 'epistemological' thinking to acquired knowledge, is assumed to be 'structurally' linked to the spatial sphere. This connection, presumed to be self-evident from the point of view of scientific discourse, is never conceptualized. Blithely indifferent to the charge of circular thinking, that discourse sets up an opposition between the status of space and the status of the 'subject', between the thinking 'I' and the object thought about. It thus rejoins the positions of the Cartesian/Western Logos, which some of its exponents indeed claim to have 'closed'.[6] Epistemological thought, in concert with the linguists' theoretical efforts, has reached a curious conclusion. It has eliminated the 'collective subject', the people as creator of a particular language, as carrier of specific etymological sequences. It has set aside the concrete subject, that subject which took over from a name-giving god. It has promoted the impersonal pronoun 'one' as creator of language in general, as creator of the system. It has failed, however, to eliminate the need for a subject of some kind. Hence the re-emergence of the abstract subject, the *cogito* of the philosophers. Hence the new lease on life of traditional philosophy in 'neo-' forms: neo-Hegelian, neo-Kantian, neo-Cartesian. This revival has profited much from the help of Husserl, whose none-too-scrupulous postulation of a (quasi-tautologous) identity of knowing Subject and conceived Essence – an identity inherent to a 'flux' (of lived experience) – underpins an almost 'pure' identity of formal and practical knowledge.[7] Nor should we be surprised to find the eminent linguist Noam Chomsky reinstating the Cartesian *cogito* or subject,[8] especially in view of the fact that he has posited the existence

---

[5] *L'archéologie du savoir* (Paris: Gallimard, 1969), p. 238. Elsewhere in the same work, Foucault speaks of 'the trajectory of a meaning' (*le parcours d'un sens*) (p. 196), of 'space of dissensions' (p. 200), etc. Eng. tr. by A. M. Sheridan Smith: *The Archaeology of Knowledge* (London: Tavistock, 1972), pp. 182, 150, 152 respectively.

[6] See Jacques Derrida, *Le vivre et le phénomène* (Paris: Presses Universitaires de France, 1967).

[7] See Michel Clouscard's critical remarks in the introduction to his *L'être et le code*. Lenin resolved this problem by brutally suppressing it: in *Materialism and Empirio-Criticism*, he argues that the thought of space reflects objective space, like a copy or photograph.

[8] See his *Cartesian Linguistics: A Chapter in the History of Rationalist Thought* (New York: Harper and Row, 1966).

of a linguistic level at which 'it will not be the case that each sentence is represented simply as a finite sequence of elements of some sort, generated from left to right by some simple device'; instead, argues Chomsky, we should expect to find 'a *finite* set of levels ordered from high to low'.[9] The fact is that Chomsky unhesitatingly postulates a mental space endowed with specific properties – with orientations and symmetries. He completely ignores the yawning gap that separates this linguistic mental space from that social space wherein language becomes practice. Similarly, J. M. Rey writes that 'Meaning presents itself as the legal authority to interchange signified elements along a single horizontal chain, within the confines [*l'espace*] of a coherent system regulated and calculated in advance.'[10] These authors, and many others, for all that they lay claim to absolute logical rigour, commit what is in fact, from the logico-mathematical point of view, the perfect paralogism: they leap over an entire area, ignoring the need for any logical links, and justify this in the vaguest possible manner by invoking, as the need arises, some such notion as *coupure* or rupture or break. They thus interrupt the continuity of their argument in the name of a discontinuity which their own methodology ought logically to prohibit. The width of the gap created in this way, and the extent of its impact, may of course vary from one author to another, or from one area of specialization to another. My criticism certainly applies in full force, however, to Julia Kristeva's σημειωτικὴ, to Jacques Derrida's 'grammatology', and to Roland Barthes's general semiology.[11] This school, whose growing renown may have something to do with its growing dogmatism, is forever promoting the basic sophistry whereby the philosophico-epistemological notion of space is fetishized and the mental realm comes to envelop the social and physical ones. Although a few of these authors suspect the existence of, or the need of, some mediation,[12] most of them

---

[9] Noam Chomsky, *Syntactic Structures* (The Hague: Mouton, 1957), pp. 24–5.

[10] J. M. Rey, *L'enjeu des signes* (Paris: Seuil, 1971), p. 13.

[11] And it extends to others, whether on their own account or via those mentioned here. Thus Barthes on Jacques Lacan: 'His topology does not concern *within* and *without*, even less *above* and *below*; it concerns, rather, a reverse and an obverse in constant motion – a front and back forever changing places as they revolve around something which is in the process of transformation, and which indeed, to begin with, *is not*' – *Critique et vérité* (Paris: Seuil, 1966), p. 27.

[12] This is certainly not true of Claude Lévi-Strauss, the whole of whose work implies that from the earliest manifestations of social life mental and social were conflated by virtue of the nomenclature of the relationships of exchange. By contrast, when Derrida gives precedence to the 'graphic' over the 'phonic', to writing over speech, or when Kristeva brings the body to the fore, clearly some search is being made for a transition or articulation between, on the one hand, the mental space previously posited (i.e. presupposed) by these authors, and, on the other hand, physical/social space.

spring without the slightest hesitation from mental to social.

What is happening here is that a powerful ideological tendency, one much attached to its own would-be scientific credentials, is expressing, in an admirably unconscious manner, those dominant ideas which are perforce the ideas of the dominant class. To some degree, perhaps, these ideas are deformed or diverted in the process, but the net result is that a particular 'theoretical practice' produces a *mental space* which is apparently, but only apparently, extra-ideological. In an inevitably circular manner, this mental space then becomes the locus of a 'theoretical practice' which is separated from social practice and which sets itself up as the axis, pivot or central reference point of Knowledge.[13] The established 'culture' reaps a double benefit from this manoeuvre: in the first place, the impression is given that the truth is tolerated, or even promoted, by that 'culture'; secondly, a multitude of small events occur within this mental space which can be exploited for useful or polemical ends. I shall return later to the peculiar kinship between this mental space and the one inhabited by the technocrats in their silent offices.[14] As for Knowledge thus defined on the basis of epistemology, and more or less clearly distinguished from ideology or from evolving science, is it not directly descended from the union between the Hegelian Concept and that scion of the great Cartesian family known as Subjectivity?

The quasi-logical presupposition of an identity between mental space (the space of the philosophers and epistemologists) and real space creates an abyss between the mental sphere on one side and the physical and social spheres on the other. From time to time some intrepid funambulist will set off to cross the void, giving a great show and sending a delightful shudder through the onlookers. By and large, however, so-called philosophical thinking recoils at the mere suggestion of any such *salto mortale*. If they still see the abyss at all, the professional philosophers avert their gaze. No matter how relevant, the problem of knowledge and the 'theory of knowledge' have been abandoned in favour of a reductionistic return to an absolute – or supposedly absolute – knowledge, namely the knowledge of the history of philosophy and the history of science. Such a knowledge can only be conceived of as separate from both ideology and non-knowledge (i.e. from lived experience). Although any separation of that kind is in fact impossible, to evoke one poses no threat to – and indeed tends to reinforce – a banal 'consensus'. After

---

[13] This pretension is to be met with in every single chapter of the *Panorama des sciences humaines* (above, note 4).

[14] See also my *Vers le cybernanthrope* (Paris: Denoël-Gonthier, 1971).

all, who is going to take issue with the True? By contrast, we all know, or think we know, where discussions of truth, illusion, lies, and appearance-*versus*-reality are liable to lead.

IV

Epistemologico-philosophical thinking has failed to furnish the basis for a science which has been struggling to emerge for a very long time, as witness an immense accumulation of research and publication. That science is – or would be – a *science of space*. To date, work in this area has produced either mere descriptions which never achieve analytical, much less theoretical, status, or else fragments and cross-sections of space. There are plenty of reasons for thinking that descriptions and cross-sections of this kind, though they may well supply inventories of what *exists in* space, or even generate a *discourse on* space, cannot ever give rise to a *knowledge of* space. And, without such a knowledge, we are bound to transfer onto the level of discourse, of language *per se* – i.e. the level of mental space – a large portion of the attributes and 'properties' of what is actually social space.

Semiology raises difficult questions precisely because it is an incomplete body of knowledge which is expanding without any sense of its own limitations; its very dynamism creates a need for such limits to be set, as difficult as that may be. When codes worked up from literary texts are applied to spaces – to urban spaces, say – we remain, as may easily be shown, on the purely descriptive level. Any attempt to use such codes as a means of deciphering social space must surely reduce that space itself to the status of a *message*, and the inhabiting of it to the status of a *reading*. This is to evade both history and practice. Yet did there not at one time, between the sixteenth century (the Renaissance – and the Renaissance city) and the nineteenth century, exist a code at once architectural, urbanistic and political, constituting a language common to country people and townspeople, to the authorities and to artists – a code which allowed space not only to be 'read' but also to be constructed? If indeed there was such a code, how did it come into being? And when, how and why did it disappear? These are all questions that I hope to answer in what follows.

As for the above-mentioned sections and fragments, they range from the ill-defined to the undefined – and thence, for that matter, to the undefinable. Indeed, talk of cross-sectioning, suggesting as it does a scientific technique (or 'theoretical practice') designed to help clarify and distinguish 'elements' within the chaotic flux of phenomena, merely adds to the muddle. Leaving aside for the moment the application of mathematical topologies to other realms, consider how fond the cognoscenti are of talk of pictural space, Picasso's space, the space of *Les demoiselles d'Avignon* or the space of *Guernica*. Elsewhere we are forever hearing of architectural, plastic or literary 'spaces'; the term is used much as one might speak of a particular writer's or artist's 'world'. Specialized works keep their audience abreast of all sorts of equally specialized spaces: leisure, work, play, transportation, public facilities – all are spoken of in spatial terms.[15] Even illness and madness are supposed by some specialists to have their own peculiar space. We are thus confronted by an indefinite multitude of spaces, each one piled upon, or perhaps contained within, the next: geographical, economic, demographic, sociological, ecological, political, commercial, national, continental, global. Not to mention nature's (physical) space, the space of (energy) flows, and so on.

Before any specific and detailed attempt is made to refute one or other of these approaches, along with whatever claim it may have to scientific status, it should be pointed out that the very multiplicity of these descriptions and sectionings makes them suspect. The fact is that all these efforts exemplify a very strong – perhaps even the dominant – tendency within present-day society and its mode of production. Under this mode of production, intellectual labour, like material labour, is subject to endless division. In addition, spatial practice consists in a projection onto a (spatial) field of all aspects, elements and moments of social practice. In the process these are separated from one another, though this does not mean that overall control is relinquished even for a moment: society as a whole continues in subjection to political practice – that is, to state power. This praxis implies and aggravates more than one contradiction, and I shall be dealing with them later. Suffice it to say at this juncture that if my analysis turns out to be correct it will be possible to claim of the sought-for 'science of space' that

> 1 it represents the political (in the case of the West, the 'neocapitalist') use of knowledge. Remember that knowledge under this system is integrated in a more or less 'immediate'

[15] [English-speaking experts tend perhaps not to use the word 'space' with quite the same facility as their French-speaking counterparts use the word *espace*, but they do have a corresponding fondness for such spatial terms as 'sector' and 'sphere' – *Translator.*]

way into the forces of production, and in a 'mediate' way into the social relations of production.

2 it implies an ideology designed to conceal that use, along with the conflicts intrinsic to the highly interested employment of a supposedly disinterested knowledge. This ideology carries no flag, and for those who accept the practice of which it is a part it is indistinguishable from knowledge.

3 it embodies at best a technological utopia, a sort of computer simulation of the future, or of the possible, within the framework of the real – the framework of the existing mode of production. The starting-point here is a knowledge which is at once integrated into, and integrative with respect to, the mode of production. The technological utopia in question is a common feature not just of many science-fiction novels, but also of all kinds of projects concerned with space, be they those of architecture, urbanism or social planning.

The above propositions need, of course, to be expounded, supported by logical arguments and shown to be true. But, if they can indeed be verified, it will be in the first place because there is a *truth of space*, an overall truth generated by analysis-followed-by-exposition, and not because a *true space* can be constituted or constructed, whether a general space as the epistemologists and philosophers believe, or a particular one as proposed by specialists in some scientific discipline or other which has a concern with space. In the second place, confirmation of these theses will imply the necessity of reversing the dominant trend towards fragmentation, separation and disintegration, a trend subordinated to a centre or to a centralized power and advanced by a knowledge which works as power's proxy. Such a reversal could not be effected without great difficulty; nor would it suffice, in order to carry it through, to replace local or 'punctual' concerns by global ones. One must assume that it would require the mobilization of a great many forces, and that in the actual course of its execution there would be a continuing need, stage by stage, for motivation and orientation.

V

Few people today would reject the idea that capital and capitalism 'influence' practical matters relating to space, from the construction of buildings to the distribution of investments and the worldwide division of labour. But it is not so clear what is meant exactly by 'capitalism' and 'influence'. What some have in mind is 'money' and its powers of intervention, or commercial exchange, the commodity and its generalization, in that 'everything' can be bought and sold. Others are concerned rather with the actors in these dramas: companies national and multinational, banks, financiers, government agencies, and so on. In either case both the unity and the diversity – and hence the contradictions – of capitalism are put in brackets. It is seen either as a mere aggregate of separate activities or else as an already constituted and closed system which derives its coherence from the fact that it endures – and solely from that fact. Actually capitalism has many facets: landed capital, commercial capital, finance capital – all play a part in practice according to their varying capabilities, and as opportunity affords; conflicts between capitalists of the same kind, or of different kinds, are an inevitable part of the process. These diverse breeds of capital, and of capitalists, along with a variety of overlapping markets – commodities, labour, knowledge, capital itself, land – are what together constitute capitalism.

Many people are inclined to forget that capitalism has yet another aspect, one which is certainly bound up with the functioning of money, with the various markets, and with the social relations of production, but which is distinct from these precisely because it is dominant. This aspect is the *hegemony* of one class. The concept of hegemony was introduced by Gramsci in order to describe the future role of the working class in the building of a new society, but it is also useful for analysing the action of the bourgeoisie, especially in relation to space. The notion is a refinement of the somewhat cruder concept of the 'dictatorship' first of the bourgeoisie and then of the proletariat. Hegemony implies more than an influence, more even than the permanent use of repressive violence. It is exercised over society as a whole, culture and knowledge included, and generally via human mediation: policies, political leaders, parties, as also a good many intellectuals and experts. It is exercised, therefore, over both institutions and ideas. The ruling class seeks to maintain its hegemony by all available means, and knowledge is one such means. The connection between knowledge (*savoir*) and power is thus made manifest, although this in no way interdicts a critical and subversive form of knowledge (*connaissance*); on the contrary, it points up the antagonism between a knowledge which serves power and a form of knowing which refuses to acknowledge power.[16]

---

[16] This is an antagonistic and hence *differentiating* distinction, a fact which Michel Foucault evades in his *Archéologie du savoir* by distinguishing between *savoir* and *con-*

Is it conceivable that the exercise of hegemony might leave space untouched? Could space be nothing more than the passive locus of social relations, the milieu in which their combination takes on body, or the aggregate of the procedures employed in their removal? The answer must be no. Later on I shall demonstrate the active – the operational or instrumental – role of space, as knowledge and action, in the existing mode of production. I shall show how space serves, and how hegemony makes use of it, in the establishment, on the basis of an underlying logic and with the help of knowledge and technical expertise, of a 'system'. Does this imply the coming into being of a clearly defined space – a capitalist space (the world market) thoroughly purged of contradictions? Once again, the answer is no. Otherwise, the 'system' would have a legitimate claim to immortality. Some over-systematic thinkers oscillate between loud denunciations of capitalism and the bourgeoisie and their repressive institutions on the one hand, and fascination and unrestrained admiration on the other. They make society into the 'object' of a systematization which must be 'closed' to be complete; they thus bestow a cohesiveness it utterly lacks upon a totality which is in fact decidedly open – so open, indeed, that it must rely on violence to endure. The position of these systematizers is in any case self-contradictory: even if their claims had some validity they would be reduced to nonsense by the fact that the terms and concepts used to define the system must necessarily be mere tools of that system itself.

VI

The theory we need, which fails to come together because the necessary critical moment does not occur, and which therefore falls back into the state of mere bits and pieces of knowledge, might well be called, by analogy, a 'unitary theory': the aim is to discover or construct a theoretical unity between 'fields' which are apprehended separately, just as molecular, electromagnetic and gravitational forces are in physics. The fields we are concerned with are, first, the *physical* – nature, the Cosmos; secondly, the *mental*, including logical and formal abstractions; and, thirdly, the *social*. In other words, we are concerned with logico-epis-

---

naissance only within the context of an *espace du jeu* or 'space of interplay' (Fr. edn, p. 241; Eng. tr., p. 185), and on the basis of chronology or 'distribution in time' (Fr. edn, p. 244; Eng. tr., p. 187). [The *savoir/connaissance* distinction cannot be conveniently expressed in English. Its significance should be clear from the discussion here; see also below pp. 367–8. Wherever the needs of clarity seemed to call for it, I have indicated in parentheses whether 'knowledge' renders *savoir* or *connaissance* – *Translator.*]

temological space, the space of social practice, the space occupied by sensory phenomena, including products of the imagination such as projects and projections, symbols and utopias.

The need for unity may be expressed in other ways too, ways that serve to underscore its importance. Reflection sometimes conflates and sometimes draws distinctions between those 'levels' which social practice establishes, in the process raising the question of their interrelationships. Thus housing, habitation – the human 'habitat', so to speak – are the concern of architecture. Towns, cities – urban space – are the bailiwick of the discipline of urbanism. As for larger, territorial spaces, regional, national, continental or worldwide, these are the responsibility of planners and economists. At times these 'specializations' are telescoped into one another under the auspices of that privileged actor, the politician. At other times their respective domains fail to overlap at all, so that neither common projects nor theoretical continuity are possible.

This state of affairs, of which the foregoing remarks do not claim to be a full critical analysis, would be brought to an end if a truly unitary theory were to be developed.

Our knowledge of the material world is based on concepts defined in terms of the broadest generality and the greatest scientific (i.e. having a content) abstraction. Even if the links between these concepts and the physical realities to which they correspond are not always clearly established, we do know that such links exist, and that the concepts or theories they imply – energy, space, time – can be neither conflated nor separated from one another. What common parlance refers to as 'matter', 'nature' or 'physical reality' – that reality within which even the crudest analysis must discern and separate different moments – has thus obviously achieved a certain unity. The 'substance' (to use the old vocabulary of philosophy) of this cosmos or 'world', to which humanity with its consciousness belongs, has properties that can be adequately summed up by means of the three terms mentioned above. When we evoke 'energy', we must immediately note that energy has to be deployed within a space. When we evoke 'space', we must immediately indicate what occupies that space and how it does so: the deployment of energy in relation to 'points' and within a time frame. When we evoke 'time', we must immediately say what it is that moves or changes therein. Space considered in isolation is an empty abstraction; likewise energy and time. Although in one sense this 'substance' is hard to conceive of, most of all at the cosmic level, it is also true to say that evidence of its existence stares us in the face: our senses and our thoughts apprehend nothing else.

Might it not be possible, then, to found our knowledge of social practice, and the general science of so-called human reality, on a model borrowed from physics? Unfortunately not. For one thing, this kind of approach has always failed in the past.[17] Secondly, following the physical model would prevent a theory of societies from using a number of useful procedures, notably the separation of levels, domains and regions. Physical theory's search for unity puts all the emphasis on the bringing-together of disparate elements. It might therefore serve as a guardrail, but never as a paradigm.

The search for a unitary theory in no way rules out conflicts within knowledge itself, and controversy and polemics are inevitable. This goes for physics, and mathematics too, for that matter; sciences that philosophers deem 'pure' precisely because they have purged them of dialectical moments are not thereby immunized against internal conflicts.

It seems to be well established that physical space has no 'reality' without the energy that is deployed within it. The modalities of this deployment, however, along with the physical relationships between central points, nuclei or condensations on the one hand and peripheries on the other are still matters for conjecture. A simple expanding-universe theory assumes an original dense core of matter and a primordial explosion. This notion of an original unity of the cosmos has given rise to many objections by reason of its quasi-theological or theogonic character. In opposition to it, Fred Hoyle has proposed a much more complex theory, according to which energy, whether at the level of the ultra-small or at that of the ultra-large, travels in every direction. On this view a single centre of the universe, whether original or final, is inconceivable. Energy/space–time condenses at an indefinite number of points (local space–times).[18]

To the extent that the theory of supposedly human space can be linked at all to a physical theory, perhaps Hoyle's is the one which best fits the bill. Hoyle looks upon space as the product of energy. Energy cannot therefore be compared to a content filling an empty container. Causalism and teleology, inevitably shot through with metaphysical abstraction, are both ruled out. The universe is seen as offering a multiplicity of particular spaces, yet this diversity is accounted for by a unitary theory, namely cosmology.

This analogy has its limits, however. There is no reason to assume an

[17] Including Claude Lévi-Strauss's attempts to draw for models on Mendeleev's classification of the elements and on general combinatorial mathematics.
[18] See Fred Hoyle, *Frontiers of Astronomy* (New York: Harper and Brothers, 1955).

isomorphism between social energies and physical energies, or between 'human' and physical fields of force. This is one form of reductionism among others which I shall have occasion explicitly to reject. All the same, human societies, like living organisms human or extra-human, cannot be conceived of independently of the universe (or of the 'world'); nor may cosmology, which cannot annex knowledge of those societies, leave them out of its picture altogether, like a state within the state.

VII

What term should be used to describe the division which keeps the various types of space away from each other, so that physical space, mental space and social space do not overlap? Distortion? Disjunction? Schism? Break? As a matter of fact the term used is far less important than the distance that separates 'ideal' space, which has to do with mental (logico-mathematical) categories, from 'real' space, which is the space of social practice. In actuality each of these two kinds of space involves, underpins and presupposes the other.

What should be the starting-point for any theoretical attempt to account for this situation and transcend it in the process? Not philosophy, certainly, for philosophy is an active and interested party in the matter. Philosophers have themselves helped bring about the schism with which we are concerned by developing abstract (metaphysical) representations of space, among them the Cartesian notion of space as absolute, infinite *res extensa*, a divine property which may be grasped in a single act of intuition because of its homogeneous (isotropic) character. This is all the more regrettable in view of the fact that the beginnings of philosophy were closely bound up with the 'real' space of the Greek city. This connection was severed later in philosophy's development. Not that we can have no recourse to philosophy, to its concepts or conceptions. But it cannot be our point of departure. What about literature? Clearly literary authors have written much of relevance, especially descriptions of places and sites. But what criteria would make certain texts more relevant than others? Céline uses everyday language to great effect to evoke the space of Paris, of the Parisian *banlieue*, or of Africa. Plato, in the *Critias* and elsewhere, offers marvellous descriptions of cosmic space, and of the space of the city as a reflection of the Cosmos. The inspired De Quincey pursuing the shadow of the woman of his dreams through the streets of London, or Baudelaire in his *Tableaux parisiens*, offer us accounts of urban space rivalling those of

Victor Hugo and Lautréamont. The problem is that any search for space in literary texts will find it everywhere and in every guise: enclosed, described, projected, dreamt of, speculated about. What texts can be considered special enough to provide the basis for a 'textual' analysis? Inasmuch as they deal with socially 'real' space, one might suppose on first consideration that architecture and texts relating to architecture would be a better choice than literary texts proper. Unfortunately, any definition of architecture itself requires a prior analysis and exposition of the concept of space.

Another possibility would be to take *general* scientific notions as a basis, notions as general as that of text, like those of information and communication, of message and code, and of sets of signs – all notions which are still being developed. The danger here is that the analysis of space might become enclosed within a single area of specialization, which, so far from helping us account for the dissociations mentioned above, would merely exacerbate them. This leaves only *universal* notions, which seemingly belong to philosophy but not to any particular specialization. Do such notions exist? Does what Hegel called the concrete universal still have any meaning? I hope to show that it does. What can be said without further ado is that the concepts of *production* and of the *act of producing* do have a certain abstract universality. Though developed by philosophers, these concepts extend beyond philosophy. They were taken over in the past, admittedly, by specialized disciplines, especially by political economy; yet they have survived that annexation. By retrieving something of the broad sense that they had in certain of Marx's writings, they have shed a good deal of the illusory precision with which the economists had endowed them. This is not to say that it will be easy to recover these concepts and put them back to work. To speak of 'producing space' sounds bizarre, so great is the sway still held by the idea that empty space is prior to whatever ends up filling it. Questions immediately arise here: what spaces? and what does it mean to speak of 'producing space'? We are confronted by the problem of how to bring concepts that have already been worked out and formalized into conjunction with this new content without falling back on mere illustration and example – notorious occasions for sophistry. What is called for, therefore, is a thoroughgoing exposition of these concepts, and of their relations, on the one hand with the extreme formal abstraction of logico-mathematical space, and on the other hand with the practico-sensory realm of social space. To proceed otherwise would result in a new fragmentation of the concrete universal into its original Hegelian moments: the *particular* (in this case descriptions or cross-sections of social space); the *general* (logical and mathematical); and the *singular* (i.e. 'places' considered as natural, in their merely physical or sensory reality).

VIII

Everyone knows what is meant when we speak of a 'room' in an apartment, the 'corner' of the street, a 'marketplace', a shopping or cultural 'centre', a public 'place', and so on. These terms of everyday discourse serve to distinguish, but not to isolate, particular spaces, and in general to describe a social space. They correspond to a specific use of that space, and hence to a spatial practice that they express and constitute. Their interrelationships are ordered in a specific way. Might it not be a good idea, therefore, first to make an inventory of them,[19] and then to try and ascertain what paradigm gives them their meaning, what syntax governs their organization?

There are two possibilities here: either these words make up an unrecognized code which we can reconstitute and explain by means of thought; alternatively, reflection will enable us, on the basis of the words themselves and the operations that are performed upon them, to construct a *spatial code*. In either event, the result of our thinking would be the construction of a 'system of space'. Now, we know from precise scientific experiments that a system of this kind is applicable only indirectly to its 'object', and indeed that it really only applies to a *discourse* on that object. The project I am outlining, however, does not aim to produce a (or *the*) discourse on space, but rather to expose the actual production of space by bringing the various kinds of space and the modalities of their genesis together within a single theory.

These brief remarks can only hint at a solution to a problem that we shall have to examine carefully later on in order to determine whether it is a *bona fide* issue or merely the expression of an obscure question about origins. This problem is: does language – logically, epistemologically or genetically speaking – precede, accompany or follow social space? Is it a precondition of social space or merely a formulation of it? The priority-of-language thesis has certainly not been established. Indeed, a good case can be made for according logical and epistemological precedence over highly articulated languages with strict rules to those

---

[19] Cf. Georges Matoré, *L'espace humain* (Paris: La Colombe, 1962), including the lexicographical index.

activities which mark the earth, leaving traces and organizing gestures and work performed in common. Perhaps what have to be uncovered are as-yet concealed relations between space and language: perhaps the 'logicalness' intrinsic to articulated language operated from the start as a spatiality capable of bringing order to the qualitative chaos (the practico-sensory realm) presented by the perception of things.

To what extent may *a* space be read or decoded? A satisfactory answer to this question is certainly not just around the corner. As I noted earlier, without as yet adducing supporting arguments or proof, the notions of message, code, information and so on cannot help us trace the genesis of a space; the fact remains, however, that an already produced space can be decoded, can be *read*. Such a space implies a process of signification. And even if there is no general code of space, inherent to language or to all languages, there may have existed specific codes, established at specific historical periods and varying in their effects. If so, interested 'subjects', as members of a particular society, would have acceded by this means at once to *their* space and to their status as 'subjects' acting within that space and (in the broadest sense of the word) comprehending it.

If, roughly from the sixteenth century to the nineteenth, a coded language may be said to have existed on the practical basis of a specific relationship between town, country and political territory, a language founded on classical perspective and Euclidean space, why and how did this coded system collapse? Should an attempt be made to reconstruct that language, which was common to the various groups making up the society – to users and inhabitants, to the authorities and to the technicians (architects, urbanists, planners)?

A theory can only take form, and be formulated, at the level of a 'supercode'. Knowledge cannot rightly be assimilated to a 'well-designed' language, because it operates at the conceptual level. It is thus not a privileged language, nor a metalanguage, even if these notions may be appropriate for the 'science of language' as such. Knowledge of space cannot be limited from the outset by categories of this kind. Are we looking, then, for a 'code of codes'? Perhaps so, but this 'meta' function of theory does not in itself explain a great deal. If indeed spatial codes have existed, each characterizing a particular spatial/social practice, and if these codifications have been *produced* along with the space corresponding to them, then the job of theory is to elucidate their rise, their role, and their demise. The shift I am proposing in analytic orientation relative to the work of specialists in this area ought by now to be clear: instead of emphasizing the rigorously formal aspect of codes,

I shall instead be putting the stress on their dialectical character. Codes will be seen as part of a practical relationship, as part of an interaction between 'subjects' and their space and surroundings. I shall attempt to trace the coming-into-being and disappearance of codings/decodings. My aim will be to highlight *contents* – i.e. the social (spatial) practices inherent to the forms under consideration.

IX

Surrealism appears quite otherwise today than it did half a century ago. A number of its pretensions have faded away, among them the substitution of poetry for politics, the politicization of poetry and the search for a transcendent revelation. All the same, though a literary movement, it cannot be reduced to the level of mere literature (which surrealism initially despised), and hence to the status of a literary event, bound up with the exploration of the unconscious (automatic writing), which had a subversive character to begin with but which was subsequently co-opted by every means available – glosses, exegeses, commentaries, fame, publicity, and so on.

The leading surrealists sought to decode inner space and illuminate the nature of the transition from this subjective space to the material realm of the body and the outside world, and thence to social life. Consequently surrealism has a theoretical import which was not originally recognized. The surrealists' effort to find a unity of this kind initiated a search which later went astray. It is discernible, for example, in André Breton's *L'amour fou*, where the introduction of imaginary and magical elements, though perhaps strange, detracts in no way from the annunciatory value of the work:

> Sometimes, for example, wishing for the visit of a particular woman, I have found myself opening a door, then shutting it, then opening it again; if this device proved inadequate to the task, I might slip the blade of a knife randomly between the pages of a book, having previously decided that a certain line on the left-hand or right-hand page would inform me more or less indirectly as to her inclinations and tell me whether to expect her soon or not at all; then I would start moving things around once more,

scrutinizing their positions relative to each other and rearranging them in unusual ways.[20]

Still, the scale of the failure of surrealism's poetic project should also be pointed out. Not that surrealist poetry lacked an accompanying conceptual apparatus designed to explain its orientation; indeed, so numerous are the movement's theoretical texts – manifestoes and others – that one might well ask what would remain of surrealism were they left out of consideration. The intrinsic shortcomings of the poetry run deeper, however: it prefers the *visual* to the act of seeing, rarely adopts a 'listening' posture, and curiously neglects the musical both in its mode of expression and, even more, in its central 'vision'. 'It was as though the deep night of human existence had suddenly been pierced', writes Breton, 'as though natural necessity had consented to become one with logical necessity and so plunged all things into a state of total transparency.'[21]

As Breton himself acknowledges,[22] a project of Hegelian derivation was to be pursued solely via an affective, and hence subjective, overburdening of the (loved) 'object' by means of a hyper-exaltation of symbols. Thus the surrealists, proclaiming – though none too loudly and certainly without any supporting evidence – that the Hegelian 'end of history' lay within, and would be advanced by, their poetry, succeeded only in producing a lyrical metalanguage of history, an illusory fusing of subject with object in a transcendental metabolism. Their purely verbal metamorphosis, anamorphosis or anaphorization of the relationship between 'subjects' (people) and things (the realm of everyday life) overloaded meaning – and changed nothing. There was simply no way, by virtue of language alone, to make the leap from *exchange* (of goods) to *use*.

Like that of the surrealists, the work of Georges Bataille now has a meaning somewhat different from the one it had originally. Bataille too sought (among other things) a junction between the space of inner experience on the one hand, and, on the other, the space of physical nature (below the level of consciousness: tree, sex, acephal) and social space (communication, speech). Like the surrealists – though not, like them, on the trail of an imagined synthesis – Bataille left his mark everywhere between real, infra-real and supra-real. His way was Nietzsche's – eruptive and disruptive. He accentuates divisions and widens

---

[20] André Breton, *L'amour fou* (Paris: Gallimard, 1937), p. 23. The same might be said, despite the passing of so many years, of much of Eluard's poetry.
[21] Ibid., p. 6.
[22] Ibid., p. 61.

immediacy of consciousness; hence Husserlian phenomenology with its
'Heraclitean' flux of phenomena and subjectivity of the ego; and hence,
later, a whole philosophical tradition.[27]

In Georg Lukács's anti-Hegelian Hegelianism, space serves to define
reification, as also false consciousness. Rediscovered time, under the
direction of a class consciousness elevated to the sublime level at which
it can survey history's twists and turns at a glance, breaks the primacy
of the spatial.[28]

Only Nietzsche, since Hegel, has maintained the primordiality of
space and concerned himself with the spatial problematic – with the
repetitiveness, the circularity, the simultaneity of that which seems
diverse in the temporal context and which arises at different times. In
the realm of becoming, but standing against the flux of time, every
defined form, whether physical, mental or social, struggles to establish
and maintain itself. Yet Nietzschean space preserves not a single feature
of the Hegelian view of space as product and residue of historical time.
'I believe in absolute space as the substratum of force: the latter limits
and forms', writes Nietzsche.[29] Cosmic space contains energy, contains
forces, and proceeds from them. The same goes for terrestrial and social
space: 'Where there is space there is being.' The relationships between
force (energy), time and space are problematical. For example, one can
neither conceive of a beginning (an origin) nor yet do without such an
idea. As soon as that (albeit essential) activity which discerns and
marks distinctions is removed from the picture, 'The interrupted and the
successive are concordant.' An energy or force can only be identified by
means of its effects in space, even if forces 'in themselves' are distinct
from their effects (and how can any 'reality' – energy, space or time –
be grasped 'in itself' by intellectual analysis?). Just as Nietzschean space
has nothing in common with Hegelian space, so Nietzschean time, as
theatre of universal tragedy, as the cyclical, repetitious space–time of
death and of life, has nothing in common with Marxist time – that is,
historicity driven forward by the forces of production and adequately

---

[27] A tradition to which both Maurice Merleau-Ponty and Gilles Deleuze belong. Cf.
Gilles Deleuze and Félix Guattari, *L'anti-Oedipe*, rev. edn (Paris: Editions de Minuit,
1973), p. 114.

[28] See Jean Gabel, *La fausse conscience* (Paris: Editions de Minuit, 1962), pp. 193ff.
Eng. tr. by M. A. and K. A. Thompson: *False Consciousness* (New York: Harper and
Row, 1975), pp. 253 ff. Also, of course, Lukács's *History and Class Consciousness*, tr.
Rodney Livingstone (London: Merlin Press, 1971; Cambridge, Mass.: MIT Press, 1971).

[29] See the collection entitled – mistakenly – *The Will to Power*, fragment 545. Eng. edn,
ed. and tr. Walter Kaufmann (New York: Random House, 1967), p. 293.

(to be optimistic) oriented by industrial, proletarian and revolutionary rationality.

This is perhaps a convenient moment to consider what has been happening in the second half of the twentieth century, the period to which 'we' are witnesses.

1 The state is consolidating on a world scale. It weighs down on society (on all societies) in full force; it plans and organizes society 'rationally', with the help of knowledge and technology, imposing analogous, if not homologous, measures irrespective of political ideology, historical background, or the class origins of those in power. The state crushes time by reducing differences to repetitions or circularities) dubbed 'equilibrium', 'feedback', 'self-regulation', and so on). Space in its Hegelian form comes back into its own. This modern state promotes and imposes itself as the stable centre – definitively – of (national) societies and spaces. As both the end and the meaning of history – just as Hegel had forecast – it flattens the social and 'cultural' spheres. It enforces a logic that puts an end to conflicts and contradictions. It neutralizes whatever resists it by castration or crushing. Is this social entropy? Or is it a monstrous excrescence transformed into normality? Whatever the answer, the results lie before us.

2 In this same space there are, however, other forces on the boil, because the rationality of the state, of its techniques, plans and programmes, provokes opposition. The violence of power is answered by the violence of subversion. With its wars and revolutions, defeats and victories, confrontation and turbulence, the modern world corresponds precisely to Nietzsche's tragic vision. State-imposed normality makes permanent transgression inevitable. As for time and negativity, whenever they re-emerge, as they must, they do so explosively. This is a new negativity, a tragic negativity which manifests itself as incessant violence. These seething forces are still capable of rattling the lid of the cauldron of the state and its space, for differences can never be totally quieted. Though defeated, they live on, and from time to time they begin fighting ferociously to reassert themselves and transform themselves through struggle.

3 Nor has the working class said its last word. It continues on its way, sometimes underground, sometimes in the light of day. It is not an easy matter to get rid of the class struggle, which has taken myriad forms not accounted for by the impoverished schema usually so referred to – a schema which is nowhere to be found in Marx even if its devotees

claim to be Marxists. It may be that a fatal balance of power has now been reached which will prevent the working class's opposition to the bourgeoisie from ever becoming an open antagonism, so that society totters while the state rots in place or reasserts itself in convulsive fashion. It may be that world revolution will break out after a period of latency. Or perhaps world war will circle the planet in the wake of the world market. At all events, everything suggests at present that the workers in the industrialized countries are opting neither for indefinite growth and accumulation nor for violent revolution leading to the disappearance of the state, but rather for the withering away of work itself. Merely to consider the possibilities is to realize that Marxist thought has not disappeared, and indeed that it cannot disappear.

Confrontation of the theses and hypotheses of Hegel, Marx and Nietzsche is just beginning – and only with great difficulty at that. As for philosophical thought and thought about space and time, it is split. On the one hand we have the philosophy of time, of duration, itself broken up into partial considerations and emphases: historical time, social time, mental time, and so on. On the other hand we have epistemological thought, which constructs an abstract space and cogitates about abstract (logico-mathematical) spaces. Most if not all authors ensconce themselves comfortably enough within the terms of mental (and therefore neo-Kantian or neo-Cartesian) space, thereby demonstrating that 'theoretical practice' is already nothing more than the egocentric thinking of specialized Western intellectuals – and indeed may soon be nothing more than an entirely separated, schizoid consciousness.

The aim of this book is to detonate this state of affairs. More specifically, apropos of space, it aims to foster confrontation between those ideas and propositions which illuminate the modern world even if they do not govern it, treating them not as isolated theses or hypotheses, as 'thoughts' to be put under the microscope, but rather as prefigurations lying at the threshold of modernity.[30]

---

[30] Here, without further ado – and I hope without too much irony – are some of the sources I have in mind: the works of Charles Dodgson / Lewis Carroll (but with the emphasis on the author of *Symbolic Logic* and *Logic without Tears* rather than on the author of the Alice books); Hermann Hesse's *Das Glasperlenspiel* (1943), tr. by Mervyn Savill as *Magister Ludi* (London: Aldus, 1949 and New York: Henry Holt, 1949) and by Richard and Clara Winston as *The Glass Bead Game* (New York: Holt, Rinehart and Winston, 1969), especially the passage on the theory of the game and its relationship with language and with space – the space of the game itself and the space in which the game is played, namely Castalia; Hermann Weyl's *Symmetry* (Princeton, NJ: Princeton University Press, 1952); and Nietzsche – especially, in *Das Philosophenbuch/Le Livre du philosophe* (Paris: Aubier-Flammarion, 1969), the fragments on language and the 'theoreti-

XI

This aim does not imply the elaboration of a *critical theory* of existing space designed as a substitute for the descriptions and cross-sections that accept that space or for other critical theories that deal with society in general, with political economy, with culture, and so on. The substitution of a negative and critical utopia of space (or of 'man' or 'society') for the dominant technological utopia is no longer sufficient. Critical theory, after being driven into practical opposition – and even into the most radical form of it, whether 'punctual' (i.e. attacking particularly vulnerable points) or global – has had its day.

It might be supposed that our first priority should be the methodical destruction of the *codes* relating to space. Nothing could be further from the case, however, because the codes inherent to knowledge and social practice have been in dissolution for a very long time already. All that remains of them are relics: words, images, metaphors. This is the outcome of an epoch-making event so generally ignored that we have to be reminded of it at every moment. The fact is that around 1910 a certain space was shattered. It was the space of common sense, of knowledge (*savoir*), of social practice, of political power, a space thitherto enshrined in everyday discourse, just as in abstract thought, as the environment of and channel for communications; the space, too, of classical perspective and geometry, developed from the Renaissance onwards on the basis of the Greek tradition (Euclid, logic) and bodied forth in Western art and philosophy, as in the form of the city and town. Such were the shocks and onslaughts suffered by this space that today it retains but a feeble pedagogical reality, and then only with great difficulty, within a conservative educational system. Euclidean and perspectivist space have disappeared as systems of reference, along with other former 'commonplaces' such as the town, history, paternity, the tonal system in music, traditional morality, and so forth. This was truly a crucial moment. Naturally, 'common-sense' space, Euclidean space and perspectivist space did not disappear in a puff of smoke without leaving any trace in our consciousness, knowledge or educational methods; they could no more have done so than elementary algebra and

---

cal introduction on truth and lies'.

It should be borne in mind that the works cited here, like those mentioned elsewhere in this book, are meant to be placed in the context of our discussion – in the context of spatial practice and its levels (planning, 'urbanism', architecture).

arithmetic, or grammar, or Newtonian physics. The fact remains that it is too late for destroying codes in the name of a critical theory; our task, rather, is to describe their already completed destruction, to measure its effects, and (perhaps) to construct a new code by means of theoretical 'supercoding'.

It must be stressed that what is needed is not a replacement for the dominant tendency, however desirable that may once have been, but instead a reversal of that tendency. As I shall attempt at some length to show, even if absolute proof is impossible, such a reversal or inversion would consist, as in Marx's time, in a movement from *products* (whether studied in general or in particular, described or enumerated) to *production*.

This reversal of tendency and of meaning has nothing to do with the conversion of signified elements into signifiers, as practised under the banner of an intellectualizing concern for 'pure' theory. The elimination of the signified element, the putting-in-brackets of the 'expressive', the exclusive appeal to formal signifiers – these operations *precede* the reversal of tendency which leads from products to productive activity; they merely simulate that reversal by reducing it to a sequence of abstract interventions performed upon language (and essentially upon literature).

XII

(Social) space is a (social) product. This proposition might appear to border on the tautologous, and hence on the obvious. There is good reason, however, to examine it carefully, to consider its implications and consequences before accepting it. Many people will find it hard to endorse the notion that space has taken on, within the present mode of production, within society as it actually is, a sort of reality of its own, a reality clearly distinct from, yet much like, those assumed in the same global process by commodities, money and capital. Many people, finding this claim paradoxical, will want proof. The more so in view of the further claim that the space thus produced also serves as a tool of thought and of action; that in addition to being a means of production it is also a means of control, and hence of domination, of power; yet that, as such, it escapes in part from those who would make use of it. The social and political (state) forces which engendered this space now seek, but fail, to master it completely; the very agency that has forced spatial reality towards a sort of uncontrollable autonomy now strives to run it into the ground, then shackle and enslave it. Is this space an

abstract one? Yes, but it is also 'real' in the sense in which concrete abstractions such as commodities and money are real. Is it then concrete? Yes, though not in the sense that an object or product is concrete. Is it instrumental? Undoubtedly, but, like knowledge, it extends beyond instrumentality. Can it be reduced to a projection – to an 'objectification' of knowledge? Yes and no: knowledge objectified in a product is no longer coextensive with knowledge in its theoretical state. If space embodies social relationships, how and why does it do so? And what relationships are they?

It is because of all these questions that a thoroughgoing analysis and a full overall exposition are called for. This must involve the introduction of new ideas – in the first place the idea of a diversity or multiplicity of spaces quite distinct from that multiplicity which results from seg-menting and cross-sectioning space *ad infinitum*. Such new ideas must then be inserted into the context of what is generally known as 'history', which will consequently itself emerge in a new light.

Social space will be revealed in its particularity to the extent that it ceases to be indistinguishable from mental space (as defined by the philosophers and mathematicians) on the one hand, and physical space (as defined by practico-sensory activity and the perception of 'nature') on the other. What I shall be seeking to demonstrate is that such a social space is constituted neither by a collection of things or an aggregate of (sensory) data, nor by a void packed like a parcel with various contents, and that it is irreducible to a 'form' imposed upon phenomena, upon things, upon physical materiality. If I am successful, the social character of space, here posited as a preliminary hypothesis, will be confirmed as we go along.

XIII

If it is true that (social) space is a (social) product, how is this fact concealed? The answer is: by a double illusion, each side of which refers back to the other, reinforces the other, and hides behind the other. These two aspects are the illusion of transparency on the one hand and the illusion of opacity, or 'realistic' illusion, on the other.

1 *The illusion of transparency*   Here space appears as luminous, as intelligible, as giving action free rein. What happens in space lends a miraculous quality to thought, which becomes incarnate by means of a *design* (in both senses of the word). The design serves as a mediator –

itself of great fidelity – between mental activity (invention) and social activity (realization); and it is deployed in space. The illusion of transparency goes hand in hand with a view of space as innocent, as free of traps or secret places. Anything hidden or dissimulated – and hence dangerous – is antagonistic to transparency, under whose reign everything can be taken in by a single glance from that mental eye which illuminates whatever it contemplates. Comprehension is thus supposed, without meeting any insurmountable obstacles, to conduct what is perceived, i.e. its object, from the shadows into the light; it is supposed to effect this displacement of the object either by piercing it with a ray or by converting it, after certain precautions have been taken, from a murky to a luminous state. Hence a rough coincidence is assumed to exist between social space on the one hand and mental space – the (topological) space of thoughts and utterances – on the other. By what path, and by means of what magic, is this thought to come about? The presumption is that an encrypted reality becomes readily decipherable thanks to the intervention first of speech and then of writing. It is said, and believed, that this decipherment is effected solely through transposition and through the illumination that such a strictly topological change brings about.

What justification is there for thus claiming that within the spatial realm the known and the transparent are one and the same thing? The fact is that this claim is a basic postulate of a diffuse ideology which dates back to classical philosophy. Closely bound up with Western 'culture', this ideology stresses speech, and overemphasizes the written word, to the detriment of a social practice which it is indeed designed to conceal. The fetishism of the spoken word, or ideology of speech, is reinforced by the fetishism and ideology of writing. For some, whether explicitly or implicitly, speech achieves a total clarity of communication, flushing out whatever is obscure and either forcing it to reveal itself or destroying it by sheer force of anathema. Others feel that speech alone does not suffice, and that the test and action of the written word, as agent of both malediction and sanctification, must also be brought into play. The act of writing is supposed, beyond its immediate effects, to imply a discipline that facilitates the grasping of the 'object' by the writing and speaking 'subject'. In any event, the spoken and written word are taken for (social) practice; it is assumed that absurdity and obscurity, which are treated as aspects of the same thing, may be dissipated without any corresponding disappearance of the 'object'. Thus communication brings the non-communicated into the realm of the communicated – the *incommunicable* having no existence beyond that of an ever-pursued residue. Such are the assumptions of an ideology which, in positing the transparency of space, identifies knowledge, information and communication. It was on the basis of this ideology that people believed for quite a time that a revolutionary social transformation could be brought about by means of communication alone. 'Everything must be said! No time limit on speech! Everything must be written! Writing transforms language, therefore writing transforms society! Writing is a signifying practice!' Such agendas succeed only in conflating revolution and transparency.

The illusion of transparency turns out (to revert for a moment to the old terminology of the philosophers) to be a transcendental illusion: a trap, operating on the basis of its own quasi-magical power, but by the same token referring back immediately to other traps — traps which are its alibis, its masks.

2 *The realistic illusion* This is the illusion of natural simplicity — the product of a naïve attitude long ago rejected by philosophers and theorists of language, on various grounds and under various names, but chiefly because of its appeal to naturalness, to substantiality. According to the philosophers of the good old idealist school, the credulity peculiar to common sense leads to the mistaken belief that 'things' have more of an existence than the 'subject', his thought and his desires. To reject this illusion thus implies an adherence to 'pure' thought, to Mind or Desire. Which amounts to abandoning the realistic illusion only to fall back into the embrace of the illusion of transparency.

Among linguists, semanticists and semiologists one encounters a primary (and indeed an ultimate) naïvety which asserts that language, rather than being defined by its form, enjoys a 'substantial reality'. On this view language resembles a 'bag of words' from which the proper and adequate word for each thing or 'object' may be picked. In the course of any reading, the imaginary and the symbolic dimensions, the landscape and the horizon which line the reader's path, are all taken as 'real', because the *true* characteristics of the text — its signifying form as much as its symbolic content — are a blank page to the *naif* in his unconsciousness. (It is worth noting *en passant* that his illusions provide the *naif* with pleasures which knowledge is bound to abolish along with those illusions themselves. Science, moreover, though it may replace the innocent delights of naturalness with more refined and sophisticated pleasures, can in no wise guarantee that these will be any more delectable.)

The illusion of substantiality, naturalness and spatial opacity nurtures its own mythology. One thinks of the space-oriented artist, at work in a hard or dense reality delivered direct from the domain of Mother Nature. More likely a sculptor than a painter, an architect sooner than a musician or poet, such an artist tends to work with materials that resist or evade his efforts. When space is not being overseen by the geometer, it is liable to take on the physical qualities and properties of the earth.

The illusion of transparency has a kinship with philosophical idealism; the realistic illusion is closer to (naturalistic and mechanistic) material-ism. Yet these two illusions do not enter into antagonism with each other after the fashion of philosophical systems, which armour themselves like battleships and seek to destroy one another. On the contrary, each illusion embodies and nourishes the other. The shifting back and forth between the two, and the flickering or oscillatory effect that it produces, are thus just as important as either of the illusions considered in isolation. Symbolisms deriving from nature can obscure the rational lucidity which the West has inherited from its history and from its successful domi-nation of nature. The apparent translucency taken on by obscure histori-cal and political forces in decline (the state, nationalism) can enlist images having their source in the earth or in nature, in paternity or in maternity. The rational is thus naturalized, while nature cloaks itself in nostalgias which supplant rationality.

XIV

As a programmatic foretaste of the topics I shall be dealing with later, I shall now review some of the implications and consequences of our initial proposition – namely, that (social) space is a (social) product.

The first implication is that (physical) natural space is disappearing. Granted, natural space was – and it remains – the common point of departure: the origin, and the original model, of the social process – perhaps even the basis of all 'originality'. Granted, too, that natural space has not vanished purely and simply from the scene. It is still the background of the picture; as decor, and more than decor, it persists everywhere, and every natural detail, every natural object is valued even more as it takes on symbolic weight (the most insignificant animal, trees, grass, and so on). As source and as resource, nature obsesses us, as do childhood and spontaneity, via the filter of memory. Everyone wants to protect and save nature; nobody wants to stand in the way of an attempt to retrieve its authenticity. Yet at the same time everything conspires to harm it. The fact is that natural space will soon be lost to view. Anyone so inclined may look over their shoulder and see it sinking below the horizon behind us. Nature is also becoming lost to *thought*. For what is nature? How can we form a picture of it as it was before the intervention of humans with their ravaging tools? Even the powerful myth of nature is being transformed into a mere fiction, a negative utopia: nature is now seen as merely the raw material out of which the productive forces of a variety of social systems have forged their particular spaces. True, nature is resistant, and infinite in its depth, but it has been defeated, and now waits only for its ultimate voidance and destruction.

XV

A second implication is that every society – and hence every mode of production with its subvariants (i.e. all those societies which exemplify the general concept – produces a space, its own space. The city of the ancient world cannot be understood as a collection of people and things in space; nor can it be visualized solely on the basis of a number of texts and treatises on the subject of space, even though some of these, as for example Plato's *Critias* and *Timaeus* or Aristotle's *Metaphysics* A, may be irreplaceable sources of knowledge. For the ancient city had its own spatial practice: it forged its own – *appropriated* – space. Whence the need for a study of that space which is able to apprehend it as such, in its genesis and its form, with its own specific time or times (the rhythm of daily life), and its particular centres and polycentrism (agora, temple, stadium, etc.).

The Greek city is cited here only as an example – as one step along the way. Schematically speaking, each society offers up its own peculiar space, as it were, as an 'object' for analysis and overall theoretical explication. I say each society, but it would be more accurate to say each mode of production, along with its specific relations of production; any such mode of production may subsume significant variant forms, and this makes for a number of theoretical difficulties, many of which we shall run into later in the shape of inconsistencies, gaps and blanks in our general picture. How much can we really learn, for instance, confined as we are to Western conceptual tools, about the Asiatic mode of production, its space, its towns, or the relationship it embodies between town and country – a relationship reputedly represented figuratively or ideographically by the Chinese characters?

More generally, the very notion of social space resists analysis because of its novelty and because of the real and formal complexity that it connotes. Social space contains – and assigns (more or less) appropriate places to – (1) the *social relations of reproduction*, i.e. the bio-physiological relations between the sexes and between age groups, along with the specific organization of the family; and (2) the *relations of production*, i.e. the division of labour and its organization in the form of hierarchical social functions. These two sets of relations, production and reproduction, are inextricably bound up with one another: the division of labour has repercussions upon the family and is of a piece with it; conversely, the organization of the family interferes with the division of labour. Yet social space must discriminate between the two – not always successfully, be it said – in order to 'localize' them.

To refine this scheme somewhat, it should be pointed out that in precapitalist societies the two interlocking levels of biological reproduction and socio-economic production together constituted social reproduction – that is to say, the reproduction of society as it perpetuated itself generation after generation, conflict, feud, strife, crisis and war notwithstanding. That a decisive part is played by space in this continuity is something I shall be attempting to demonstrate below.

The advent of capitalism, and more particularly 'modern' neocapitalism, has rendered this state of affairs considerably more complex. Here *three* interrelated levels must be taken into account: (1) *biological reproduction* (the family); (2) the *reproduction of labour power* (the working class *per se*); and (3) the *reproduction of the social relations of production* – that is, of those relations which are constitutive of capitalism and which are increasingly (and increasingly effectively) sought and imposed as such. The role of space in this tripartite ordering of things will need to be examined in its specificity.

To make things even more complicated, social space also contains specific representations of this double or triple interaction between the social relations of production and reproduction. Symbolic representation serves to maintain these social relations in a state of coexistence and cohesion. It displays them while displacing them – and thus concealing them in symbolic fashion – with the help of, and onto the backdrop of, nature. Representations of the relations of reproduction are sexual symbols, symbols of male and female, sometimes accompanied, sometimes not, by symbols of age – of youth and of old age. This is a symbolism which conceals more than it reveals, the more so since the relations of reproduction are divided into frontal, public, overt – and hence coded – relations on the one hand, and, on the other, covert, clandestine and repressed relations which, precisely because they are repressed, characterize transgressions related not so much to sex *per se* as to sexual pleasure, its preconditions and consequences.

Thus space may be said to embrace a multitude of intersections, each with its assigned location. As for representations of the relations of production, which subsume power relations, these too occur in space: space contains them in the form of buildings, monuments and works of art. Such frontal (and hence brutal) expressions of these relations do not completely crowd out their more clandestine or underground aspects; all power must have its accomplices – and its police.

A conceptual triad has now emerged from our discussion, a triad to which we shall be returning over and over again.

1 *Spatial practice*, which embraces production and reproduction, and the particular locations and spatial sets characteristic of each social formation. Spatial practice ensures continuity and some degree of cohesion. In terms of social space, and of each member of a given society's relationship to that space, this cohesion implies a guaranteed level of *competence* and a specific level of *performance*.[31]

2 *Representations of space*, which are tied to the relations of production and to the 'order' which those relations impose, and hence to knowledge, to signs, to codes, and to 'frontal' relations.

3 *Representational spaces*, embodying complex symbolisms, sometimes coded, sometimes not, linked to the clandestine or underground side of social life, as also to art (which may come eventually to be defined less as a code of space than as a code of representational spaces).

XVI

In reality, social space 'incorporates' social actions, the actions of subjects both individual and collective who are born and who die, who suffer and who act. From the point of view of these subjects, the

---

[31] These terms are borrowed from Noam Chomsky, but this should not be taken as implying any subordination of the theory of space to linguistics.

behaviour of their space is at once vital and mortal: within it they develop, give expression to themselves, and encounter prohibitions; then they perish, and that same space contains their graves. From the point of view of knowing (*connaissance*), social space works (along with its concept) as a tool for the analysis of society. To accept this much is at once to eliminate the simplistic model of a one-to-one or 'punctual' correspondence between social actions and social locations, between spatial functions and spatial forms. Precisely because of its crudeness, however, this 'structural' schema continues to haunt our consciousness and knowledge (*savoir*).

It is not the work of a moment for a society to generate (produce) an appropriated social space in which it can achieve a form by means of self-presentation and self-representation – a social space to which that society is not identical, and which indeed is its tomb as well as its cradle. This act of creation is, in fact, a *process*. For it to occur, it is necessary (and this necessity is precisely what has to be explained) for the society's practical capabilities and sovereign powers to have at their disposal special places: religious and political sites. In the case of precapitalist societies, more readily comprehensible to anthropology, ethnology and sociology than to political economy, such sites are needed for symbolic sexual unions and murders, as places where the principle of fertility (the Mother) may undergo renewal and where fathers, chiefs, kings, priests and sometimes gods may be put to death. Thus space emerges consecrated – yet at the same time protected from the forces of good and evil: it retains the aspect of those forces which facilitates social continuity, but bears no trace of their other, dangerous side.

A further necessity is that space – natural and social, practical and symbolic – should come into being inhabited by a (signifying and signified) higher 'reality'. By Light, for instance – the light of sun, moon or stars as opposed to the shadows, the night, and hence death; light identified with the True, with life, and hence with thought and knowledge and, ultimately, by virtue of mediations not immediately apparent, with established authority. So much is intimated by myths, whether Western or Oriental, but it is only actualized in and through (religio-political) space. Like all social practice, spatial practice is lived directly before it is conceptualized; but the speculative primacy of the conceived over the lived causes practice to disappear along with life, and so does very little justice to the 'unconscious' level of lived experience *per se*.

Yet another requirement is that the family (long very large, but never unlimited in size) be rejected as sole centre or focus of social practice, for such a state of affairs would entail the dissolution of society; but at the same time that it be retained and maintained as the 'basis' of personal and direct relationships which are bound to nature, to the earth, to procreation, and thus to reproduction.

Lastly, death must be both represented and rejected. Death too has a 'location', but that location lies below or above appropriated social space; death is relegated to the infinite realm so as to disenthral (or purify) the finiteness in which social practice occurs, in which the law that that practice has established holds sway. Social space thus remains the space of society, of social life. Man does not live by words alone; all 'subjects' are situated in a space in which they must either recognize themselves or lose themselves, a space which they may both enjoy and modify. In order to accede to this space, individuals (children, adolescents) who are, paradoxically, already within it, must pass tests. This has the effect of setting up reserved spaces, such as places of initiation, within social space. All holy or cursed places, places characterized by the presence or absence of gods, associated with the death of gods, or with hidden powers and their exorcism – all such places qualify as special preserves. Hence in absolute space the absolute has no place, for otherwise it would be a 'non-place'; and religio-political space has a rather strange composition, being made up of areas set apart, reserved – and so mysterious.

As for magic and sorcery, they too have their own spaces, opposed to (but presupposing) religio-political space; also set apart and reserved, such spaces are cursed rather than blessed and beneficent. By contrast, certain ludic spaces, devoted for their part to religious dances, music, and so on, were always felt to be beneficent rather than baleful.

Some would doubtless argue that the ultimate foundation of social space is *prohibition*, adducing in support of this thesis the unsaid in communication between the members of a society; the gulf between them, their bodies and consciousnesses, and the difficulties of social intercourse; the dislocation of their most immediate relationships (such as the child's with its mother), and even the dislocation of their bodily integrity; and, lastly, the never fully achieved restoration of these relations in an 'environment' made up of a series of zones defined by interdictions and bans.

Along the same lines, one might go so far as to explain social space in terms of a dual prohibition: the prohibition which separates the (male) child from his mother because incest is forbidden, and the prohibition which separates the child from its body because language in constituting consciousness breaks down the unmediated unity of the body – because, in other words, the (male) child suffers symbolic castration and his own phallus is objectified for him as part of outside reality. Hence the Mother, her sex and her blood, are relegated to the realm of the cursed and the sacred – along with sexual pleasure, which is thus rendered both fascinating and inaccessible.

The trouble with this thesis [32] is that it assumes the logical, epistemological and anthropological priority of language over space. By the same token, it puts prohibitions – among them that against incest – and not productive activity, at the origin of society. The pre-existence of an objective, neutral and empty space is simply taken as read, and only the space of speech (and writing) is dealt with as something that must be created. These assumptions obviously cannot become the basis for an adequate account of social/spatial practice. They apply only to an imaginary society, an ideal type or model of society which this ideology dreams up and then arbitrarily identifies with all 'real' societies. All the same, the existence within space of *phallic verticality*, which has a long history but which at present is becoming more prevalent, cries out for explanation. The same might be said apropos of the general fact that walls, enclosures and façades serve to define both a *scene* (where something takes place) and an *obscene* area to which everything that cannot or may not happen on the scene is relegated: whatever is inadmissible, be it malefic or forbidden, thus has its own hidden space on the near or the far side of a frontier. It is true that explaining everything in psychoanalytic terms, in terms of the unconscious, can only lead to an intolerable reductionism and dogmatism; the same goes for the overestimation of the 'structural'. Yet structures do exist, and there is such a thing as the 'unconscious'. Such little-understood aspects of consciousness would provide sufficient justification in themselves for research in this area. If it turned out, for instance, that every society, and particularly (for our purposes) the city, had an underground and repressed life, and hence an 'unconscious' of its own, there can be no doubt that interest in psychoanalysis, at present on the decline, would get a new lease on life.

XVII

The third implication of our initial hypothesis will take an even greater effort to elaborate on. If space is a product, our knowledge of it must be expected to reproduce and expound the process of production. The

---

[32] A thesis basic to the approach of Jacques Lacan and his followers.

'object' of interest must be expected to shift from *things in space* to the actual *production of space*, but this formulation itself calls for much additional explanation. Both partial products located *in space* – that is, things – and discourse *on space* can henceforth do no more than supply clues to, and testimony about, this productive process – a process which subsumes signifying processes without being reducible to them. It is no longer a matter of the space of this or the space of that: rather, it is space in its totality or global aspect that needs not only to be subjected to analytic scrutiny (a procedure which is liable to furnish merely an infinite series of fragments and cross-sections subordinate to the analytic project), but also to be *engendered* by and within theoretical understanding. Theory *reproduces* the generative process – by means of a concatenation of concepts, to be sure, but in a very strong sense of the word: from within, not just from without (descriptively), and globally – that is, moving continually back and forth between past and present. The historical and its consequences, the 'diachronic', the 'etymology' of locations in the sense of what happened at a particular spot or place and thereby changed it – all of this becomes inscribed in space. The past leaves its traces; time has its own script. Yet this space is always, now and formerly, a *present* space, given as an immediate whole, complete with its associations and connections in their actuality. Thus production process and product present themselves as two inseparable aspects, not as two separable ideas.

It might be objected that at such and such a period, in such and such a society (ancient/slave, medieval/feudal, etc.), the active groups did not 'produce' space in the sense in which a vase, a piece of furniture, a house, or a fruit tree is 'produced'. So how exactly did those groups contrive to produce their space? The question is a highly pertinent one and covers all 'fields' under consideration. Even neocapitalism or 'organized' capitalism, even technocratic planners and programmers, cannot produce a space with a perfectly clear understanding of cause and effect, motive and implication.

Specialists in a number of 'disciplines' might answer or try to answer the question. Ecologists, for example, would very likely take natural ecosystems as a point of departure. They would show how the actions of human groups upset the balance of these systems, and how in most cases, where 'pre-technological' or 'archaeo-technological' societies are concerned, the balance is subsequently restored. They would then examine the development of the relationship between town and country, the perturbing effects of the town, and the possibility or impossibility of a new balance being established. Then, from their point of view, they would adequately have clarified and even explained the genesis of modern social space. Historians, for their part, would doubtless take a different approach, or rather a number of different approaches according to the individual's method or orientation. Those who concern themselves chiefly with events might be inclined to establish a chronology of decisions affecting the relations between cities and their territorial dependencies, or to study the construction of monumental buildings. Others might seek to reconstitute the rise and fall of the institutions which underwrote those monuments. Still others would lean toward an economic study of exchange between city and territory, town and town, state and town, and so on.

To follow this up further, let us return to the three concepts introduced earlier.

1 *Spatial practice* The spatial practice of a society secretes that society's space; it propounds and presupposes it, in a dialectical interaction; it produces it slowly and surely as it masters and appropriates it. From the analytic standpoint, the spatial practice of a society is revealed through the deciphering of its space.

What is spatial practice under neocapitalism? It embodies a close association, within perceived space, between daily reality (daily routine) and urban reality (the routes and networks which link up the places set aside for work, 'private' life and leisure). This association is a paradoxical one, because it includes the most extreme separation between the places it links together. The specific spatial competence and performance of every society member can only be evaluated empirically. 'Modern' spatial practice might thus be defined – to take an extreme but significant case – by the daily life of a tenant in a government-subsidized high-rise housing project. Which should not be taken to mean that motorways or the politics of air transport can be left out of the picture. A spatial practice must have a certain cohesiveness, but this does not imply that it is coherent (in the sense of intellectually worked out or logically conceived).

2 *Representations of space*: conceptualized space, the space of scientists, planners, urbanists, technocratic subdividers and social engineers, as of a certain type of artist with a scientific bent – all of whom identify what is lived and what is perceived with what is conceived. (Arcane speculation about Numbers, with its talk of the golden number, moduli and 'canons', tends to perpetuate this view of matters.) This is the

dominant space in any society (or mode of production). Conceptions of space tend, with certain exceptions to which I shall return, towards a system of verbal (and therefore intellectually worked out) signs.

3 *Representational spaces*: space as directly *lived* through its associated images and symbols, and hence the space of 'inhabitants' and 'users', but also of some artists and perhaps of those, such as a few writers and philosophers, who *describe* and aspire to do no more than describe. This is the dominated – and hence passively experienced – space which the imagination seeks to change and appropriate. It overlays physical space, making symbolic use of its objects. Thus representational spaces may be said, though again with certain exceptions, to tend towards more or less coherent systems of non-verbal symbols and signs.

The (relative) autonomy achieved by space *qua* 'reality' during a long process which has occurred especially under capitalism or neocapitalism has brought new contradictions into play. The contradictions within space itself will be explored later. For the moment I merely wish to point up the dialectical relationship which exists within the triad of the perceived, the conceived, and the lived.

A triad: that is, three elements and not two. Relations with two elements boil down to oppositions, contrasts or antagonisms. They are defined by significant effects: echoes, repercussions, mirror effects. Philosophy has found it very difficult to get beyond such dualisms as subject and object, Descartes's *res cogitans* and *res extensa*, and the Ego and non-Ego of the Kantians, post-Kantians and neo-Kantians. 'Binary' theories of this sort no longer have anything whatsoever in common with the Manichaean conception of a bitter struggle between two cosmic principles; their dualism is entirely mental, and strips everything which makes for living activity from life, thought and society (i.e. from the physical, mental and social, as from the lived, perceived and conceived). After the titanic effects of Hegel and Marx to free it from this straitjacket, philosophy reverted to supposedly 'relevant' dualities, drawing with it – or perhaps being drawn by – several specialized sciences, and proceeding, in the name of transparency, to define intelligibility in terms of opposites and systems of opposites. Such a system can have neither materiality nor loose ends: it is a 'perfect' system whose rationality is supposed, when subjected to mental scrutiny, to be self-evident. This paradigm apparently has the magic power to turn obscurity into transparency and to move the 'object' out of the shadows into the light merely by articulating it. In short, it has the power to *decrypt*. Thus knowledge (*savoir*), with a remarkable absence of consciousness, put itself in thrall to power, suppressing all resistance, all obscurity, in its very being.

In seeking to understand the three moments of social space, it may help to consider the *body*. All the more so inasmuch as the relationship to space of a 'subject' who is a member of a group or society implies his relationship to his own body and vice versa. Considered overall, social practice presupposes the use of the body: the use of the hands, members and sensory organs, and the gestures of work as of activity unrelated to work. This is the realm of the *perceived* (the practical basis of the perception of the outside world, to put it in psychology's terms). As for *representations of the body*, they derive from accumulated scientific knowledge, disseminated with an admixture of ideology: from knowledge of anatomy, of physiology, of sickness and its cure, and of the body's relations with nature and with its surroundings or 'milieu'. Bodily *lived* experience, for its part, maybe both highly complex and quite peculiar, because 'culture' intervenes here, with its illusory immediacy, via symbolisms and via the long Judaeo-Christian tradition, certain aspects of which are uncovered by psychoanalysis. The 'heart' as *lived* is strangely different from the heart as *thought* and *perceived*. The same holds *a fortiori* for the sexual organs. Localizations can absolutely not be taken for granted where the lived experience of the body is concerned: under the pressure of morality, it is even possible to achieve the strange result of a body without organs – a body chastised, as it were, to the point of being castrated.

The perceived–conceived–lived triad (in spatial terms: spatial practice, representations of space, representational spaces) loses all force if it is treated as an abstract 'model'. If it cannot grasp the concrete (as distinct from the 'immediate'), then its import is severely limited, amounting to no more than that of one ideological mediation among others.

That the lived, conceived and perceived realms should be interconnected, so that the 'subject', the individual member of a given social group, may move from one to another without confusion – so much is a logical necessity. Whether they constitute a coherent whole is another matter. They probably do so only in favourable circumstances, when a common language, a consensus and a code can be established. It is reasonable to assume that the Western town, from the Italian Renaissance to the nineteenth century, was fortunate enough to enjoy such auspicious conditions. During this period the representation of space tended to dominate and subordinate a representational space, of religious origin, which was now reduced to symbolic figures, to images of Heaven and Hell, of the Devil and the angels, and so on. Tuscan painters, architects and theorists developed a representation of space – perspective – on the basis of a social practice which was itself, as we shall see, the result of a historic change in the relationship between town and country. Common sense meanwhile, though more or less reduced to silence, was still preserving virtually intact a representational space, inherited from the Etruscans, which had survived all the centuries of Roman and Christian dominance. The vanishing line, the vanishing-point and the meeting of parallel lines 'at infinity' were the determinants of a representation, at once intellectual and visual, which promoted the primacy of the gaze in a kind of 'logic of visualization'. This representation, which had been in the making for centuries, now became enshrined in architectural and urbanistic practice as the *code* of linear perspective.

For the present investigation to be brought to a satisfactory conclusion, for the theory I am proposing to be confirmed as far as is possible, the distinctions drawn above would have to be generalized in their application to cover all societies, all periods, all 'modes of production'. That is too tall an order for now, however, and I shall at this point merely advance a number of preliminary arguments. I would argue, for example, that representations of space are shot through with a knowledge (*savoir*) – i.e. a mixture of understanding (*connaissance*) and ideology – which is always relative and in the process of change. Such representations are thus objective, though subject to revision. Are they then true or false? The question does not always have a clear meaning: what does it mean, for example, to ask whether perspective is true or false? Representations of space are certainly abstract, but they also play a part in social and political practice: established relations between objects and people in represented space are subordinate to a logic which will sooner or later break them up because of their lack of consistency. Representational spaces, on the other hand, need obey no rules of consistency or cohesiveness. Redolent with imaginary and symbolic elements, they have their source in history – in the history of a people as well as in the history of each individual belonging to that people. Ethnologists, anthropologists and psychoanalysts are students of such representational spaces, whether they are aware of it or not, but they nearly always forget to set them alongside those representations of space which coexist, concord or interfere with them; they even more frequently ignore social practice. By contrast, these experts have no difficulty discerning those aspects of representational spaces which interest them: childhood memories, dreams, or uterine images and symbols (holes, passages, labyrinths). Representational space is alive: it speaks. It has an affective kernel or centre: Ego, bed, bedroom, dwelling, house; or: square, church, graveyard. It embraces the loci of passion, of action and of lived situations, and thus immediately implies time. Consequently it may be qualified in various ways: it may be directional, situational or relational, because it is essentially qualitative, fluid and dynamic.

If this distinction were generally applied, we should have to look at history itself in a new light. We should have to study not only the history of space, but also the history of representations, along with that of their relationships – with each other, with practice, and with ideology. History would have to take in not only the genesis of these spaces but also, and especially, their interconnections, distortions, displacements, mutual interactions, and their links with the spatial practice of the particular society or mode of production under consideration.

We may be sure that representations of space have a practical impact, that they intervene in and modify spatial *textures* which are informed by effective knowledge and ideology. Representations of space must therefore have a substantial role and a specific influence in the production of space. Their intervention occurs by way of construction – in other words, by way of architecture, conceived of not as the building of a particular structure, palace or monument, but rather as a project embedded in a spatial context and a texture which call for 'representations' that will not vanish into the symbolic or imaginary realms.

By contrast, the only products of representational spaces are symbolic works. These are often unique; sometimes they set in train 'aesthetic' trends and, after a time, having provoked a series of manifestations and incursions into the imaginary, run out of steam.

This distinction must, however, be handled with considerable caution. For one thing, there is a danger of its introducing divisions and so defeating the object of the exercise, which is to rediscover the unity of the productive process. Furthermore, it is not at all clear *a priori* that it can legitimately be generalized. Whether the East, specifically China, has experienced a contrast between representations of space and representational spaces is doubtful in the extreme. It is indeed quite possible that the Chinese characters combine two functions in an inextricable way, that on the one hand they convey the order of the world (space–time), while on the other hand they lay hold of that concrete (practical and social) space–time wherein symbolisms hold sway, where works of art are created, and where buildings, palaces and temples are built. I shall return to this question later – although, lacking adequate knowledge of the Orient, I shall offer no definite answer to it. On the other hand, apropos of the West, and of Western practice from ancient Greece and Rome onwards, I shall be seeking to show the development of this distinction, its import and meaning. Not, be it said right away, that the distinction has necessarily remained unchanged in the West right up until the modern period, or that there have never been role reversals (representational spaces becoming responsible for productive activity, for example).

There have been societies – the Chavin of the Peruvian Andes are a case in point[33] – whose representation of space is attested to by the plans of their temples and palaces, while their representational space appears in their art works, writing-systems, fabrics, and so on. What would be the relationship between two such aspects of a particular period? A problem confronting us here is that we are endeavouring with conceptual means to reconstruct a connection which originally in no way resembled the application of a pre-existing knowledge to 'reality'. Things become very difficult for us in that symbols which we can readily conceive and intuit are inaccessible as such to our abstract knowledge – a knowledge that is bodiless and timeless, sophisticated and efficacious, yet 'unrealistic' with respect to certain 'realities'. The question is what intervenes, what occupies the interstices between representations of space and representational spaces. A culture, perhaps? Certainly – but the word has less content than it seems to have. The work of artistic creation? No doubt – but that leaves unanswered the queries 'By whom?' and 'How?' Imagination? Perhaps – but why? and for whom?

The distinction would be even more useful if it could be shown that today's theoreticians and practitioners worked either for one side of it or the other, some developing representational spaces and the remainder working out representations of space. It is arguable, for instance, that Frank Lloyd Wright endorsed a communitarian representational space deriving from a biblical and Protestant tradition, whereas Le Corbusier was working towards a technicist, scientific and intellectualized representation of space.

Perhaps we shall have to go further, and conclude that the producers of space have always acted in accordance with a representation, while the 'users' passively experienced whatever was imposed upon them inasmuch as it was more or less thoroughly inserted into, or justified

[33] See François Hébert-Stevens, *L'art de l'Amérique du Sud* (Paris: Arthaud, 1973), pp. 55ff. For a sense of medieval space – both the representation of space and representational space – see *Le Grand et le Petit Albert* (Paris: Albin Michel, 1972), particularly 'Le traité des influences astrales'. Another edn: *Le Grand et le Petit Albert: les secrets de la magie* (Paris: Belfond, 1972).

by, their representational space. How such manipulation might occur is a matter for our analysis to determine. If architects (and urban planners) do indeed have a representation of space, whence does it derive? Whose interests are served when it becomes 'operational'? As to whether or not 'inhabitants' possess a representational space, if we arrive at an affirmative answer, we shall be well on the way to dispelling a curious misunderstanding (which is not to say that this misunderstanding will disappear in social and political practice).

The fact is that the long-obsolescent notion of ideology is now truly on its last legs, even if critical theory still holds it to be necessary. At no time has this concept been clear. It has been much abused by evocations of Marxist, bourgeois, proletarian, revolutionary or socialist ideology; and by incongruous distinctions between ideology in general and specific ideologies, between 'ideological apparatuses' and institutions of knowledge, and so forth.

What is an ideology without a space to which it refers, a space which it describes, whose vocabulary and links it makes use of, and whose code it embodies? What would remain of a religious ideology – the Judaeo-Christian one, say – if it were not based on places and their names: church, confessional, altar, sanctuary, tabernacle? What would remain of the Church if there were no churches? The Christian ideology, carrier of a recognizable if disregarded Judaism (God the Father, etc.), has created the spaces which guarantee that it endures. More generally speaking, what we call ideology only achieves consistency by intervening in social space and in its production, and by thus taking on body therein. Ideology *per se* might well be said to consist primarily in a discourse upon social space.

According to a well-known formulation of Marx's, knowledge (*connaissance*) becomes a productive force immediately, and no longer through any mediation, as soon as the capitalist mode of production takes over.[34] If so, a definite change in the relationship between ideology and knowledge must occur: knowledge must replace ideology. Ideology, to the extent that it remains distinct from knowledge, is characterized by rhetoric, by metalanguage, hence by verbiage and lucubration (and no longer by philosophico-metaphysical systematizing, by 'culture' and 'values'). Ideology and logic may even become indistinguishable – at least to the extent that a stubborn demand for coherence and cohesion

[34] Karl Marx, *Grundrisse*, tr. Martin Nicolaus (Harmondsworth, Middx: Penguin, 1973).

manages to erase countervailing factors proceeding either from above (information and knowledge [*savoir*]) or from below (the space of daily life).

Representations of space have at times combined ideology and knowledge within a (social-spatial) practice. Classical perspective is the perfect illustration of this. The space of today's planners, whose system of localization assigns an exact spot to each activity, is another case in point.

The area where ideology and knowledge are barely distinguishable is subsumed under the broader notion of *representation*, which thus supplants the concept of ideology and becomes a serviceable (operational) tool for the analysis of spaces, as of those societies which have given rise to them and recognized themselves in them.

In the Middle Ages, spatial practice embraced not only the network of local roads close to peasant communities, monasteries and castles, but also the main roads between towns and the great pilgrims' and crusaders' ways. As for representations of space, these were borrowed from Aristotelian and Ptolemaic conceptions, as modified by Christianity: the Earth, the underground 'world', and the luminous Cosmos, Heaven of the just and of the angels, inhabited by God the Father, God the Son, and God the Holy Ghost. A fixed sphere within a finite space, diametrically bisected by the surface of the Earth; below this surface, the fires of Hell; above it, in the upper half of the sphere, the Firmament – a cupola bearing the fixed stars and the circling planets – and a space criss-crossed by divine messages and messengers and filled by the radiant Glory of the Trinity. Such is the conception of space found in Thomas Aquinas and in the *Divine Comedy*. Representational spaces, for their part, determined the foci of a vicinity: the village church, graveyard, hall and fields, or the square and the belfry. Such spaces were interpretations, sometimes marvellously successful ones, of cosmological representations. Thus the road to Santiago de Compostela was the equivalent, on the earth's surface, of the Way that led from Cancer to Capricorn on the vault of the heavens, a route otherwise known as the Milky Way – a trail of divine sperm where souls are born before following its downward trajectory and falling to earth, there to seek as best they may the path of redemption – namely, the pilgrimage that will bring them to Compostela ('the field of stars'). The body too, unsurprisingly, had a role in the interplay between representations relating to space. 'Taurus rules over the neck', wrote Albertus Magnus, 'Gemini over the shoulders; Cancer over the hands and arms; Leo over the breast, the heart and the diaphragm; Virgo over the stomach; Libra takes care of the second part of the back; Scorpio is responsible for those parts that belong to lust. . . .'

It is reasonable to assume that spatial practice, representations of space and representational spaces contribute in different ways to the production of space according to their qualities and attributes, according to the society or mode of production in question, and according to the historical period. Relations between the three moments of the perceived, the conceived and the lived are never either simple or stable, nor are they 'positive' in the sense in which this term might be opposed to 'negative', to the indecipherable, the unsaid, the prohibited, or the unconscious. Are these moments and their interconnections in fact conscious? Yes – but at the same time they are disregarded or misconstrued. Can they be described as 'unconscious'? Yes again, because they are generally unknown, and because analysis is able – though not always without error – to rescue them from obscurity. The fact is, however, that these relationships have always had to be given utterance, which is not the same thing as being known – even 'unconsciously'.

XVIII

If space is produced, if there is a productive process, then we are dealing with *history*; here we have the fourth implication of our hypothesis. The history of space, of its production *qua* 'reality', and of its forms and representations, is not to be confused either with the causal chain of 'historical' (i.e. dated) events, or with a sequence, whether teleological or not, of customs and laws, ideals and ideology, and socio-economic structures or institutions (superstructures). But we may be sure that the forces of production (nature; labour and the organization of labour; technology and knowledge) and, naturally, the relations of production play a part – though we have not yet defined it – in the production of space.

It should be clear from the above that the passage from one mode of production to another is of the highest theoretical importance for our purposes, for it results from contradictions in the social relations of production which cannot fail to leave their mark on space and indeed to revolutionize it. Since, *ex hypothesi*, each mode of production has its own particular space, the shift from one mode to another must entail the production of a new space. Some people claim a special status for the mode of production, which they conceive of as a finished whole or closed system; the type of thinking which is forever searching for transparency or substantiality, or both, has a natural predilection for an 'object' of this kind. Contrary to this view of matters, however, examination of the transitions between modes of production will reveal that a fresh space is indeed generated during such changes, a space which is planned and organized subsequently. Take for example the Renaissance town, the dissolution of the feudal system and the rise of merchant capitalism. This was the period during which the code already referred to above was constituted; the analysis of this code – with the accent on its paradigmatic aspects – will take up a good few pages later in the present discussion. It began forming in antiquity, in the Greek and Roman cities, as also in the works of Vitruvius and the philosophers; later it would become the language of the writer. It corresponded to spatial practice, and doubtless to the representation of space rather than to representational spaces still permeated by magic and religion. What the establishment of this code meant was that 'people' – inhabitants, builders, politicians – stopped going from urban messages to the code in order to decipher reality, to decode town and country, and began instead to go from code to messages, so as to produce a discourse and a reality adequate to the code. This code thus has a history, a history determined, in the West, by the entire history of cities. Eventually it would allow the organization of the cities, which had been several times overturned, to become knowledge and power – to become, in other words, an *institution*. This development heralded the decline and fall of the autonomy of the towns and urban systems in their historical reality. The state was built on the back of the old cities, and their structure and code were shattered in the process. Notice that a code of this kind is a superstructure, which is not true of the town itself, its space, or the 'town–country' relationship within that space. The code served to fix the alphabet and language of the town, its primary signs, their paradigm and their syntagmatic relations. To put it in less abstract terms, façades were harmonized to create perspectives; entrances and exits, doors and windows, were subordinated to façades – and hence also to perspectives; streets and squares were arranged in concord with the public buildings and palaces of political leaders and institutions (with municipal authorities still predominating). At all levels, from family dwellings to monumental edifices, from 'private' areas to the territory as a whole, the elements of this space were disposed and composed in a manner at once familiar and surprising which even in the late twentieth century has not lost its charm. It is clear, therefore, that a spatial code is not simply a means of reading or interpreting space: rather it is a means of living in that space, of understanding it, and of producing it. As such it brings together verbal signs (words and sentences, along with the meaning invested in them by a signifying process) and non-verbal signs (music, sounds, evocations, architectural constructions).

The history of space cannot be limited to the study of the special moments constituted by the formation, establishment, decline and dissolution of a given code. It must deal also with the global aspect – with modes of production as generalities covering specific societies with their particular histories and institutions. Furthermore, the history of space may be expected to periodize the development of the productive process in a way that does not correspond exactly to widely accepted periodizations.

Absolute space was made up of fragments of nature located at sites which were chosen for their intrinsic qualities (cave, mountaintop, spring, river), but whose very consecration ended up by stripping them of their natural characteristics and uniqueness. Thus natural space was soon populated by political forces. Typically, architecture picked a site in nature and transferred it to the political realm by means of a symbolic mediation; one thinks, for example, of the statues of local gods or goddesses in Greek temples, or of the Shintoist's sanctuary, empty or else containing nothing but a mirror. A sanctified inwardness set itself up in opposition to the outwardness in nature, yet at the same time it echoed and restored that outwardness. The absolute space where rites and ceremonies were performed retained a number of aspects of nature, albeit in a form modified by ceremonial requirements: age, sex, genitality (fertility) – all still had a part to play. At once civil and religious, absolute space thus preserved and incorporated bloodlines, family, unmediated relationships – but it transposed them to the city, to the political state founded on the town. The socio-political forces which occupied this space also had their administrative and military extensions: scribes and armies were very much part of the picture. Those who produced space (peasants or artisans) were not the same people as managed it, as used it to organize social production and reproduction; it was the priests, warriors, scribes and princes who possessed what others had produced, who appropriated space and became its fully entitled owners.

Absolute space, religious and political in character, was a product of the bonds of consanguinity, soil and language, but out of it evolved a space which was relativized and *historical.* Not that absolute space disappeared in the process; rather it survived as the bedrock of historical space and the basis of representational spaces (religious, magical and political symbolisms). Quickened by an internal dialectic which urged it on towards its demise though simultaneously prolonging its life, absolute space embodied an antagonism between full and empty. After the fashion of a cathedral's 'nave' or 'ship', the invisible fullness of political space (the space of the town–state's nucleus or 'city') set up its rule in the emptiness of a natural space confiscated from nature. Then the forces of history smashed naturalness forever and upon its ruins established the space of accumulation (the accumulation of all wealth and resources: knowledge, technology, money, precious objects, works of art and symbols). For the theory of this accumulation, and particularly of its primitive stage, in which the respective roles of nature and history are still hard to distinguish, we are indebted to Marx; but, inasmuch as Marx's theory is incomplete, I shall have occasion to discuss this further below. One 'subject' dominated this period: the historical town of the West, along with the countryside under its control. It was during this time that productive activity (labour) became no longer one with the process of reproduction which perpetuated social life; but, in becoming independent of that process, labour fell prey to abstraction, whence abstract social labour – and *abstract space*.

This abstract space took over from historical space, which nevertheless lived on, though gradually losing its force, as substratum or under-pinning of representational spaces. Abstract space functions 'objectally', as a set of things/signs and their formal relationships: glass and stone, concrete and steel, angles and curves, full and empty. Formal and quantitative, it erases distinctions, as much those which derive from nature and (historical) time as those which originate in the body (age, sex, ethnicity). The signification of this ensemble refers back to a sort of super-signification which escapes meaning's net: the functioning of capitalism, which contrives to be blatant and covert at one and the same time. The dominant form of space, that of the centres of wealth and power, endeavours to mould the spaces it dominates (i.e. peripheral spaces), and it seeks, often by violent means, to reduce the obstacles and resistance it encounters there. Differences, for their part, are forced into the symbolic forms of an art that is itself abstract. A symbolism derived from that mis-taking of sensory, sensual and sexual which is intrinsic to the things/signs of abstract space finds objective expression in derivative ways: monuments have a phallic aspect, towers exude arrogance, and the bureaucratic and political authoritarianism immanent to a repressive space is everywhere. All of which calls, of course, for thorough analysis. A characteristic contradiction of abstract space consists in the fact that, although it denies the sensual and the sexual, its only immediate point of reference is genitality: the family unit, the type of dwelling (apartment, bungalow, cottage, etc.), fatherhood and motherhood, and the assumption that fertility and fulfilment are identical. The reproduction of social relations is thus crudely conflated with biological reproduction, which is itself conceived of in the crudest and most simplistic way imaginable. In *spatial practice*, the reproduction of social relations is predominant. The *representation of space*, in thrall to both knowledge and power, leaves only the narrowest leeway to *representational spaces*, which are limited to works, images and memories whose content, whether sensory, sensual or sexual, is so far displaced that it barely achieves symbolic force. Perhaps young children can live in a space of this kind, with its indifference to age and sex (and even to time itself), but adolescence perforce suffers from it, for it cannot discern its own reality therein: it furnishes no male or female images nor any images of possible pleasure. Inasmuch as adolescents are unable to challenge either the dominant system's imperious architecture or its deployment of signs, it is only by way of revolt that they have any prospect of recovering the world of differences – the natural, the sensory/sensual, sexuality and pleasure.

Abstract space is not defined only by the disappearance of trees, or by the receding of nature; nor merely by the great empty spaces of the state and the military – plazas that resemble parade grounds; nor even by commercial centres packed tight with commodities, money and cars. It is not in fact defined on the basis of what is perceived. Its abstraction has nothing simple about it: it is not transparent and cannot be reduced either to a logic or to a strategy. Coinciding neither with the abstraction of the sign, nor with that of the concept, it operates *negatively*. Abstract space relates negatively to that which perceives and underpins it – namely, the historical and religio-political spheres. It also relates negatively to something which it carries within itself and which seeks to emerge from it: a differential space–time. It has nothing of a 'subject' about it, yet it acts like a subject in that it transports and maintains specific social relations, dissolves others and stands opposed to yet others. It functions *positively* vis-à-vis its own implications: technology, applied sciences, and knowledge bound to power. Abstract space may even be described as at once, and inseparably, the locus, medium and tool of this 'positivity'. How is this possible? Does it mean that this space could be defined in terms of a reifying alienation, on the assumption that the milieu of the commodity has itself become a commodity to be sold wholesale and retail? Perhaps so, yet the 'negativity' of abstract space is not negligible, and its abstraction cannot be reduced to an 'absolute thing'. A safer assumption would seem to be that the status of abstract space must henceforward be considered a highly complex one. It is true that it dissolves and incorporates such former 'subjects' as the village and the town; it is also true that it replaces them. It sets itself up as the space of power, which will (or at any rate may) eventually lead to its own dissolution on account of conflicts (contradictions) arising within it. What we seem to have, then, is an apparent subject, an impersonal pseudo-subject, the abstract 'one' of modern social space, and – hidden within it, concealed by its illusory transparency – the real 'subject', namely state (political) power. Within this space, and on the subject of this space, everything is openly declared: everything is said or written. Save for the fact that there is very little to be said – and even less to be 'lived', for lived experience is crushed, vanquished by what is 'conceived of'. History is experienced as nostalgia, and nature as regret – as a horizon fast disappearing behind us. This may explain why affectivity, which, along with the sensory/sensual realm, cannot accede to abstract space and so informs no symbolism, is referred to by a term that denotes both a subject and that subject's denial by the absurd rationality of space: that term is 'the unconscious'.

In connection with abstract space, a space which is also instrumental (i.e. manipulated by all kinds of 'authorities' of which it is the locus and milieu), a question arises whose full import will become apparent only later. It concerns the silence of the 'users' of this space. Why do they allow themselves to be manipulated in ways so damaging to their spaces and their daily life without embarking on massive revolts? Why is protest left to 'enlightened', and hence elite, groups who are in any case largely exempt from these manipulations? Such elite circles, at the margins of political life, are highly vocal, but being mere wordmills, they have little to show for it. How is it that protest is never taken up by supposedly left-wing political parties? And why do the more honest politicians pay such a high price for displaying a bare minimum of straightforwardness?[35] Has bureaucracy already achieved such power that no political force can successfully resist it? There must be many reasons for such a startlingly strong – and worldwide – trend. It is difficult to see how so odd an indifference could be maintained without diverting the attention and interest of the 'users' elsewhere, without throwing sops to them in response to their demands and proposals, or without supplying replacement fulfilments for their (albeit vital) objec-

---

[35] I am thinking, for instance, of the Parti Socialiste Unifié (PSU) and its leader Michel Rocard, defeated in the French elections of 1973, or of George McGovern's defeat in the US presidential election of 1971.

tives. Perhaps it would be true to say that the place of social space as a whole has been usurped by a part of that space endowed with an illusory special status – namely, the part which is concerned with writing and imagery, underpinned by the written text (journalism, literature), and broadcast by the media; a part, in short, that amounts to abstraction wielding awesome reductionistic force vis-à-vis 'lived' experience.

Given that abstract space is buttressed by non-critical (positive) knowledge, backed up by a frightening capacity for violence, and maintained by a bureaucracy which has laid hold of the gains of capitalism in the ascendent and turned them to its own profit, must we conclude that this space will last forever? If so, we should have to deem it the locus and milieu of the ultimate abjection, of that final stability forecast by Hegel, the end result of social entropy. To such a state of affairs our only possible response would be the spasms of what Georges Bataille calls the acephal. Whatever traces of vitality remained would have a wasteland as their only refuge.

From a less pessimistic standpoint, it can be shown that abstract space harbours specific contradictions. Such spatial contradictions derive in part from the old contradictions thrown up by historical time. These have undergone modifications, however: some are aggravated, others blunted. Amongst them, too, completely fresh contradictions have come into being which are liable eventually to precipitate the downfall of abstract space. The reproduction of the social relations of production within this space inevitably obeys two tendencies: the dissolution of old relations on the one hand and the generation of new relations on the other. Thus, despite – or rather because of – its negativity, abstract space carries within itself the seeds of a new kind of space. I shall call that new space 'differential space', because, inasmuch as abstract space tends towards homogeneity, towards the elimination of existing differences or peculiarities, a new space cannot be born (produced) unless it accentuates differences. It will also restore unity to what abstract space breaks up – to the functions, elements and moments of social practice. It will put an end to those localizations which shatter the integrity of the individual body, the social body, the corpus of human needs, and the corpus of knowledge. By contrast, it will distinguish what abstract space tends to identify – for example, social reproduction and genitality, gratification and biological fertility, social relationships and family relationships. (The persistence of abstract space notwithstanding, the pressure for these distinctions to be drawn is constantly on the increase; the space of gratification, for instance, if indeed it is ever produced, will have nothing whatsoever to do with functional spaces in general, and in particular with the space of genitality as expressed in the family cell and its insertion into the piled-up boxes of 'modern' buildings, tower blocks, 'urban complexes', and what-have-you.)

XIX

If indeed every society produces a space, its own space, this will have other consequences in addition to those we have already considered. Any 'social existence' aspiring or claiming to be 'real', but failing to produce its own space, would be a strange entity, a very peculiar kind of abstraction unable to escape from the ideological or even the 'cultural' realm. It would fall to the level of folklore and sooner or later disappear altogether, thereby immediately losing its identity, its denomination and its feeble degree of reality. This suggests a possible criterion for distinguishing between ideology and practice as well as between ideology and knowledge (or, otherwise stated, for distinguishing between the *lived* on the one hand and the *perceived* and the *conceived* on the other, and for discerning their interrelationship, their oppositions and dispositions, and what they reveal *versus* what they conceal).

There is no doubt that medieval society – that is, the feudal mode of production, with its variants and local peculiarities – created its own space. Medieval space built upon the space constituted in the preceding period, and preserved that space as a substrate and prop for its symbols; it survives in an analogous fashion itself today. Manors, monasteries, cathedrals – these were the strong points anchoring the network of lanes and main roads to a landscape transformed by peasant communities. This space was the take-off point for Western European capital accumulation, the original source and cradle of which were the towns.

Capitalism and neocapitalism have produced abstract space, which includes the 'world of commodities', its 'logic' and its worldwide strategies, as well as the power of money and that of the political state. This space is founded on the vast network of banks, business centres and major productive entities, as also on motorways, airports and information lattices. Within this space the town – once the forcing-house of accumulation, fountainhead of wealth and centre of historical space – has disintegrated.

What of socialism – or, rather, what of what is today so confusedly referred to as socialism? There is no 'communist society' in existence, and the very concept of communism has become obscure inasmuch as the notion serves chiefly to sustain two opposing yet complementary myths, the myth of anti-communism on the one hand and the myth that a communist revolution has been carried through somewhere on the other. To rephrase the question therefore: has state socialism produced a space of its own?

The question is not unimportant. A revolution that does not produce a new space has not realized its full potential; indeed it has failed in that it has not changed life itself, but has merely changed ideological superstructures, institutions or political apparatuses. A social transformation, to be truly revolutionary in character, must manifest a creative capacity in its effects on daily life, on language and on space – though its impact need not occur at the same rate, or with equal force, in each of these areas.

Which having been said, there is no easy or quick answer to the question of 'socialism's' space; much careful thought is called for here. It may be that the revolutionary period, the period of intense change, merely establishes the preconditions for a new space, and that the realization of that space calls for a rather longer period – for a period of calm. The prodigious creative ferment in Soviet Russia between 1920 and 1930 was halted even more dramatically in the fields of architecture and urbanism than it was in other areas; and those fertile years were followed by years of sterility. What is the significance of this sterile outcome? Where can an architectural production be found today that might be described as 'socialist' – or even as *new* when contrasted with the corresponding efforts of capitalist planning? In the former Stalinallee, East Berlin – now renamed Karl-Marx-Allee? In Cuba, Moscow or Peking? Just how wide by now is the rift between the 'real' society rightly or wrongly referred to as socialist and Marx and Engels' project for a new society? How is the total space of a 'socialist' society to be conceived of? How is it appropriated? In short, what do we find when we apply the yardstick of space – or, more precisely, the yardstick of spatial practice – to societies with a 'socialist' mode of production? To phrase the question even more precisely, what is the relationship between, on the one hand, the entirety of that space which falls under the sway of 'socialist' relations of production and, on the other hand, the world market, generated by the capitalist mode of production, which weighs down so heavily upon the whole planet, imposing its division of labour on a worldwide scale and so governing the specific configurations of space, of the forces of production within that space, of sources of wealth and of economic fluctuations?

So many questions to which it is difficult at the present time, for lack of information or comprehension, to give satisfactory answers. One cannot help but wonder, however, whether it is legitimate to speak of socialism where no architectural innovation has occurred, where no specific space has been created; would it not be more appropriate in that case to speak of a failed transition?

As I hope to make clear later on, there are two possible ways forward for 'socialism'. The first of these would opt for accelerating growth, whatever the costs, whether for reasons of competition, prestige or power. According to this scenario, state socialism would aim to do no more than perfect capitalist strategies of growth, relying entirely on the proven strengths of large-scale enterprise and large cities, the latter constituting at once great centres of production and great centres of political power. The inevitable consequences of this approach – namely, the aggravation of inequalities in development and the abandonment of whole regions and whole sectors of the population – are seen from this viewpoint as of negligible importance. The second strategy would be founded on small and medium-sized businesses and on towns of a size compatible with that emphasis. It would seek to carry the whole territory and the whole population forward together in a process which would not separate growth from development. The inevitable urbanization of society would not take place at the expense of whole sectors, nor would it exacerbate unevenness in growth or development; it would successfully transcend the opposition between town and country instead of degrading both by turning them into an undifferentiated mass.

As for the class struggle, its role in the production of space is a cardinal one in that this production is performed solely by classes, fractions of classes and groups representative of classes. Today more than ever, the class struggle is inscribed in space. Indeed, it is that struggle alone which prevents abstract space from taking over the whole planet and papering over all differences. Only the class struggle has the capacity to differentiate, to generate differences which are not intrinsic to economic growth *qua* strategy, 'logic' or 'system' – that is to say, differences which are neither induced by nor acceptable to that growth. The forms of the class struggle are now far more varied than formerly. Naturally, they include the political action of minorities.

During the first half of the twentieth century, agrarian reforms and peasant revolutions reshaped the surface of the planet. A large portion of these changes served the ends of abstract space, because they smoothed out and in a sense automatized the previously existing space of historic peoples and cities. In more recent times, urban guerrilla actions and the intervention of the 'masses' even in urban areas have extended this movement, particularly in Latin America. The events of

May 1968 in France, when students occupied and took charge of their own space, and the working class immediately followed suit, marked a new departure. The halting of this reappropriation of space, though doubtless only temporary, has given rise to a despairing attitude. It is argued that only bulldozers or Molotov cocktails can change the dominant organization of space, that destruction must come before reconstruction. Fair enough, but it is legitimate to ask what 'reconstruction' entails. Are the same means of production to be used to produce the same products? Or must those means be destroyed also? The problem with this posture is that it minimizes the contradictions in society and space as they actually are; although there are no good grounds for doing so, it attributes a hermetic or finished quality to the 'system'; and, in the very process of heaping invective upon this system, it comes in a sense under its spell and succeeds only in glorifying its power beyond all reasonable bounds. Schizophrenic 'leftism' of this kind secretes its own, 'unconscious', contradictions. Its appeal to an absolute spontaneity in destruction and construction necessarily implies the destruction of thought, of knowledge, and of all creative capacities, on the spurious grounds that they stand in the way of an immediate and total revolution – a revolution, incidentally, which is never defined.

All the same, there is no getting around the fact that the bourgeoisie still has the initiative in its struggle for (and in) space. Which brings us back to the question of the passivity and silence of the 'users' of space.

Abstract space works in a highly complex way. It has something of a dialogue about it, in that it implies a tacit agreement, a non-aggression pact, a contract, as it were, of non-violence. It imposes reciprocity, and a communality of use. In the street, each individual is supposed not to attack those he meets; anyone who transgresses this law is deemed guilty of a criminal act. A space of this kind presupposes the existence of a 'spatial economy' closely allied, though not identical, to the verbal economy. This economy valorizes certain relationships between people in particular places (shops, cafés, cinemas, etc.), and thus gives rise to connotative discourses concerning these places; these in turn generate 'consensuses' or conventions according to which, for example, such and such a place is supposed to be trouble-free, a quiet area where people go peacefully to have a good time, and so forth. As for denotative (i.e. descriptive) discourses in this context, they have a quasi-legal aspect which also works for consensus: there is to be no fighting over who should occupy a particular spot; spaces are to be left free, and wherever possible allowance is to be made for 'proxemics' – for the maintenance of 'respectful' distances. This attitude entails in its turn a logic and a strategy of property in space: 'places and things belonging to you do not belong to me'. The fact remains, however, that communal or shared spaces, the possession or consumption of which cannot be entirely privatized, continue to exist. Cafés, squares and monuments are cases in point. The spatial consensus I have just described in brief constitutes part of civilization much as do prohibitions against acts considered vulgar or offensive to children, women, old people or the public in general. Naturally enough, its response to class struggle, as to other forms of violence, amounts to a formal and categorical rejection.

Every space is already in place before the appearance in it of actors; these actors are collective as well as individual subjects inasmuch as the individuals are always members of groups or classes seeking to appropriate the space in question. This pre-existence of space conditions the subject's presence, action and discourse, his competence and performance; yet the subject's presence, action and discourse, at the same time as they presuppose this space, also negate it. The subject experiences space as an obstacle, as a resistant 'objectality' at times as implacably hard as a concrete wall, being not only extremely difficult to modify in any way but also hedged about by Draconian rules prohibiting any attempt at such modification. Thus the *texture* of space affords opportunities not only to social acts with no particular place in it and no particular link with it, but also to a spatial practice that it does indeed determine, namely its collective and individual use: a sequence of acts which embody a signifying practice even if they cannot be reduced to such a practice. Life and death are not merely conceptualized, simulated or given expression by these acts; rather, it is in and through them that life and death actually have their being. It is within space that time consumes or devours living beings, thus giving reality to sacrifice, pleasure and pain. Abstract space, the space of the bourgeoisie and of capitalism, bound up as it is with exchange (of goods and commodities, as of written and spoken words, etc.) depends on consensus more than any space before it. It hardly seems necessary to add that within this space violence does not always remain latent or hidden. One of its contradictions is that between the appearance of security and the constant threat, and indeed the occasional eruption, of violence.

The old class struggle between bourgeoisie and aristocracy produced a space where the signs of that struggle are still manifest. Innumerable historic towns were transformed by that conflict, whose traces and results may easily be seen. After its political triumph in France, for example, the bourgeoisie smashed the aristocratic space of the Marais district in the centre of Paris, pressing it into the service of material

production and installing workshops, shops and apartments in the luxurious mansions of the area. This space was thus both uglified and
enlivened, in characteristically bourgeois fashion, through a process of
'popularization'. Today, a second phase of bourgeoisification is proceeding apace in the Marais, as it is reclaimed for residential purposes by
the elite. This is a good example of how the bourgeoisie can retain its
initiative in a great historic city. It also keeps the initiative on a much
wider scale, of course. Consider, for instance, the way in which 'pollut-
ing' industries are beginning to be exported to less developed countries
– to Brazil in the case of America, or to Spain in the European context.
It is worth noting that such trends bring about differentiation *within* a
given mode of production.

A remarkable instance of the production of space on the basis of a
difference internal to the dominant mode of production is supplied by
the current transformation of the perimeter of the Mediterranean into
a leisure-oriented space for industrialized Europe. As such, and even in
a sense as a 'non-work' space (set aside not just for vacations but also
for convalescence, rest, retirement, and so on), this area has acquired a
specific role in the social division of labour. Economically and socially,
architecturally and urbanistically, it has been subjected to a sort of neo-
colonization. At times this space even seems to transcend the constraints
imposed by the neocapitalism which governs it: the use to which it has
been put calls for 'ecological' virtues such as an immediate access to
sun and sea and a close juxtaposition of urban centres and temporary
accommodation (hotels, villas, etc.). It has thus attained a certain qualitative distinctiveness as compared with the major industrial agglomerations, where a pure culture of the quantitative reigns supreme. If, by
abandoning all our critical faculties, we were to accept this 'distinctiveness' at face value, we would get a mental picture of a space given
over completely to unproductive expense, to a vast wastefulness, to an
intense and gigantic potlatch of surplus objects, symbols and energies,
with the accent on sports, love and reinvigoration rather than on rest
and relaxation. The quasi-cultist focus of localities based on leisure
would thus form a striking contrast to the productive focus of North
European cities. The waste and expense, meanwhile, would appear as
the end-point of a temporal sequence starting in the workplace, in
production-based space, and leading to the consumption of space, sun
and sea, and of spontaneous or induced eroticism, in a great
'vacationland festival'. Waste and expense, then, instead of occurring at
the beginning, as inaugurating events, would come at the end of the
sequence, giving it meaning and justification. What a travesty such a

picture would be, however, enshrining as it does both the illusion of transparency and the illusion of naturalness. The truth is that all this seemingly non-productive expense is planned with the greatest care: centralized, organized, hierarchized, symbolized and programmed to the nth degree, it serves the interests of the tour-operators, bankers and entrepreneurs of places such as London and Hamburg. To be more precise, and to use the terminology introduced earlier: in the spatial practice of neocapitalism (complete with air transport), representations of space facilitate the manipulation of representational spaces (sun, sea, festival, waste, expense).

There are two reasons for bringing these considerations up at this point: to make the notion of the production of space as concrete as possible right away, and to show how the class struggle is waged under the hegemony of the bourgeoisie.

XX

'Change life!' 'Change society!' These precepts mean nothing without the production of an appropriate space. A lesson to be learned from the Soviet constructivists of 1920–30, and from their failure, is that new social relationships call for a new space, and vice versa. This proposition, which is a corollary of our initial one, will need to be discussed at some length. The injunction to change life originated with the poets and philosophers, in the context of a negative utopianism, but it has recently fallen into the public (i.e. the political) domain. In the process it has degenerated into political slogans – 'Live better!', 'Live differently!', 'the quality of life', 'lifestyle' – whence it is but a short step to talk of pollution, of respect for nature and for the environment, and so forth. The pressure of the world market, the transformation of the planet, the production of a new space – all these have thus disappeared into thin air. What we are left with, so far from implying the creation, whether gradual or sudden, of a different spatial practice, is simply the return of an idea to an ideal state. So long as everyday life remains in thrall to abstract space, with its very concrete constraints; so long as the only improvements to occur are technical improvements of detail (for example, the frequency and speed of transportation, or relatively better amenities); so long, in short, as the only connection between work spaces, leisure spaces and living spaces is supplied by the agencies of political power and by their mechanisms of control – so long must the project of 'changing life' remain no more than a political rallying-cry to be taken up or abandoned according to the mood of the moment.

Such are the circumstances under which theoretical thought must labour as it attempts to negotiate the obstacles in its path. To one side, it perceives the abyss of negative utopias, the vanity of a critical theory which works only at the level of words and ideas (i.e. at the ideological level). Turning in the opposite direction, it confronts highly positive technological utopias: the realm of 'prospectivism', of social engineering and programming. Here it must of necessity take note of the application to space – and hence to existing social relationships – of cybernetics, electronics and information science, if only in order to draw lessons from these developments.

The path I shall be outlining here is thus bound up with a strategic hypothesis – that is to say, with a long-range theoretical and practical project. Are we talking about a political project? Yes and no. It certainly embodies a politics of space, but at the same time goes beyond politics inasmuch as it presupposes a critical analysis of all spatial politics as of all politics in general. By seeking to point the way towards a different space, towards the space of a different (social) life and of a different mode of production, this project straddles the breach between science and utopia, reality and ideality, conceived and lived. It aspires to surmount these oppositions by exploring the dialectical relationship between 'possible' and 'impossible', and this both objectively and subjectively.

The role of strategic hypotheses in the construction of knowledge is well established. A hypothesis of this kind serves to centre knowledge around a particular focal point, a kernel, a concept or a group of concepts. The strategy involved may succeed or fail; in any case it will last for a finite length of time, long or short, before dissolving or splitting. Thus, no matter how long it may continue to govern tactical operations in the fields of knowledge and action, it must remain essentially temporary – and hence subject to revision. It demands commitment, yet appeals to no eternal truths. Sooner or later, the basis of even the most successful strategy must crumble. At which point, the concomitant removal of the centre will topple whatever has been set in place around it.

In recent times, a series of tactical and strategic operations have been undertaken with a view to the establishment (the word is apt) of a sort of impregnable fortress of knowledge. With a curious blend of naïvety and cunning, the learned promoters of such movements always express the conviction that their claims are of an irrefutably scientific nature, while at the same time ignoring the questions raised by all such claims to scientific status, and especially the question of the justification for assigning priority to what is *known* or *seen* over what is *lived*. The most recent strategic operation of this kind has sought to centre knowledge on linguistics and its ancillary disciplines: semantics, semiology, semiotics. (Earlier efforts had given a comparable centrality to political economy, history, sociology, and so on.)

This most recent hypothesis has given rise to a great mass of research and publication. Some of this work is of great importance; some of it is no doubt over- or underestimated. Naturally all such judgements, having nothing eternal about them, are subject to revision. But, inasmuch as the hypothesis itself is based on the shaky assumption that a definite (and definitive) centre can be established, it is likely to collapse. Indeed, it is already threatened with destruction from within and from without. Internally, it raises questions that it cannot answer. The question of the subject is a case in point. The systematic study of language, and/or the study of language as a system, have eliminated the 'subject' in every sense of the term. This is the sort of situation where reflective thinking must pick up the pieces of its broken mirror. Lacking a 'subject' of its own, it seizes on the old 'subjects' of the philosophers. Thus we find Chomsky readopting Descartes's *cogito* and its unique characteristics: the unicity of the deep structures of discourse and the generality of the field of consciousness. Witness also the reappearance of the Husserlian Ego, a modernized version of the *cogito*, but one which cannot maintain its philosophical (or meta-physical) substantiality – especially in face of that unconscious which was indeed invented as a way of escaping from it.

Which brings us back to an earlier part of our discussion, for what this hypothesis does is cheerfully commandeer social space and physical space and reduce them to an epistemological (mental) space – the space of discourse and of the Cartesian *cogito*. It is conveniently forgotten that the practical 'I', which is inseparably individual and social, is in a space where it must either recognize itself or lose itself. This unconsidered leap from the mental to the social and back again effectively transfers the properties of space proper onto the level of discourse – and particularly onto the level of discourse upon space. It is true that this approach seeks to supply some mediation between mental and social by evoking the body (voice, gestures, etc.). But one may wonder what connection exists between this abstract body, understood simply as a mediation between 'subject' and 'object', and a practical and fleshy body conceived of as a totality complete with spatial qualities (symmetries, asymmetries) and energetic properties (discharges, economies, waste). In fact, as I shall show later, the moment the body is envisioned as a practico-sensory totality, a decentring and recentring of knowledge occurs.

The strategy of centring knowledge on discourse avoids the particularly scabrous topic of the relationship between knowledge and power. It is also incapable of supplying reflective thought with a satisfactory answer to a theoretical question that it raises itself: do sets of non-verbal signs and symbols, whether coded or not, systematized or not, fall into the same category as verbal sets, or are they rather irreducible to them? Among non-verbal signifying sets must be included music, painting, sculpture, architecture, and certainly theatre, which in addition to a text or pretext embraces gesture, masks, costume, a stage, a *mise-en-scène* – in short, a space. Non-verbal sets are thus characterized by a spatiality which is in fact irreducible to the mental realm. There is even a sense in which landscapes, both rural and urban, fall under this head. To underestimate, ignore and diminish space amounts to the overestimation of texts, written matter, and writing systems, along with the readable and the visible, to the point of assigning to these a monopoly on intelligibility.

Simply stated, the strategic hypothesis proposed here runs as follows.

> Theoretical and practical questions relating to space are becoming more and more important. These questions, though they do not suppress them, tend to resituate concepts and problems having to do with biological reproduction, and with the production both of the means of production themselves and of consumer goods.

A given mode of production does not disappear, according to Marx, until it has liberated the forces of production and realized its full potential. This assertion may be viewed either as a statement of the obvious or as a striking paradox. When the forces of production make a leap forward, but the capitalist relations of production remain intact, the production of space itself replaces – or, rather, is superimposed upon – the production of things in space. In a number of observable and analysable instances, at any rate, such a production of space itself is entailed by the pressure of the world market and the reproduction of the capitalist relations of production. Through their manipulation of abstract space, the bourgeoisie's enlightened despotism and the capitalist system have successfully established partial control over the commodity market. They have found it harder – witness their 'monetary' problems – to establish control over the capital market itself. The combined result of a very strong political hegemony, a surge in the forces of production, and an inadequate control of markets, is a spatial chaos experienced at the most parochial level just as on a worldwide scale. The bourgeoisie and the capitalist system thus experience great difficulty in mastering what is at once their product and the tool of their mastery, namely space. They find themselves unable to reduce practice (the practico-sensory realm, the body, social-spatial practice) to their abstract space, and hence new, spatial, contradictions arise and make themselves felt. Might not the spatial chaos engendered by capitalism, despite the power and rationality of the state, turn out to be the system's Achilles' heel?

The question naturally arises whether this strategic hypothesis can in any way influence or supplant such generally accepted political strategies as world revolution carried through politically by a single party, in a single country, under the guidance of a single doctrine, through the efforts of a single class – in a word, from a single *centre*. The crisis of all such 'monocentric' strategies cleared the way not so long ago, it will be recalled, for another strategic hypothesis, one based on the idea of a social transformation accomplished by the 'third world'.

In actuality, it cannot be a matter merely of dogmatically substituting one of these hypotheses for another, nor simply of transcending the opposition between 'monocentric' and 'polycentric'. The earthshaking transformation hallowed in common parlance by the term 'revolution' has turned out to be truly earthshaking in that it is worldwide,[36] and hence also, necessarily, manifold and multiform. It advances on the theoretical as well as the political plane, for in it theory is immanent to politics. It progresses hand in hand with technology just as with knowledge and practice. In some situations peasants will remain, as they have long been, the principal factor, active and/or passive. In others, that factor may be supplied by marginal social elements or by an advanced sector of the working class now disposing of an unprecedented range of options. There are places where the transformation of the world may take on a violent and precipitate character, while in others it will progress in subterranean fashion, way below an apparently tranquil or pacified surface. A particular ruling class may succeed in presiding over changes capable of utterly destroying its opposite numbers elsewhere.

The strategic hypothesis based on space excludes neither the role of the so-called 'underdeveloped' countries nor that of the industrialized nations and their working classes. To the contrary, its basic principle and

---

[36] This is not to say that it is reducible to what Kostas Axelos, in his long philosophical meditation in the Heraclitean mould, refers to as the 'game of the world'.

objective is the bringing-together of dissociated aspects, the unification of disparate tendencies and factors. Inasmuch as it tries to take the planetary experiment in which humanity is engaged for what it is – that is to say, a series of separate and distinct assays of the world's space – this hypothesis sets itself up in clear opposition to the homogenizing efforts of the state, of political power, of the world market, and of the commodity world – tendencies which find their practical expression through and in abstract space. It implies the mobilization of differences in a single movement (including differences of natural origin, each of which ecology tends to emphasize in isolation): differences of regime, country, location, ethnic group, natural resources, and so on.

One might suppose that little argument would be required to establish that the 'right to be different' can only have meaning when it is based on actual struggles to establish differences and that the differences generated through such theoretical and practical struggles must themselves differ both from natural distinguishing characteristics and from differentiations induced within existing abstract space. The fact remains that the differences which concern us, those differences upon whose future strength theory and action may count, can only be effectively demonstrated by dint of laborious analysis.

The reconstruction of a spatial 'code' – that is, of a language common to practice and theory, as also to inhabitants, architects and scientists – may be considered from the practical point of view to be an immediate task. The first thing such a code would do is recapture the unity of dissociated elements, breaking down such barriers as that between private and public, and identifying both confluences and oppositions in space that are at present indiscernible. It would thus bring together levels and terms which are isolated by existing spatial practice and by the ideologies underpinning it: the 'micro' or architectural level and the 'macro' level currently treated as the province of urbanists, politicians and planners; the everyday realm and the urban realm; inside and outside; work and non-work (festival); the durable and the ephemeral; and so forth. The code would therefore comprise significant oppositions (i.e. paradigmatic elements) to be found amidst seemingly disparate terms, and links (syntagmatic elements) retrieved from the seemingly homogeneous mass of politically controlled space. In this sense the code might be said to contribute to the reversal of the dominant tendency and thus to play a role in the overall project. It is vital, however, that the code itself not be mistaken for a practice. The search for a language must therefore in no circumstances be permitted to become detached from practice or from the changes wrought by practice (i.e. from the worldwide process of transformation).

The working-out of the code calls itself for an effort to stay within the paradigmatic sphere: that is, the sphere of essential, hidden, implicit and unstated oppositions – oppositions susceptible of orienting a social practice – as opposed to the sphere of explicit relations, the sphere of the operational links between terms; in short, the syntagmatic sphere of language, ordinary discourse, writing, reading, literature, and so on.

A code of this kind must be correlated with a system of knowledge. It brings an alphabet, a lexicon and a grammar together within an overall framework; and it situates itself – though not in such a way as to exclude it – vis-à-vis non-knowledge (ignorance or misunderstanding); in other words, vis-à-vis the *lived* and the *perceived*. Such a knowledge is conscious of its own approximativeness: it is at once certain and uncertain. It announces its own relativity at each step, undertaking (or at least seeking to undertake) self-criticism, yet never allowing itself to become dissipated in apologias for non-knowledge, absolute spontaneity or 'pure' violence. This knowledge must find a middle path between dogmatism on the one hand and the abdication of understanding on the other.

XXI

The approach taken here may be described as 'regressive–progressive'. It takes as its starting-point the realities of the present: the forward leap of productive forces, and the new technical and scientific capacity to transform natural space so radically that it threatens nature itself. The effects of this destructive and constructive power are to be felt on all sides; they enter into combinations, often in alarming ways, with the pressures of the world market. Within this global framework, as might be expected, the Leninist principle of uneven development applies in full force: some countries are still in the earliest stages of the production of things (goods) in space, and only the most industrialized and urbanized ones can exploit to the full the new possibilities opened up by technology and knowledge. The production of space, having attained the conceptual and linguistic level, acts retroactively upon the past, disclosing aspects and moments of it hitherto uncomprehended. The past appears in a different light, and hence the process whereby that past becomes the present also takes on another aspect.

This *modus operandi* is also the one which Marx proposed in his chief 'methodological' text. The categories (concepts) which express social relationships in the most advanced society, namely bourgeois society, writes Marx, also allow 'insights into the structure and the relations of production of all the vanished social formations out of whose ruins and elements [bourgeois society] built itself up, whose partly still unconquered remnants are carried along with it, whose mere nuances have developed explicit significance within it'.[37]

Though it may seem paradoxical at first sight, this method appears on closer inspection to be fairly sensible. For how *could* we come to understand a genesis, the genesis of the present, along with the preconditions and processes involved, other than by starting from that present, working our way back to the past and then retracing our steps? Surely this must be the method adopted by any historian, economist or sociologist – assuming, of course, that such specialists aspire to any methodology at all.

Though perfectly clear in its formulation and application, Marx's approach does have its problems, and they become apparent as soon as he applies his method to the concept and reality of *labour*. The main difficulty arises from the fact that the 'regressive' and the 'progressive' movements become intertwined both in the exposition and in the research procedure itself. There is a constant risk of the regressive phase telescoping into the progressive one, so interrupting or obscuring it. The beginning might then appear at the end, and the outcome might emerge at the outset. All of which serves to add an extra level of complexity to the uncovering of those contradictions which drive every historical process forward – and thus (according to Marx) towards its end.

This is indeed the very problem which confronts us in the present context. A new concept, that of the production of space, appears at the start; it must 'operate' or 'work' in such a way as to shed light on processes from which it cannot separate itself because it is a product of them. Our task, therefore, is to employ this concept by giving it free rein without for all that according it, after the fashion of the Hegelians,

---

[37] Marx, *Grundrisse*, p. 105. This is an appropriate moment to point out a serious blunder in *Panorama des sciences sociales* (see above, note 4), where the method here discussed is attributed to Jean-Paul Sartre. Sartre's own discussion of method, however, explicitly cites Henri Lefebvre, 'Perspectives', *Cahiers internationaux de sociologie* (1953) – an article reprinted in my *Du rural à l'urbain* (Paris: Anthropos, 1970); see Sartre, *Critique de la raison dialectique* (Paris: Gallimard, 1960), pp. 41 and 42, and *Panorama*, pp. 89ff. *Panorama* is thus wrong on two counts, for what is involved here is actually the trajectory of Marxist thought itself.

a life and strength of its own *qua* concept – without, in other words, according an autonomous reality to knowledge. Ultimately, once it has illuminated and thereby validated its own coming-into-being, the production of space (as theoretical concept and practical reality in indissoluble conjunction) will become clear, and our demonstration will be over: we shall have arrived at a truth 'in itself and for itself', complete and yet relative.

In this way the method can become progressively more dialectical without posing a threat to logic and consistency. Not that there is no danger of falling into obscurity or, especially, into repetitiousness. Marx certainly failed to avoid such risks completely. And he was very aware of them: witness the fact that the exposition in *Capital* by no means follows exactly the method set forth in the *Grundrisse*; Marx's great doctrinal dissertation starts off from a form, that of exchange value, and not from the concepts brought to the fore in the earlier work, namely production and labour. On the other hand, the approach adumbrated in the *Grundrisse* is taken up again apropos of the accumulation of capital: in England, studying the most advanced form of capitalism in order to understand the system in other countries and the process of its actual growth, Marx cleaved firmly to his initial methodological precepts.

2

Social Space

I

Our project calls for a very careful examination of the notions and terminology involved, especially since the expression 'the production of space' comprises two terms neither of which has ever been properly clarified.

In Hegelianism, 'production' has a cardinal role: first, the (absolute) Idea produces the world; next, nature produces the human being; and the human being in turn, by dint of struggle and labour, produces at once history, knowledge and self-consciousness – and hence that Mind which reproduces the initial and ultimate Idea.

For Marx and Engels, the concept of production never emerges from the ambiguity which makes it such a fertile idea. It has two senses, one very broad, the other restrictive and precise. In its broad sense, humans as social beings are said to produce their own life, their own consciousness, their own world. There is nothing, in history or in society, which does not have to be achieved and produced. 'Nature' itself, as apprehended in social life by the sense organs, has been modified and therefore in a sense produced. Human beings have produced juridical, political, religious, artistic and philosophical forms. Thus production in the broad sense of the term embraces a multiplicity of works and a great diversity of forms, even forms that do not bear the stamp of the producer or of the production process (as is the case with the logical form: an abstract form which can easily be perceived as atemporal and therefore non-produced – that is, metaphysical).

Neither Marx nor Engels leaves the concept of production in an indeterminate state of this kind. They narrow it down, but with the result that works in the broad sense are no longer part of the picture;

what they have in mind is things only: *products*. This narrowing of the concept brings it closer to its everyday, and hence banal, sense – the sense it has for the economists. As for the question of who does the producing, and how they do it, the more restricted the notion becomes the less it connotes creativity, inventiveness or imagination; rather, it tends to refer solely to labour. 'It was an immense step forward for Adam Smith to throw out every limiting specification of wealth-creating activity [and to consider only] labour in general. . . . With the abstract universality of wealth-creating activity we now have the universality of the object defined as wealth, the product as such or again labour as such. . . .'[1] Production, product, labour: these three concepts, which emerge simultaneously and lay the foundation for political economy, are abstractions with a special status, *concrete* abstractions that make possible the relations of production. So far as the concept of production is concerned, it does not become fully concrete or take on a true content until replies have been given to the questions that it makes possible: 'Who produces?', 'What?', 'How?', 'Why and for whom?' Outside the context of these questions and their answers, the concept of production remains purely abstract. In Marx, as in Engels, the concept never attains concreteness. (It is true that, very late on, Engels at his most economistic sought to confine the notion to its narrowest possible meaning: 'the *ultimately* determining element in history is the production and repro-duction of real life', he wrote in a letter to Bloch on 21 September 1890. This sentence is at once dogmatic and vague: production is said to subsume biological, economic and social reproduction, and no further clarification is forthcoming.)

What constitutes the forces of production, according to Marx and Engels? Nature, first of all, plays a part, then labour, hence the organiz-ation (or division) of labour, and hence also the instruments of labour, including technology and, ultimately, knowledge.

Since the time of Marx and Engels the concept of production has come to be used so very loosely that it has lost practically all definition. We speak of the production of knowledge, or ideologies, or writings and meanings, of images, of discourses, of language, of signs and sym-bols; and, similarly, of 'dream-work' or of the work of 'operational' concepts, and so on. Such is the extension of these concepts that their comprehension has been seriously eroded. What makes matters worse is that the authors of such extensions of meaning quite consciously

[1] Karl Marx, *Grundrisse*, tr. Martin Nicolaus (Harmondsworth, Middx.: Penguin, 1973), p. 104.

abuse a procedure which Marx and Engels used ingenuously, endowing the broad or philosophical sense of the concepts with a positivity properly belonging to the narrow or scientific (economic) sense.

There is thus every reason to take up these concepts once more, to try and restore their value and to render them dialectical, while attempting to define with some degree of rigour the relationship between 'production' and 'product', as likewise those between 'works' and 'products' and 'nature' and 'production'. It may be pointed out right away that, whereas a *work* has something irreplaceable and unique about it, a *product* can be reproduced exactly, and is in fact the result of repetitive acts and gestures. Nature creates and does not produce; it provides resources for a creative and productive activity on the part of social humanity; but it supplies only *use value*, and every use value – that is to say, any product inasmuch as it is not exchangeable – either returns to nature or serves as a natural good. The earth and nature cannot, of course, be divorced from each other.

Why do I say that nature does not produce? The original meaning of the word suggests the contrary: to lead out and forward, to bring forth from the depths. And yet, nature does not labour: it is even one of its defining characteristics that it *creates*. What it creates, namely individual 'beings', simply surges forth, simply appears. Nature knows nothing of these creations – unless one is prepared to postulate the existence within it of a calculating god or providence. A tree, a flower or a fruit is not a 'product' – even if it is in a garden. A rose has no why or wherefore; it blooms because it blooms. In the words of Angelus Silesius, it 'cares not whether it is seen'. It does not know that it is beautiful, that it smells good, that it embodies a symmetry of the nth order. It is surely almost impossible not to pursue further or to return to such questions. 'Nature' cannot operate according to the same teleology as human beings. The 'beings' it creates are works; and each has 'something' unique about it even if it belongs to a genus and a species: a tree is a particular tree, a rose a particular rose, a horse a particular horse. Nature appears as the vast territory of births. 'Things' are born, grow and ripen, then wither and die. The reality behind these words is infinite. As it deploys its forces, nature is violent, generous, niggardly, bountiful, and above all open. Nature's space is not staged. To ask why this is so is a strictly meaningless question: a flower does not know that it is a flower any more than death knows upon whom it is visited. If we are to believe the word 'nature', with its ancient metaphysical and theological credentials, what is essential occurs in the depths. To say 'natural' is to say spontaneous. But today nature is drawing away from us, to say the very least. It is becoming impossible to escape the notion that nature is being murdered by 'anti-nature' – by abstraction, by signs and images, by discourse, as also by labour and its products. Along with God, nature is dying. 'Humanity' is killing both of them – and perhaps committing suicide into the bargain.

Humanity, which is to say social practice, creates works and produces things. In either case labour is called for, but in the case of works the part played by labour (and by the creator *qua* labourer) seems secondary, whereas in the manufacture of products it predominates.

In clarifying the philosophical (Hegelian) concept of production, and calling for this purpose upon the economists and political economy, Marx was seeking a rationality immanent to that concept and to its content (i.e. activity). A rationality so conceived would release him from any need to evoke a pre-existing reason of divine or 'ideal' (hence theological and metaphysical) origin. It would also eliminate any suggestion of a goal governing productive activity and conceived of as preceding and outlasting that activity. Production in the Marxist sense transcends the philosophical opposition between 'subject' and 'object', along with all the relationships constructed by the philosophers on the basis of that opposition. How, then, is the rationality immanent to production to be defined? By the fact, first of all, that it organizes a sequence of actions with a certain 'objective' (i.e. the object to be produced) in view. It imposes a temporal and spatial order upon related operations whose results are coextensive. From the start of an activity so oriented towards an objective, spatial elements – the body, limbs, eyes – are mobilized, including both *materials* (stone, wood, bone, leather, etc.) and *matériel* (tools, arms, language, instructions and agendas). Relations based on an order to be followed – that is to say, on simultaneity and synchronicity – are thus set up, by means of intellectual activity, between the component elements of the action undertaken on the physical plane. All productive activity is defined less by invariable or constant factors than by the incessant to-and-fro between temporality (succession, concatenation) and spatiality (simultaneity, synchronicity). This form is inseparable from orientation towards a goal – and thus also from functionality (the end and meaning of the action, the energy utilized for the satisfaction of a 'need') and from the structure set in motion (know-how, skills, gestures and co-operation in work, etc.). The formal relationships which allow separate actions to form a coherent whole cannot be detached from the material preconditions of individual and collective activity; and this holds true whether the aim is to move a rock, to hunt game, or to make a simple or complex object. The rationality of space, according to this analysis, is not the outcome of a quality or property of human action in general, or human labour as such, of 'man', or of social organization. On the contrary, it is itself the origin and source – not distantly but immediately, or rather inherently – of the rationality of activity; an origin which is concealed by, yet at the same time implicit in, the inevitable empiricism of those who use their hands and tools, who adjust and combine their gestures and direct their energies as a function of specific tasks.

By and large, the concept of production is still that same 'concrete universal' which Marx described on the basis of Hegel's thinking, although it has since been somewhat obscured and watered down. This fact has indeed been the justification offered for a number of critical appraisals. Only a very slight effort is made, however, to veil the tactical aim of such criticisms: the liquidation of this concept, of Marxist concepts in general, and hence of the *concrete universal* as such, in favour of the generalization of the abstract and the unrealistic in a sort of wilful dalliance with nihilism.[2]

On the right, so to speak, the concept of production can scarcely be separated out from the ideology of productivism, from a crude and brutal economism whose aim is to annex it for its own purposes. On the other hand, it must be said, in response to the left-wing or 'leftist' notion that words, dreams, texts and concepts labour and produce on their own account, that this leaves us with a curious image of labour without labourers, products without a production process or production without products, and works without creators (no 'subject' – and no 'object' either!). The phrase 'production of knowledge' does make a certain amount of sense so far as the development of concepts is concerned: every concept must come into being and must mature. But without the facts, and without the discourse of social beings or 'subjects', who could be said to produce concepts? There is a point beyond which reliance on such formulas as 'the production of knowledge' leads onto very treacherous ground: knowledge may be conceived of on the model of industrial production, with the result that the existing division of labour and use of machines, especially cybernetic machines, is uncritically accepted; alternatively, the concept of production as well as the concept of knowledge may be deprived of all specific content, and this from the point of view of the 'object' as well as from that of the

---

[2] See Jean Baudrillard, *Le miroir de la production* (Tournai: Casterman, 1973). Eng. tr. by Mark Poster: *The Mirror of Production* (St Louis: Telos Press, 1975).

'subject' – which is to give *carte blanche* to wild speculation and pure irrationalism.

(Social) space is not a thing among other things, nor a product among other products: rather, it subsumes things produced, and encompasses their interrelationships in their coexistence and simultaneity – their (relative) order and/or (relative) disorder. It is the outcome of a sequence and set of operations, and thus cannot be reduced to the rank of a simple object. At the same time there is nothing imagined, unreal or 'ideal' about it as compared, for example, with science, representations, ideas or dreams. Itself the outcome of past actions, social space is what permits fresh actions to occur, while suggesting others and prohibiting yet others. Among these actions, some serve production, others consumption (i.e. the enjoyment of the fruits of production). Social space implies a great diversity of knowledge. What then is its exact status? And what is the nature of its relationship to production?

'To produce space': this combination of words would have meant strictly nothing when the philosophers exercised all power over concepts. The space of the philosophers could be created only by God, as his first work; this is as true for the God of the Cartesians (Descartes, Malebranche, Spinoza, Leibniz) as for the Absolute of the post-Kantians (Schelling, Fichte, Hegel). Although, later on, space began to appear as a mere degradation of 'being' as it unfolded in a temporal continuum, this pejorative view made no basic difference: though relativized and devalued, space continued to depend on the absolute, or upon duration in the Bergsonian sense.

Consider the case of a city – a space which is fashioned, shaped and invested by social activities during a finite historical period. Is this city a *work* or a *product*? Take Venice, for instance. If we define works as unique, original and primordial, as occupying a space yet associated with a particular time, a time of maturity between rise and decline, then Venice can only be described as a work. It is a space just as highly expressive and significant, just as unique and unified as a painting or a sculpture. But what – and whom – does it express and signify? These questions can give rise to interminable discussion, for here content and meaning have no limits. Happily, one does not have to know the answers, or to be a 'connoisseur', in order to experience Venice as festival. Who conceived the architectural and monumental unity which extends from each palazzo to the city as a whole? The truth is that no one did – even though Venice, more than any other place, bears witness to the existence, from the sixteenth century on, of a unitary code or common language of the city. This unity goes deeper, and in a sense higher, than the spectacle Venice offers the tourist. It combines the city's reality with its ideality, embracing the practical, the symbolic and the imaginary. In Venice, the *representation of space* (the sea at once dominated and exalted) and *representational space* (exquisite lines, refined pleasures, the sumptuous and cruel dissipation of wealth accumulated by any and every means) are mutually reinforcing. Something similar may be said of the space of the canals and streets, where water and stone create a texture founded on reciprocal reflection. Here everyday life and its functions are coextensive with, and utterly transformed by, a theatricality as sophisticated as it is unsought, a sort of involuntary *mise-en-scène*. There is even a touch of madness added for good measure.

But the moment of creation is past; indeed, the city's disappearance is already imminent. Precisely because it is still full of life, though threatened with extinction, this work deeply affects anyone who uses it as a source of pleasure and in so doing contributes in however small a measure to its demise. The same thing goes for a village, or for a fine vase. These 'objects' occupy a space which is not produced as such. Think now of a flower. 'A rose does not know that it is a rose.'[3] Obviously, a city does not present itself in the same way as a flower, ignorant of its own beauty. It has, after all, been 'composed' by people, by well-defined groups. All the same, it has none of the intentional character of an 'art object'. For many people, to describe something as a work of art is simply the highest praise imaginable. And yet, what a distance there is between a work of nature and art's intentionality! What exactly were the great cathedrals? The answer is that they were political acts. The ancient function of statues was to immortalize the dead so that they would not harm the living. Fabrics or vases served a purpose. One is tempted to say, in fact, that the appearance of art, a short time prior to the appearance of its concept, implies the degeneration of works: that no work has ever been created as a work of art, and hence that art – especially the art of writing, or literature – merely heralds that decline. Could it be that art, as a specialized activity, has destroyed works and replaced them, slowly but implacably, by products destined to be exchanged, traded and reproduced *ad infinitum*? Could it be that the space of the finest cities came into being after the fashion of plants and flowers in a garden – after the fashion, in other words, of works of nature, just as unique as they, albeit fashioned by highly civilized people?

---

[3] Cf. Heidegger's commentary on Angelus Silesius's diptych in *Der Satz vom Grund* (Pfullingen: Neske, 1957), pp. 68–71.

The question is an important one. Can works really be said to stand in a transcendent relationship to products? Can the historical spaces of village and city be adequately dealt with solely by reference to the notion of a work? Are we concerned here with collectivities still so close to nature that the concepts of production and product, and hence any idea of a 'production of space', are largely irrelevant to our understanding of them? Is there not a danger here too of fetishizing the notion of the work, and so erecting unjustified barriers between creation and production, nature and labour, festival and toil, the unique and the reproducible, difference and repetition, and, ultimately, the living and the dead?

Another result of such an approach would be to force a radical break between the historical and economic realms. There is no need to subject modern towns, their outskirts and new buildings, to careful scrutiny in order to reach the conclusion that everything here resembles everything else. The more or less accentuated split between what is known as 'architecture' and what is known as 'urbanism' – that is to say, between the 'micro' and 'macro' levels, and between these two areas of concern and the two professions concerned – has not resulted in an increased diversity. On the contrary. It is obvious, sad to say, that repetition has everywhere defeated uniqueness, that the artificial and contrived have driven all spontaneity and naturalness from the field, and, in short, that products have vanquished works. Repetitious spaces are the outcome of repetitive gestures (those of the workers) associated with instruments which are both duplicatable and designed to duplicate: machines, bull-dozers, concrete-mixers, cranes, pneumatic drills, and so on. Are these spaces interchangeable because they are homologous? Or are they homogeneous so that they can be exchanged, bought and sold, with the only differences between them being those assessable in money – i.e. quantifiable – terms (as volumes, distances, etc.)? At all events, repetition reigns supreme. Can a space of this kind really still be described as a 'work'? There is an overwhelming case for saying that it is a product *strictu sensu*: it is reproducible and it is the result of repetitive actions. Thus space is undoubtedly produced even when the scale is not that of major highways, airports or public works. A further important aspect of spaces of this kind is their increasingly pronounced visual character. They are made with the visible in mind: the visibility of people and things, of spaces and of whatever is contained by them. The predominance of visualization (more important than 'spectacularization', which is in any case subsumed by it) serves to conceal repetitiveness. People *look*, and take sight, take seeing, for life itself. We build on the basis of papers and plans. We buy on the basis of images. Sight and seeing, which in the Western tradition once epitomized intelligibility, have turned into a trap: the means whereby, in social space, diversity may be simulated and a travesty of enlightenment and intelligibility ensconced under the sign of transparency.

Let us return now to the exemplary case of Venice. Venice is indeed a unique space, a true marvel. But is it a work of art? No, because it was not planned in advance. It was born of the sea, but gradually, and not, like Aphrodite, in an instant. To begin with, there was a challenge (to nature, to enemies) and an aim (trade). The space of the settlement on the lagoon, encompassing swamps, shallows and outlets to the open sea, cannot be separated from a vaster space, that of a system of commercial exchange which was not yet worldwide but which took in the Mediterranean and the Orient. Another prerequisite of Venice's development was the continuity ensured by a grand design, by an ongoing practical project, and by the dominance of a political caste, by the 'thalassocracy' of a merchant oligarchy. Beginning with the very first piles driven into the mud of the lagoon, every single site in the city had of course to be planned and realized by people – by political 'chiefs', by groups supporting them, and by those who performed the work of construction itself. Closely behind practical responses to the challenge of the sea (the port, navigable channels) came public gatherings, festivals, grandiose ceremonies (such as the marriage of the Doge and the sea) and architectural inventiveness. Here we can see the relationship between a place built by collective will and collective thought on the one hand, and the productive forces of the period on the other. For this is a place that has been *laboured on*. Sinking pilings, building docks and harbourside installations, erecting palaces – these tasks also constituted social labour, a labour carried out under difficult conditions and under the constraint of decisions made by a caste destined to profit from it in every way. Behind Venice the work, then, there assuredly lay production. Had not the emergence of social surplus production – a form preceding capitalist surplus value – already heralded this state of things? In the case of Venice, a rider must be added to the effect that the surplus labour and the social surplus production were not only realized but also for the most part expanded on the spot – that is to say, in the city of Venice. The fact that this surplus production was put to an aesthetically satisfying use, in accordance with the tastes of people who were prodigiously gifted, and highly civilized for all their ruthlessness, can in no way conceal its origins. All Venice's now-declining splendour reposes after its fashion on oft-repeated gestures on the part of carpenters and masons, sailors and stevedores; as also, of course, on those of patricians managing their affairs from day to day. All the same, every bit of Venice is part of a great hymn to diversity in pleasure and inventiveness in celebration, revelry and sumptuous ritual. If indeed there is a need at all to preserve the distinction between works and products, its import must be quite relative. Perhaps we shall discover a subtler relationship between these two terms than either identity or opposition. Each work occupies a space; it also engenders and fashions that space. Each product too occupies a space, and circulates within it. The question is therefore what relationship might exist between these two modalities of occupied space.

Even in Venice, social space is produced and reproduced in connection with the forces of production (and with the relations of production). And these forces, as they develop, are not taking over a pre-existing, empty or neutral space, or a space determined solely by geography, climate, anthropology, or some other comparable consideration. There is thus no good reason for positing such a radical separation between works of art and products as to imply the work's total transcendence of the product. The benefit to be derived from this conclusion is that it leaves us some prospect of discovering a dialectical relationship in which works are in a sense inherent in products, while products do not press all creativity into the service of repetition.

A social space cannot be adequately accounted for either by nature (climate, site) or by its previous history. Nor does the growth of the forces of production give rise in any direct causal fashion to a particular space or a particular time. Mediations, and mediators, have to be taken into consideration: the action of groups, factors within knowledge, within ideology, or within the domain of representations. Social space contains a great diversity of objects, both natural and social, including the networks and pathways which facilitate the exchange of material things and information. Such 'objects' are thus not only things but also relations. As objects, they possess discernible peculiarities, contour and form. Social labour transforms them, rearranging their positions within spatio-temporal configurations without necessarily affecting their materiality, their natural state (as in the case, for instance, of an island, gulf, river or mountain).

Let us turn now to another example: Tuscany. Another *Italian* example, be it noted, and no doubt this is because in Italy the history of precapitalism is especially rich in meaning and the growth leading up to the industrial era particularly rapid, even if this progress was to be offset during the eighteenth and nineteenth centuries by slowdown and relative retardation.

From about the thirteenth century, the Tuscan urban oligarchy of merchants and burghers began transforming lordly domains or latifundia that they had inherited or acquired by establishing the *métayage* system (or *colonat partiaire*) on these lands: serfs gave way to *métayers*. A *métayer* was supposed to receive a share of what he produced and hence, unlike a slave or a serf, he had a vested interest in production. The trend thus set in train, which gave rise to a new social reality, was based neither on the towns alone, nor on the country alone, but rather on their (dialectical) relationship in space, a space which had its own basis in their history. The urban bourgeoisie needed at once to feed the town-dwellers, invest in agriculture, and draw upon the territory as a whole as it supplied the markets that it controlled with cereals, wool, leather, and so on. Confronted by these requirements, the bourgeoisie transformed the country, and the countryside, according to a preconceived plan, according to a model. The houses of the *métayers*, known as *poderi*, were arranged in a circle around the mansion where the proprietor would come to stay from time to time, and where his stewards lived on a permanent basis. Between *poderi* and mansion ran alleys of cypresses. Symbol of property, immortality and perpetuity, the cypress thus inscribed itself upon the countryside, imbuing it with depth and meaning. These trees, the criss-crossing of these alleys, sectioned and organized the land. Their arrangement was evocative of the laws of perspective, whose fullest realization was simultaneously appearing in the shape of the urban piazza in its architectural setting. Town and country – and the relationship between them – had given birth to a space which it would fall to the painters, and first among them in Italy to the Siena school, to identify, formulate and develop.

In Tuscany, as elsewhere during the same period (including France, which we shall have occasion to discuss later in connection with the 'history of space'), it was not simply a matter of material production and the consequent appearance of social forms, or even of a social production of material realities. The new social forms were not 'inscribed' in a pre-existing space. Rather, a space was produced that was neither rural nor urban, but the result of a newly engendered spatial relationship between the two.

The cause of, and reason for, this transformation was the growth of productive forces – of crafts, of early industry, and of agriculture. But growth could only occur via the town–country relationship, and hence via those groups which were the motor of development: the urban oligarchy and a portion of the peasantry. The result was an increase in wealth, hence also an increase in surplus production, and this in turn had a retroactive effect on the initial conditions. Luxurious spending on the construction of palaces and monuments gave artists, and primarily painters, a chance to express, after their own fashion, what was happening, to display what they perceived. These artists 'discovered' perspective and developed the theory of it because a space in perspective lay before them, because such a space had already been produced. Work and product are only distinguishable here with the benefit of analytic hindsight. To separate them completely, to posit a radical fissure between them, would be tantamount to destroying the movement that brought both into being – or, rather, since it is all that remains to us, to destroy the concept of that movement. The growth I have been describing, and the development that went hand in hand with it, did not take place without many conflicts, without class struggle between the aristocracy and the rising bourgeoisie, between *populo minuto* and *populo grosso* in the towns, between townspeople and country people, and so on. The sequence of events corresponds in large measure to the *révolution communale* that took place in a part of France and elsewhere in Europe, but the links between the various aspects of the overall process are better known for Tuscany than for other regions, and indeed they are more marked there, and their effects more striking.

Out of this process emerged, then, a new representation of space: the visual perspective shown in the works of painters and given form first by architects and later by geometers. Knowledge emerged from a practice, and elaborated upon it by means of formalization and the application of a logical order.

This is not to say that during this period in Italy, even in Tuscany around Florence and Siena, townspeople and villagers did not continue to experience space in the traditional emotional and religious manner – that is to say, by means of the representation of an interplay between good and evil forces at war throughout the world, and especially in and around those places which were of special significance for each individual: his body, his house, his land, as also his church and the graveyard which received his dead. Indeed this *representational space* continued to figure in many works of painters and architects. The point is merely that *some* artists and men of learning arrived at a very different *representation of space*: a homogeneous, clearly demarcated space complete with horizon and vanishing-point.

II

Towards the middle of the nineteenth century, in a few 'advanced' countries, a new reality began to agitate populations and exercise minds because it posed a multitude of problems to which no solutions were as yet apparent. This 'reality' – to use a conventional and rather crude term – did not offer itself either to analysis or to action in a clear and distinct way. In the practical realm, it was known as 'industry'; for theoretical thought, it was 'political economy'; and the two went hand in hand. Industrial practice brought a set of new concepts and questions into play; reflection on this practice, in conjunction with reflection on the past (history) and with the critical evaluation of innovations (sociology), gave birth to a science that would soon come to predominate, namely political economy.

How did the people of that time actually proceed, whether those who laid claim to responsibilities in connection with knowledge (philosophers, scholars, and especially 'economists') or those who did so in the sphere of action (politicians, of course, but also capitalist entrepreneurs)? They proceeded, certainly, in a fashion which to them seemed solid, irrefutable and 'positive' (cf. the emergence of positivism at the same period).

Some people counted things, objects. Some, such as the inspired Charles Babbage, described machines; others described the products of machinery, with the emphasis on the needs that the things thus produced fulfilled, and on the markets open to them. With a few exceptions, these people became lost in detail, swamped by mere facts; although the ground seemed firm at the outset – as indeed it was – their efforts simply missed the mark. This was no impediment, however, in extreme cases, to the passing-off of the description of some mechanical device, or of some selling-technique, as knowledge in the highest sense of the term. (It scarcely needs pointing out how little has changed in this respect in the last century or more.)

Things and products that are measured, that is to say reduced to the common measure of money, do not speak the truth about themselves. On the contrary, it is in their nature as things and products to conceal that truth. Not that they do not speak at all: they use their own language, the language of things and products, to tout the satisfaction they can supply and the needs they can meet; they use it too to lie, to dissimulate not only the amount of social labour that they contain, not only the productive labour that they embody, but also the social relationships of

exploitation and domination on which they are founded. Like all languages, the language of things is as useful for lying as it is for telling the
truth. Things lie, and when, having become commodities, they lie in
order to conceal their origin, namely social labour, they tend to set
themselves up as absolutes. Products and the circuits they establish (in
space) are fetishized and so become more 'real' than reality itself – that is,
than productive activity itself, which they thus take over. This tendency
achieves its ultimate expression, of course, in the world market. Objects
hide something very important, and they do so all the more effectively
inasmuch as we (i.e. the 'subject') cannot do without them; inasmuch,
too, as they do give us pleasure, be it illusory or real (and how can
illusion and reality be distinguished in the realm of pleasure?). But
appearances and illusion are located not in the use made of things or
in the pleasure derived from them, but rather within things themselves,
for things are the substrate of mendacious signs and meanings. The
successful unmasking of *things* in order to reveal (social) relationships
– such was Marx's great achievement, and, whatever political tendencies
may call themselves Marxist, it remains the most durable accomplishment of Marxist thought. A rock on a mountainside, a cloud, a blue
sky, a bird on a tree – none of these, of course, can be said to lie.
Nature presents itself as it is, now cruel, now generous. It does not seek
to deceive; it may reserve many an unpleasant surprise for us, but it
never lies. So-called social reality is dual, multiple, plural. To what
extent, then, does it furnish a reality at all? If reality is taken in the
sense of materiality, social reality no longer *has* reality, nor *is* it reality.
On the other hand, it contains and implies some terribly concrete
abstractions (including, as cannot be too often emphasized, money,
commodities and the exchange of material goods), as well as 'pure'
forms: exchange, language, signs, equivalences, reciprocities, contracts,
and so on.

According to Marx (and no one who has considered the matter at all
has managed to demolish this basic analytical premise), merely to note
the existence of things, whether specific objects or 'the object' in general,
is to ignore what things at once embody and dissimulate, namely social
relations and the forms of those relations. When no heed is paid to the
relations that inhere in social facts, knowledge misses its target; our
understanding is reduced to a confirmation of the undefined and indefinable multiplicity of things, and gets lost in classifications, descriptions
and segmentations.

In order to arrive at an inversion and revolution of meaning that
would reveal authentic meaning, Marx had to overthrow the certainties

of an epoch; the nineteenth century's confident faith in things, in reality, had to go by the board. The 'positive' and the 'real' have never lacked for justifications or for strong supporting arguments from the standpoint of common sense and of everyday life, so Marx had his work cut out when it fell to him to demolish such claims. Admittedly, a fair part of the job had already been done by the philosophers, who had considerably eroded the calm self-assurance of common sense. But it was still up to Marx to smash such philosophical abstractions as the appeal to transcendence, to conscience, to Mind or to Man: he still had to transcend philosophy and preserve the truth at the same time.

To the present-day reader, Marx's work may seem peppered with polemics that were flogged to death long ago. Yet, despite the superfluity, these discussions have not lost all their significance (no thanks, be it said, to the far more superfluous commentaries of the orthodox Marxists). Already in Marx's time there were plenty of people ready to sing paeans to the progress achieved through economic, social or political rationality. They readily envisaged such a rationality as the way forward to a 'better' reality. To them, Marx responded by showing that what they took for progress was merely a growth in the productive forces, which, so far from solving so-called 'social' and 'political' problems, was bound to exacerbate them. On the other hand, to those who lamented the passing of an earlier era, this same Marx pointed out the new possibilities opened up by the growing forces of production. To revolutionaries raring for immediate all-out action, Marx offered concepts; to fact-collectors, he offered theories whose 'operational' import would only become apparent later on: theories of the organization of production as such, theories of planning.

On the one hand, Marx retrieved the contents which the predominant tendency – the tendency of the ruling class, though not so perceived – sought to avoid at all costs. Specifically, these contents were productive labour, the productive forces, and the relations and mode of production. At the same time, countering the tendency to fragment reality, to break it down into 'facts' and statistics, Marx identified the most general form of social relations, namely the form of exchange (exchange value). (Not their sole form, it must be emphasized, but rather the form in its generality.)

Now let us consider for a moment any given space, any 'interval' provided that it is not empty. Such a space contains things yet is not itself a thing or material 'object'. Is it then a floating 'medium', a simple abstraction, or a 'pure' form? No – precisely because it has a content. We have already been led to the conclusion that any space implies, contains and dissimulates social relationships – and this despite the fact that a space is not a thing but rather a set of relations between things (objects and products). Might we say that it is or tends to become the absolute Thing? The answer must be affirmative to the extent that every thing which achieves autonomy through the process of exchange (i.e. attains the status of a commodity) tends to become absolute – a tendency, in fact, that defines Marx's concept of fetishism (practical alienation under capitalism). The Thing, however, never quite becomes absolute, never quite emancipates itself from activity, from use, from need, from 'social being'. What are the implications of this for space? That is the key question.

When we contemplate a field of wheat or maize, we are well aware that the furrows, the pattern of sowing, and the boundaries, be they hedges or wire fences, designate relations of production and property. We also realize that this is much less true of uncultivated land, heath or forest. The more a space partakes of nature, the less it enters into the social relations of production. There is nothing surprising about this; the same holds true after all for a rock or a tree. On the other hand, spaces of this type, spaces with predominantly natural traits or containing objects with predominantly natural traits, are, like nature itself, on the decline. Take national or regional 'nature parks', for instance: it is not at all easy to decide whether such places are natural or artificial. The fact is that the once-prevalent characteristic 'natural' has grown indistinct and become a subordinate feature. Inversely, the social character of space – those social relations that it implies, contains and dissimulates – has begun *visibly* to dominate. This typical quality of visibility does not, however, imply decipherability of the inherent social relations. On the contrary, the analysis of these relations has become harder and more paradoxical.

What can be said, for example, of a peasant dwelling? It embodies and implies particular social relations; it shelters a family – a particular family belonging to a particular country, a particular region, a particular soil; and it is a component part of a particular site and a particular countryside. No matter how prosperous or humble such a dwelling may be, it is as much a work as it is a product, even though it is invariably representative of a type. It remains, to a greater or lesser degree, part of nature. It is an object intermediate between work and product, between nature and labour, between the realm of symbols and the realm of signs. Does it engender a space? Yes. Is that space natural or cultural? Is it immediate or mediated – and, if the latter, mediated by whom and to what purpose? Is it a given or is it artificial? The answer to such questions must be: 'Both.' The answer is ambiguous because the questions are too simple: between 'nature' and 'culture', as between work and product, complex relationships (mediations) already obtain. The same goes for time and for the 'object' in space.

To compare different maps of a region or country – say France – is to be struck by the remarkable diversity among them. Some, such as maps that show 'beauty spots' and historical sites and monuments to the accompaniment of an appropriate rhetoric, aim to mystify in fairly obvious ways. This kind of map designates places where a ravenous consumption picks over the last remnants of nature and of the past in search of whatever nourishment may be obtained from the *signs* of anything historical or original. If the maps and guides are to be believed, a veritable feast of authenticity awaits the tourist. The conventional signs used on these documents constitute a code even more deceptive than the things themselves, for they are at one more remove from reality. Next, consider an ordinary map of roads and other communications in France. What such a map reveals, its meaning – not, perhaps, to the most ingenuous inspection, but certainly to an intelligent perusal with even minimal preparation – is at once clear and hard to decipher. A diagonal band traverses the supposedly one and indivisible Republic like a bandolier. From Berre-l'Etang to Le Havre via the valleys of the Rhône (the great Delta), the Saône and the Seine, this stripe represents a narrow over-industrialized and over-urbanized zone which relegates the rest of our dear old France to the realm of underdevelopment and 'touristic potential'. Until only recently this state of affairs was a sort of official secret, a project known only to a few technocrats. Today (summer 1973) it is common knowledge – a banality. Perhaps not so banal, though, if one turns from tourist maps to a map of operational and projected military installations in southern France. It will readily be seen that this vast area, which has been earmarked, except for certain well-defined areas, for tourism, for national parks – that is, for economic and social decline – is also destined for heavy use by a military which finds such peripheral regions ideal for its diverse purposes.

These spaces are *produced*. The 'raw material' from which they are produced is nature. They are products of an activity which involves the economic and technical realms but which extends well beyond them, for these are also political products, and strategic spaces. The term 'strategy' connotes a great variety of products and actions: it combines peace with war, the arms trade with deterrence in the event of crisis, and the use of resources from *peripheral* spaces with the use of riches from industrial, urban, state-dominated centres.

Space is never produced in the sense that a kilogram of sugar or a yard of cloth is produced. Nor is it an aggregate of the places or locations of such products as sugar, wheat or cloth. Does it then come into being after the fashion of a superstructure? Again, no. It would be more accurate to say that it is at once a precondition and a result of social superstructures. The state and each of its constituent institutions call for spaces – but spaces which they can then organize according to their specific requirements; so there is no sense in which space can be treated solely as an *a priori* condition of these institutions and the state which presides over them. Is space a social relationship? Certainly – but one which is inherent to property relationships (especially the ownership of the earth, of land) and also closely bound up with the forces of production (which impose a form on that earth or land); here we see the polyvalence of social space, its 'reality' at once formal and material. Though a *product* to be used, to be consumed, it is also a *means of production*; networks of exchange and flows of raw materials and energy fashion space and are determined by it. Thus this means of production, produced as such, cannot be separated either from the productive forces, including technology and knowledge, or from the social division of labour which shapes it, or from the state and the superstructures of society.

III

As it develops, then, the concept of social space becomes broader. It infiltrates, even invades, the concept of production, becoming part – perhaps the essential part – of its content. Thence it sets a very specific dialectic in motion, which, while it does not abolish the production–consumption relationship as this applies to things (goods, commodities, objects of exchange), certainly does modify it by widening it. Here a unity transpires between levels which analysis often keeps separate from one another: the forces of production and their component elements (nature, labour, technology, knowledge); structures (property relations); superstructures (institutions and the state itself).

How many maps, in the descriptive or geographical sense, might be needed to deal exhaustively with a given space, to code and decode all its meanings and contents? It is doubtful whether a finite number can ever be given in answer to this sort of question. What we are most likely confronted with here is a sort of instant infinity, a situation reminiscent of a Mondrian painting. It is not only the codes – the map's legend, the conventional signs of map-making and map-reading – that are liable to change, but also the objects represented, the lens through which they are viewed, and the scale used. The idea that a small number of maps or even a single (and singular) map might be sufficient can only apply in a specialized area of study whose own self-affirmation depends on isolation from its context.

There are data of the greatest relevance today, furthermore, that it would be very difficult, if not impossible, to map at all. For example, where, how, by whom, and to what purpose is information stored and processed? How is computer technology deployed and whom does it serve? We know enough in this area to suspect the existence of a space peculiar to information science, but not enough to describe that space, much less to claim close acquaintanceship with it.

We are confronted not by one social space but by many – indeed, by an unlimited multiplicity or uncountable set of social spaces which we refer to generically as 'social space'. No space disappears in the course of growth and development: the *worldwide does not abolish the local*. This is not a consequence of the law of uneven development, but a law in its own right. The intertwinement of social spaces is also a law. Considered in isolation, such spaces are mere abstractions. As concrete abstractions, however, they attain 'real' existence by virtue of networks and pathways, by virtue of bunches or clusters of relationships. Instances of this are the worldwide networks of communication, exchange and information. It is important to note that such newly developed networks do not eradicate from their social context those earlier ones, superimposed upon one another over the years, which constitute the various *markets*: local, regional, national and international markets; the market in commodities, the money or capital market, the labour market, and the market in works, symbols and signs; and lastly – the most recently created – the market in spaces themselves. Each market, over the centuries, has been consolidated and has attained concrete form by means of a network: a network of buying- and selling-points in the case of the exchange of commodities, of banks and stock exchanges in the case of the circulation of capital, of labour exchanges in the case of the labour market, and so on. The corresponding buildings, in the towns, bear material testimony to this evolution. Thus social space, and especially urban space, emerged in all its diversity – and with a structure far more reminiscent of flaky *mille-feuille* pastry than of the homogeneous and isotropic space of classical (Euclidean/Cartesian) mathematics.

Social spaces interpenetrate one another and/or superimpose themselves upon one another. They are not *things*, which have mutually limiting boundaries and which collide because of their contours or as a result of inertia. Figurative terms such as 'sheet' and 'stratum' have serious drawbacks: being metaphorical rather than conceptual, they assimilate space to things and thus relegate its concept to the realm of abstraction. Visible boundaries, such as walls or enclosures in general, give rise for their part to an appearance of separation between spaces where in fact what exists is an ambiguous continuity. The space of a room, bedroom, house or garden may be cut off in a sense from social space by barriers and walls, by all the signs of private property, yet still remain fundamentally part of that space. Nor can such spaces be considered empty 'mediums', in the sense of containers distinct from their contents. Produced over time, distinguishable yet not separable, they can be compared neither to those local spaces evoked by astronomers such as Hoyle, nor to sedimentary substrata, although this last comparison is certainly more defensible than any to be derived from mathematics. A much more fruitful analogy, it seems to me, may be found in hydrodynamics, where the principle of the superimposition of small movements teaches us the importance of the roles played by scale, dimension and rhythm. Great movements, vast rhythms, immense waves – these all collide and 'interfere' with one another; lesser movements, on the other hand, interpenetrate. If we were to follow this model, we would say that any social locus could only be properly understood by taking two kinds of determinations into account: on the one hand, that locus would be mobilized, carried forward and sometimes smashed apart by major tendencies, those tendencies which 'interfere' with one another; on the other hand, it would be penetrated by, and shot through with, the weaker tendencies characteristic of networks and pathways.

This does not, of course, explain what it is that produces these various movements, rhythms and frequencies; nor how they are sustained; nor, again, how precarious hierarchical relationships are preserved between major and minor tendencies, between the strategic and tactical levels, or between networks and locations. A further problem with the metaphor of the dynamics of fluids is that it suggests a particular analysis and explication; if taken too far, that analysis could lead us into serious error. Even if a viable parallel may be drawn with physical phenomena (waves, types of waves, their associated 'quanta' – the classification of radiation in terms of wavelengths), this analogy might guide our analysis, but must not be allowed to govern the theory as a whole. A paradoxical implication of this paradigm is that the shorter the wavelength the greater the relative quantum of energy attaching to each discrete element. Is there anything in social space comparable to this law of physical space? Perhaps so, inasmuch, at any rate, as the practical and social 'base' may be said to preserve a concrete existence, inasmuch as the counter-violence which arises in response to a given major strategic trend invariably has a specific and local source, namely the energy of an 'element' at the base – the energy, as it were, of 'elemental' movement.

Be that as it may, the *places* of social space are very different from those of natural space in that they are not simply juxtaposed: they may be intercalated, combined, superimposed – they may even sometimes collide. Consequently the local (or 'punctual', in the sense of 'determined by a particular "point"') does not disappear, for it is never absorbed by the regional, national or even worldwide level. The national and regional levels take in innumerable 'places'; national space embraces the regions; and world space does not merely subsume national spaces, but even (for the time being at least) precipitates the formation of new national spaces through a remarkable process of fission. All these spaces, meanwhile, are traversed by myriad currents. The hypercomplexity of social space should by now be apparent, embracing as it does individual entities and peculiarities, relatively fixed points, movements, and flows and waves – some interpenetrating, others in conflict, and so on.

The principle of the interpenetration and superimposition of social spaces has one very helpful result, for it means that each fragment of space subjected to analysis masks not just one social relationship but a host of them that analysis can potentially disclose. It will be recalled that the same goes for *objects*: corresponding to needs, they result from a division of labour, enter into the circuits of exchange, and so forth.

Our initial hypothesis having now been considerably expanded, a number of remarks are called for.

1 There is a certain similarity between the present situation, in both its practical and its theoretical aspects, and the one which came to prevail in the middle of the nineteenth century. A fresh set of questions – a fresh 'problematic' as the philosophers say – is in the process of usurping the position of the old problems, substituting itself for them and superimposing itself upon them without for all that abolishing them completely.

The most 'orthodox' among the Marxists will doubtless wish to deny this state of affairs. They are firmly and exclusively committed to the study of production in the usual sense of the production of things, of 'goods', of commodities. They are even reluctant to acknowledge that, inasmuch as the 'city' constitutes a means of production (inasmuch as it amounts to something more than the sum of the 'productive factors' that it embodies), there is a conflict between the social character of this production and the private ownership of its location. This attitude trivializes thought in general and critical thought in particular. There are even some people, seemingly, who go so far as to claim that any discussion of space, of the city, of the earth and urban sphere, tends only to obscure 'class consciousness' and thus help demobilize the workers so far as class struggle is concerned. One should not have to waste time on such asininity but, sad to say, we shall be obliged to come back to this complaint later on.

2 Our chief concern is with space. The problematic of space, which subsumes the problems of the urban sphere (the city and its extensions) and of everyday life (programmed consumption), has displaced the problematic of industrialization. It has not, however, destroyed that earlier set of problems: the social relationships that obtained previously still obtain; the new problem is, precisely, the problem of their *reproduction*.

3 In Marx's time, economic science (or, rather, attempts to elevate political economy to the rank of a science) became swallowed up in the enumeration and description of products (objects, things) – in the application to them of the methods of book-keeping. Already at that time there were specialists waiting to divide up these tasks, and to perform them with the help of concepts or pseudo-concepts which were not yet referred to as 'operational' but which were already an effective means for classifying and counting and mentally pigeonholing 'things'. Marx replaced this study of things taken 'in themselves', in isolation from one another, with a critical analysis of productive activity itself (social labour; the relations and mode of production). Resuming and renewing the initiatives of the founders of so-called economic science (Smith, Ricardo), he combined these with a fundamental critique of capitalism, so achieving a higher level of knowledge.

4 A comparable approach is called for today, an approach which would analyse not things in space but space itself, with a view to uncovering the social relationships embedded in it. The dominant tendency fragments space and cuts it up into pieces. It enumerates the things, the various objects, that space contains. Specializations divide space among them and act upon its truncated parts, setting up mental barriers and practico-social frontiers. Thus architects are assigned architectural space as their (private) property, economists come into possession of economic space, geographers get their own 'place in the sun', and so on. The *ideologically* dominant tendency divides space up into parts and parcels

in accordance with the social division of labour. It bases its image of the forces occupying space on the idea that space is a passive receptacle. Thus, instead of uncovering the social relationships (including class relationships) that are latent in spaces, instead of concentrating our attention on the production of space and the social relationships inherent to it – relationships which introduce specific contradictions into production, so echoing the contradiction between the private ownership of the means of production and the social character of the productive forces – we fall into the trap of treating space as space 'in itself', as space as such. We come to think in terms of spatiality, and so to fetishize space in a way reminiscent of the old fetishism of commodities, where the trap lay in exchange, and the error was to consider 'things' in isolation, as 'things in themselves'.

5 There can be no doubt that the problematic of space results from a growth in the forces of production. (Talk of 'growth' *tout court* is better avoided, since this abstraction is forever being used in an ideological manner.) The forces of production and technology now permit of intervention at every level of space: local, regional, national, worldwide. Space as a whole, geographical or historical space, is thus modified, but without any concomitant abolition of its underpinnings – those initial 'points', those first foci or nexuses, those 'places' (localities, regions, countries) lying at different levels of a social space in which nature's space has been replaced by a space-*qua*-product. In this way reflexive thought passes from produced space, from the space of production (the production of things in space) to the production of space as such, which occurs on account of the (relatively) continuous growth of the productive forces but which is confined within the (relatively) discontinuous frameworks of the dominant relations and mode of production. Consequently, before the concept of the production of space can fully be grasped, it will be necessary to dispel ideologies which serve to conceal the use of the productive forces within modes of production in general, and within the dominant mode of production in particular. The ideologies which have to be destroyed for our immediate purposes are those which promote (abstract) spatiality and segmented representations of space. Naturally, such ideologies do not present themselves for what they are; instead, they pass themselves off as established knowledge. The difficulty and complexity of our critical task derives from the fact that it applies at once to the (mental) forms and practical (social) contents of space.

6 The search for a science of space has been going on for years, and this from many angles of approach: philosophy, epistemology, ecology,

geopolitics, systems theory (decision-making systems; cognitive systems), anthropology, ethnology, and so on. Yet such a science, forever teetering on the brink of existence, has yet to come into being. This situation is truly tantalizing for workers in these fields, but the reason for it is not far to seek. Knowledge of spaces wavers between description and dissection. Things in space, or pieces of space, are described. Part-spaces are carved out for inspection from social space as a whole. Thus we are offered a geographical space, an ethnological space, a demographic space, a space peculiar to the information sciences, and so on *ad infinitum*. Elsewhere we hear of pictural, musical or plastic spaces. What is always overlooked is the fact that this sort of fragmentation tallies not only with the tendency of language itself, not only with the wishes of specialists of all kinds, but also with the goals of existing society, which, within the overall framework of a strictly controlled and thus homogeneous totality, splits itself up into the most heterogeneous spaces: housing, labour, leisure, sport, tourism, astronautics, and so on. The result is that all focus is lost as the emphasis shifts either to what exists in space (things considered on their own, in reference to themselves, their past, or their names), or else to space emptied, and thus detached from what it contains: either objects in space or else a space without objects, a neutral space. So it is indeed because of its predilection for partial representations that this search for knowledge is confounded, integrated unintentionally into existing society and forced to operate within that society's framework. It is continually abandoning any global perspective, accepting fragmentation and so coming up with mere shards of knowledge. From time to time it makes an arbitrary 'totalization' on the basis of some issue or other, thus creating yet another 'area of specialization'. What is urgently required here is a clear distinction between an imagined or sought-after 'science of space' on the one hand and real knowledge of the production of space on the other. Such a knowledge, in contrast to the dissection, interpretations and representations of a would-be science of space, may be expected to rediscover *time* (and in the first place the time of production) in and through space.

7 The real knowledge that we hope to attain would have a retrospective as well as a prospective import. Its implications for history, for example, and for our understanding of time, will become apparent if our hypothesis turns out to be correct. It will help us to grasp how societies generate their (social) space and time – their representational spaces and their representations of space. It should also allow us, not to foresee the future, but to bring relevant factors to bear on the future in prospect –

on the *project*, in other words, of another space and another time in
another (possible or impossible) society.

IV

To suggest out of the blue that there is a need for a 'critique of space'
is liable to seem paradoxical or even intellectually outrageous. In the
first place, it may well be asked what such an expression might mean;
one normally criticizes a person or a thing – and space is neither. In
philosophical terms, space is neither subject nor object. How can it be
effectively grasped? It is inaccessible to the so-called critical spirit (a
spirit which apparently reached its apogee in the watered-down Marxism
of 'critical theory'). Perhaps this difficulty explains why there is no
architectural or urbanistic criticism on a par with the criticism of art,
literature, music and theatre. There would certainly seem to be a need
for such criticism: its 'object' is at least as important and interesting as
the aesthetic objects of everyday consumption. We are talking, after all,
of the setting in which we live. Criticism of literature, art or drama is
concerned with people and institutions: with painters, dealers, galleries,
shows, museums, or else with publishers, authors and the culture market.
Architectural and urbanistic space seems, by contrast, out of range. On
the mental level, it is evoked in daunting terms: readability, visibility,
intelligibility. Socially, it appears as the intangible outcome of history,
society and culture, all of which are supposedly combined within it.
Should we conclude that the absence of a criticism of space is simply
the result of a lack of an appropriate terminology? Perhaps – but, if so,
the reasons for this lack themselves need explaining.

At all events, a criticism of space is certainly called for inasmuch as
spaces cannot be adequately explained on the basis either of the mythical
image of pure transparency or of its opposite, the myth of the opacity
of nature; inasmuch, too, as spaces conceal their contents by means of
meanings, by means of an absence of meaning or by means of an
overload of meaning; and inasmuch, lastly, as spaces sometimes lie just
as things lie, even though they are not themselves things.

Eventually, moreover, it would also fall to a critique of this kind to
rip aside appearances which have nothing particularly mendacious about
them. Consider a house, and a street, for example. The house has six
storeys and an air of stability about it. One might almost see it as the
epitome of immovability, with its concrete and its stark, cold and rigid

outlines. (Built around 1950: no metal or plate glass yet.) Now, a critical analysis would doubtless destroy the appearance of solidity of this house, stripping it, as it were, of its concrete slabs and its thin non-load-bearing walls, which are really glorified screens, and uncovering a very different picture. In the light of this imaginary analysis, our house would emerge as permeated from every direction by streams of energy which run in and out of it by every imaginable route: water, gas, electricity, telephone lines, radio and television signals, and so on. Its image of immobility would then be replaced by an image of a complex of mobilities, a nexus of in and out conduits. By depicting this convergence of waves and currents, this new image, much more accurately than any drawing or photograph, would at the same time disclose the fact that this piece of 'immovable property' is actually a two-faceted machine analogous to an active body: at once a machine calling for massive energy supplies, and an information-based machine with low energy requirements. The occupants of the house perceive, receive and manipulate the energies which the house itself consumes on a massive scale (for the lift, kitchen, bathroom, etc.).

Comparable observations, of course, might be made apropos of the whole street, a network of ducts constituting a structure, having a global form, fulfilling functions, and so on. Or apropos of the city, which consumes (in both senses of the word) truly colossal quantities of energy, both physical and human, and which is in effect a constantly burning, blazing bonfire. Thus as exact a picture as possible of this space would differ considerably from the one embodied in the representational space which its inhabitants have in their minds, and which for all its inaccuracy plays an integral role in social practice.

The error – or illusion – generated here consists in the fact that, when social space is placed beyond our range of vision in this way, its practical character vanishes and it is transformed in philosophical fashion into a kind of absolute. In face of this fetishized abstraction, 'users' spontaneously turn themselves, their presence, their 'lived experience' and their bodies into abstractions too. Fetishized abstract space thus gives rise to two practical abstractions: 'users' who cannot recognize themselves within it, and a thought which cannot conceive of adopting a critical stance towards it. If this state of affairs were to be successfully reversed, it would become clear that the critical analysis of space as directly experienced poses more serious problems than any partial activity, no matter how important, including literature, reading and writing, art, music, and the rest. Vis-à-vis lived experience, space is neither a mere 'frame', after the fashion of the frame of a painting, nor a form or container of a virtually neutral kind, designed simply to receive whatever is poured into it. Space is social morphology: it is to lived experience what form itself is to the living organism, and just as intimately bound up with function and structure. To picture space as a 'frame' or container into which nothing can be put unless it is smaller than the recipient, and to imagine that this container has no other purpose than to preserve what has been put in it – this is probably the initial error. But is it error, or is it ideology? The latter, more than likely. If so, who promotes it? Who exploits it? And why and how do they do so?

The *theoretical* error is to be content to see a space without conceiving of it, without concentrating discrete perceptions by means of a mental act, without assembling details into a whole 'reality', without apprehending contents in terms of their interrelationships within the containing forms. The rectification of this error would very likely lead to the dissolution of not a few major ideological illusions. This has been the thrust of the preceding remarks, in which I have sought to show that a space that is apparently 'neutral', 'objective', fixed, transparent, innocent or indifferent implies more than the convenient establishment of an inoperative system of knowledge, more than an error that can be avoided by evoking the 'environment', ecology, nature and anti-nature, culture, and so forth. Rather, it is a whole set of errors, a complex of illusions, which can even cause us to forget completely that there is a total subject which acts continually to maintain and reproduce its own conditions of existence, namely the state (along with its foundation in specific social classes and fractions of classes). We also forget that there is a total object, namely absolute political space – that strategic space which seeks to impose itself as reality despite the fact that it is an abstraction, albeit one endowed with enormous powers because it is the locus and medium of Power. Whence the abstraction of the 'user' and of that so-called critical thinking which loses all its critical capacities when confronted by the great Fetishes.

There are many lines of approach to this truth. The important thing, however, is to take one or other of them instead of making excuses or simply taking flight (even if it is forward flight). In the ordinary way, the study of 'real' (i.e. social) space is referred to specialists and their respective specialities – to geographers, town-planners, sociologists, *et alii*. As for knowledge of 'true' (i.e. mental) space, it is supposed to fall within the province of the mathematicians and philosophers. Here we have a double or even multiple error. To begin with, the split between 'real' and 'true' serves only to avoid any confrontation between practice and theory, between lived experience and concepts, so that both sides of these dualities are distorted from the outset. Another trap is the resort to specialities which antedate 'modernity', which are themselves older than capitalism's absorption of the entirety of space for its own purposes, older than the actual possibility, thanks to science and technology, of producing space. Surely it is the supreme illusion to defer to architects, urbanists or planners as being experts or ultimate authorities in matters relating to space. What the 'interested parties' here fail to appreciate is that they are bending their *demands* (from below) to suit *commands* (from above), and that this unforced renunciation on their part actually runs ahead of the wishes of the manipulators of consciousness. The real task, by contrast, is to uncover and stimulate demands *even at the risk* of their wavering in face of the imposition of oppressive and repressive commands. It is, one suspects, the ideological error *par excellence* to go instead in search of specialists of 'lived experience' and of the morphology of everyday life.

Let everyone look at the space around them. What do they see? Do they see *time*? They live time, after all; they are *in* time. Yet all anyone sees is movements. In nature, time is apprehended within space – in the very heart of space: the hour of the day, the season, the elevation of the sun above the horizon, the position of the moon and stars in the heavens, the cold and the heat, the age of each natural being, and so on. Until nature became *localized* in underdevelopment, each place showed its age and, like a tree trunk, bore the mark of the years it had taken it to grow. Time was thus inscribed in space, and natural space was merely the lyrical and tragic script of natural time. (Let us not follow the bad example of those philosophers who speak in this connection merely of the degradation of duration or of the outcome of 'evolution'.) With the advent of modernity time has vanished from social space. It is recorded solely on measuring-instruments, on clocks, that are as isolated and functionally specialized as this time itself. Lived time loses its form and its social interest – with the exception, that is, of time spent working. Economic space subordinates time to itself; political space expels it as threatening and dangerous (to power). The primacy of the economic and above all of the political implies the supremacy of space over time. It is thus possible that the error concerning space that we have been discussing actually concerns time more directly, more intimately, than it does space, time being even closer to us, and more fundamental. Our time, then, this most essential part of lived experience, this greatest good of all goods, is no longer visible to us, no longer intelligible. It cannot be constructed. It is consumed, exhausted, and that is all. It leaves no traces. It is concealed in space, hidden under a pile of debris to be disposed of as soon as possible; after all, rubbish is a pollutant.

This manifest expulsion of time is arguably one of the hallmarks of modernity. It must surely have more far-reaching implications than the simple effacement of marks or the erasing of words from a sheet of paper. Since time can apparently be assessed in terms of money, however, since it can be bought and sold just like any object ('time is money'), little wonder that it disappears after the fashion of an object. At which point it is no longer even a dimension of space, but merely an incomprehensible scribble or scrawl that a moment's work can completely rub out. It is reasonable to ask if this expulsion or erasure of time is directed at historical time. The answer is: certainly, but only for symbolic purposes. It is, rather, the time needed for living, time as an irreducible good, which eludes the logic of visualization and spatialization (if indeed one may speak of logic in this context). Time may have been promoted to the level of ontology by the philosophers, but it has been murdered by society.

How could so disturbing, so outrageous an operation have been carried out without causing an outcry? How can it have been passed off as 'normal'? The fact is that it has been made part and parcel of social norms, of normative activity. One wonders just how many errors, or worse, how many lies, have their roots in the modernist trio, triad or trinity of readability–visibility–intelligibility.

We may seem by now to have left the practico-social realm far behind and to be back once more amidst some very old distinctions: appearance *versus* reality, truth *versus* lies, illusion *versus* revelation. Back, in short, in philosophy. And that is true, certainly, inasmuch as our analysis is an extension of the philosophical project; this, I hope, has already been made abundantly clear. On the other hand, the 'object' of criticism has shifted: we are concerned with practical and social activities which are supposed to embody and 'show' the truth, but which actually comminute space and 'show' nothing besides the deceptive fragments thus produced. The claim is that space can be shown by means of space itself. Such a procedure (also known as tautology) uses and abuses a familiar technique that is indeed as easy to abuse as it is to use – namely, a shift from the part to the whole: metonymy. Take images, for example: photographs, advertisements, films. Can images of this kind really be expected to expose errors concerning space? Hardly. Where there is error or illusion, the image is more likely to secrete it and reinforce it than to reveal it. No matter how 'beautiful' they may be, such images belong to an incriminated 'medium'. Where the error consists in a segmentation of space, moreover – and where the illusion consists in the failure to perceive this dismemberment – there is simply no possibility of any image rectifying the mistake. On the contrary, images fragment; they are themselves fragments of space. Cutting things up and rearranging them, *découpage* and *montage* – these are the alpha and omega of the art of image-making. As for error and illusion, they reside already in the artist's eye and gaze, in the photographer's lens, in the draftsman's pencil and on his blank sheet of paper. Error insinuates itself into the very objects that the artist discerns, as into the sets of objects that he selects. Wherever there is illusion, the optical and visual world plays an integral and integrative, active and passive, part in it. It fetishizes abstraction and imposes it as the norm. It detaches the pure form from its impure content – from lived time, everyday time, and from bodies with their opacity and solidity, their warmth, their life and their death. After its fashion, the image kills. In this it is like all signs. Occasionally, however, an artist's tenderness or cruelty transgresses the limits of the image. Something else altogether may then emerge, a truth and a reality answering to criteria quite different from those of exactitude, clarity, readability and plasticity. If this is true of images, moreover, it must apply equally well to sounds, to words, to bricks and mortar, and indeed to signs in general.[4]

Our space has strange effects. For one thing, it unleashes desire. It presents desire with a 'transparency' which encourages it to surge forth in an attempt to lay claim to an apparently clear field. Of course this foray comes to naught, for desire encounters no object, nothing desirable, and no work results from its action. Searching in vain for plenitude, desire must make do with words, with the rhetoric of desire. Disillusion leaves space empty – an emptiness that words convey. Spaces are devastated – and devastating; incomprehensibly so (without prolonged reflection at least). 'Nothing is allowed. Nothing is forbidden', in the words of one inhabitant. Spaces are strange: homogeneous, rationalized, and as such constraining; yet at the same time utterly dislocated. Formal boundaries are gone between town and country, between centre and periphery, between suburbs and city centres, between the domain of automobiles and the domain of people. Between happiness and unhappiness, for that matter. And yet everything ('public facilities', blocks of flats, 'environments for living') is separated, assigned in isolated fashion

---

[4] See for example a photographic feature by Henri Cartier-Bresson in *Politique-Hebdo*, 29 June 1972.

to unconnected 'sites' and 'tracts'; the spaces themselves are specialized just as operations are in the social and technical division of labour.

It may be said of this space that it presupposes and implies a logic of visualization. Whenever a 'logic' governs an operational sequence, a strategy, whether conscious or unconscious, is necessarily involved. So, if there is a 'logic of visualization' here, we need to understand how it is formed and how applied. The arrogant verticality of skyscrapers, and especially of public and state buildings, introduces a phallic or more precisely a phallocratic element into the visual realm; the purpose of this display, of this need to impress, is to convey an impression of authority to each spectator. Verticality and great height have ever been the spatial expression of potentially violent power. This very particular type of spatialization, though it may seem 'normal' or even 'natural' to many people, embodies a twofold 'logic', which is to say a twofold strategy, in respect of the spectator. On the one hand, it embodies a metonymic logic consisting in a continual to-and-fro movement – enforced with carrot and stick – between the part and the whole. In an apartment building comprising stack after stack of 'boxes for living in', for example, the spectators-*cum*-tenants grasp the relationship between part and whole directly; furthermore, they recognize themselves in that relationship. By constantly expanding the scale of things, this movement serves to compensate for the pathetically small size of each set of living-quarters; it posits, presupposes and imposes homogeneity in the subdivision of space; and, ultimately, it takes on the aspect of pure logic – and hence of tautology: space contains space, the visible contains the visible – and boxes fit into boxes.

The second 'logic' embodied in this spatialization is a logic (and strategy) of metaphor – or, rather, of constant metaphorization. Living bodies, the bodies of 'users' – are caught up not only in the toils of parcellized space, but also in the web of what philosophers call 'analogons': images , signs and symbols. These bodies are transported out of themselves, transferred and emptied out, as it were, via the eyes: every kind of appeal, incitement and seduction is mobilized to tempt them with doubles of themselves in prettified, smiling and happy poses; and this campaign to void them succeeds exactly to the degree that the images proposed correspond to 'needs' that those same images have helped fashion. So it is that a massive influx of information, of messages, runs head on into an inverse flow constituted by the evacuation from the innermost body of all life and desire. Even cars may fulfil the function of analogons, for they are at once extensions of the body and mobile homes, so to speak, fully equipped to receive these wandering bodies. Were it not for the eyes and the dominant form of space, words and dispersed fragments of discourse would be quite incapable of ensuring this 'transfer' of bodies.

Metaphor and metonymy, then. These familiar concepts are borrowed, of course, from linguistics. Inasmuch, however, as we are concerned not with words but rather with space and spatial practice, such conceptual borrowing has to be underwritten by a careful examination of the relationship between space and language.

Any determinate and hence demarcated space necessarily embraces some things and excludes others; what it rejects may be relegated to nostalgia or it may be simply forbidden. Such a space asserts, negates and denies. It has some characteristics of a 'subject', and some of an 'object'. Consider the great power of a façade, for example. A façade admits certain acts to the realm of what is visible, whether they occur on the façade itself (on balconies, window ledges, etc.) or are to be seen *from* the façade (processions in the street, for example). Many other acts, by contrast, it condemns to obscenity: these occur *behind* the façade. All of which already seems to suggest a 'psychoanalysis of space'.

In connection with the city and its extensions (outskirts, suburbs), one occasionally hears talk of a 'pathology of space', of 'ailing neighbourhoods', and so on. This kind of phraseology makes it easy for people who use it – architects, urbanists or planners – to suggest the idea that they are, in effect, 'doctors of space'. This is to promote the spread of some particularly mystifying notions, and especially the idea that the modern city is a product not of the capitalist or neocapitalist system but rather of some putative 'sickness' of society. Such formulations serve to divert attention from the criticism of space and to replace critical analysis by schemata that are at once not very rational and very reactionary. Taken to their logical limits, these theses can deem society as a whole and 'man' as a social being to be sicknesses of nature. Not that such a position is utterly indefensible from a strictly philosophical viewpoint: one is at liberty to hold that 'man' is a monster, a mistake, a failed species on a failed planet. My point is merely that this philosophical view, like many others, leads *necessarily* to nihilism.

V

Perhaps it would make sense to decide without further ado to seek inspiration in Marx's *Capital* – not in the sense of sifting it for quotations nor in the sense of subjecting it to the 'ultimate exegesis', but in the sense of following *Capital*'s plan in dealing with space. There are several good arguments in favour of doing so, including the parallels I mentioned earlier between the set of problems with which we are concerned and the set which existed in Marx's time. In view of the fact that there are plenty of 'Marxists' who think that discussing problems related to space (problems of cities or of the management of the land) merely serves to obfuscate the real political problems, such an association between the study of space and Marx's work might also help dispel some gross misunderstandings.

The plan of *Capital*, as it has emerged from the many commentaries on and rereadings of the book (the most literal-minded of which seem, incidentally, to be the best), itself constitutes a strong argument in favour of proceeding in this way. In his work preparatory to *Capital*, Marx was able to develop such essential concepts as that of (social) labour. Labour has existed in all societies, as have representations of it (pain, punishment, etc.), but only in the eighteenth century did the concept itself emerge. Marx shows how and why this was so, and then, having dealt with these preliminaries, he proceeds to the essential, which is neither a substance nor a 'reality', but rather a *form*. Initially, and centrally, Marx uncovers an (almost) pure form, that of the circulation of material goods, or *exchange*. This is a quasi-logical form similar to, and indeed bound up with, other 'pure' forms (identity and difference, equivalence, consistency, reciprocity, recurrence, and repetition). The circulation and exchange of material goods are distinct but not separate from the circulation and exchange of signs (language, discourse). The 'pure' form here has a bipolar structure (use value *versus* exchange value), and it has functions which *Capital* sets forth. As a *concrete abstraction*, it is developed by thought – just as it developed in time and space – until it reaches the level of social practice: via money, and via labour and its determinants (i.e. its dialectic: individual *versus* social, divided *versus* global, particular *versus* mean, qualitative *versus* quantitative). This kind of development is more fruitful conceptually than classical deduction, and suppler than induction or construction. In this case, of course, it culminates in the notion of surplus value. The pivot, however, remains unchanged: by virtue of a dialectical paradox, that pivot is a quasi-void, a near-absence – namely the form of exchange, which governs social practice.

Now, as for the form of social space, we are acquainted with it; it has already been identified. Another concrete abstraction, it has emerged in several stages (in certain philosophies and major scientific theories) from representations of space and from representational spaces. This has occurred quite recently. Like that of exchange, the form of social space has an affinity with logical forms: it calls for a content and cannot be conceived of as having no content; but, thanks to abstraction, it is in fact conceived of, precisely, as independent of any specific content. Similarly, the form of material exchange does not determine what is exchanged: it merely stipulates that *something*, which has a use, is also an object of exchange. So too with the form of non-material communication, which does not determine what sign is to be communicated, but simply that there must be a stock of distinct signs, a message, a channel and a code. Nor, finally, does a logical form decide what is consistent, or what is thought, although it does prescribe the necessity, if thought is to exist, for formal consistency.

The form of social space is encounter, assembly, simultaneity. But what assembles, or what is assembled? The answer is: everything that there is *in space*, everything that is produced either by nature or by society, either through their co-operation or through their conflicts. Everything: living beings, things, objects, works, signs and symbols. Natural space juxtaposes – and thus disperses: it puts places and that which occupies them side by side. It particularizes. By contrast, social space implies actual or potential assembly at a single point, or around that point. It implies, therefore, the possibility of accumulation (a possibility that is realized under specific conditions). Evidence in support of this proposition is supplied by the space of the village, by the space of the dwelling; it is overwhelmingly confirmed by urban space, which clearly reveals many basic aspects of social space that are still hard to discern in villages. Urban space gathers crowds, products in the markets, acts and symbols. It concentrates all these, and accumulates them. To say 'urban space' is to say centre and centrality, and it does not matter whether these are actual or merely possible, saturated, broken up or under fire, for we are speaking here of a dialectical centrality.

It would thus be quite possible to elaborate on this form, to illuminate its structures (centre/periphery), its social functions, its relationship to labour (the various markets) and hence to production and reproduction, its connections with precapitalist and capitalist production relations, the roles of historic cities and of the modern urban fabric, and so on. One might also go into the dialectical processes bound up with this relationship between a form and its contents: the explosions, the saturation points, the challenges arising from internal contradictions, the assaults mounted by contents being pushed out towards the periphery, and so forth. In and of itself, social space does not have all of the characteristics of 'things' as opposed to creative activity. Social space *per se* is at once *work* and *product* – a materialization of 'social being'. In specific sets of circumstances, however, it may take on fetishized and autonomous characteristics of things (of commodities and money).

There is thus no lack of arguments for undertaking the ambitious project we have been discussing. A number of objections may also be reasonably raised, however – quite aside from those based on the very immensity of the task.

In the first place, the plan of *Capital* is not the only one Marx ever formulated. Its aims concern exposition rather than content; it envisages a strict formal structure, but one which impoverishes because of its reductionism. In the *Grundrisse* we find a different project, another plan and a more fruitful one. Whereas *Capital* stresses a homogenizing rationality founded on the quasi-'pure' form, that of (exchange) value, the *Grundrisse* insists at all levels on difference. Not that the *Grundrisse* leaves form out of the picture; rather, it goes from one content to the next and generates forms on the basis of these contents. Less rigour, less emphasis on logical consistency, and hence a less elaborate formaliz-ation or axiomatization – all leave the door open to more concrete themes, especially in connection with the (dialectical) relations between town and country, between natural reality and social reality. In the *Grundrisse* Marx takes all the historical mediations into consideration, including the village community, the family, and so on.[5] The 'world of the commodity' is less far removed from its historical context and practical conditions, matters which are only taken up in the concluding (and unfinished) portion of *Capital*.

Secondly, there have after all been some changes and new develop-ments in the last hundred years. Even if we want to keep Marx's concepts and categories (including the concept of production) in their central theoretical position, it is still necessary to incorporate a number of categories that Marx considered only at the end of his life. A case in point is the reproduction of the relations of production, which superim-poses itself upon the reproduction of the means of production, and upon the (quantitatively) expanded reproduction of products, but which remains distinct from these. When reproduction is treated as a concept, however, it brings other concepts in its wake: repetition, reproducibility, and so on. Such ideas had no more place in Marx's work than did the terms 'urban', 'everyday life' or 'space'.

If the *production of space* does indeed correspond to a leap forward in the productive forces (in technology, in knowledge, in the domination

[5] See my *La pensée marxiste et la ville* (Tournai: Casterman, 1972).

of nature), and if therefore this tendency, when pushed to its limit – or, better, when it has overcome its limits – must eventually give rise to a *new mode of production* which is neither state capitalism nor state socialism, but the collective management of space, the social management of nature, and the transcendence of the contradiction between nature and anti-nature, then clearly we cannot rely solely on the application of the 'classical' categories of Marxist thought.

Thirdly (though what I am about to say actually takes in and extends the first two points), another new development since Marx's time is the emergence of a plethora of disciplines known as 'social' or 'human' sciences. Their vicissitudes – for each has had its own particular ups and downs – have occasioned not a little anxious inquiry concerning disparities of development, crises, sudden expansions followed by equally sudden declines, and so on. The specialists and specialized institutions naturally seek to deny, combat or silence whatever is liable to damage their reputation, but their efforts in this direction have been largely in vain. Resounding failures and catastrophic collapses have been frequent. The early economists, for example, deluded themselves into thinking that they could safely ignore the Marxist injunctions to give critical thought priority over model-building, and to treat political economy as the science of poverty. Their consequent humiliation was an eminently public event, all their attempts to prevent this notwithstanding. As for linguistics, the illusions and the failure here could scarcely be more obvious, especially in view of the fact that, following the earlier examples of history and political economy, this specialization set itself up as the epitome of science – as the 'science of sciences', so to speak. In actuality linguistics can legitimately concern itself only with the deciphering of texts and messages, with coding and decoding. After all, 'man' does not live by words alone. In recent decades, linguistics has become a metalanguage, and an analysis of metalanguages; an analysis, consequently, of social repetitiveness, one which allows us – no more and no less – to apprehend the enormous redundancy of past writings and discourse.

Despite the uneven character and vicissitudes of their development, the existence of these sciences cannot be denied. In Marx's time, by contrast, they did not exist, or existed only in virtual or embryonic form; their degree of specialization was negligible and their future expansionist ambitions were as yet inconceivable.

These areas of specialized knowledge, at once isolated and imperialistic – the two are surely connected – have specific relationships with mental and social space. Some groups of scholars have simply sliced off their share, so to speak – staking out and enclosing their particular 'field'. Others, following the example of the mathematicians, have constructed a mental space so designed as to facilitate the interpretation, according to their particular principles, of theoretical and practical (social) history; in this way they have arrived at specific representations of space. Architecture offers plenty of instances of procedures of this kind, which are essentially circular in form. Architects have a trade. They raise the question of architecture's 'specificity' because they want to establish that trade's claim to legitimacy. Some of them then draw the conclusion that there are such things as 'architectural space' and 'architectural production' (specific, of course). Whereupon they close their case. This relationship between cutting-up and representation, as it refers to space, has already found its place in the order (and the disorder) of the connections we have been examining.

Sections and interpretations of this kind can be understood and taken up not as a function of some 'science of space', or of some totalizing concept of 'spatiality', but rather from the standpoint of *productive* activity. Specialists have already inventoried the objects in space, some of them cataloguing those that come from nature, others those that are produced. When knowledge *of* space (as a product, and not as an aggregate of objects produced) is substituted for knowledge of things *in* space, such enumerations and descriptions take on another meaning. It is possible to conceive of a 'political economy of space' which would go back to the old political economy and rescue it from bankruptcy, as it were, by offering it a new object: the production of space. If the critique of political economy (which was for Marx identical with knowledge of the economic realm) were then to be resumed, it would no doubt demonstrate how that political economy of space corresponded exactly to the self-presentation of space as the worldwide medium of the definitive installation of capitalism. A similar approach might well be adopted towards history, psychology, anthropology, and so on – perhaps even towards psychoanalysis.

This orientation calls for thoroughly clarified distinctions to be drawn between thought and discourse *in* space (i.e. in one particular space, dated and located), thought and discourse *about* space (i.e. restricted to words and signs, images and symbols), and thought *adequate to the understanding of* space (i.e. grounded in developed concepts). These distinctions are themselves founded on a more fundamental one: they presuppose careful critical attention, on the one hand, to the *materials* used (words, images, symbols, concepts), and, on the other hand, to the *matériel* used (collection procedures, tools for cutting-up and reassembling, etc.) – all this within the framework of the scientific division of labour.

The distinction between materials and *matériel*, though originally developed in other conceptual contexts, is in fact well worth borrowing for our purposes. Materials are indispensable and durable: stone, brick, cement and concrete, for example – or, in the musical sphere, scales, modes and tones. *Matériel*, by contrast, is quickly used up; it must be replaced often; it is comprised of tools and directions for their use; and its adaptative capability is limited: when new needs arise, new *matériel* must be invented to meet them. Instances of *matériel* in music would be the piano, the saxophone or the lute. In the construction industry, new techniques and equipment fall under this rubric. This distinction may achieve a certain 'operational' force inasmuch as it can be used to discriminate between what is ephemeral and what is more permanent: to decide what, in a particular scientific discipline, is worth preserving or reassigning to new tasks, and what deserves only to be rejected or relegated to a subsidiary role. For obsolete *matériel* can have only marginal applications; it often ends up, for example, in the realm of pedagogy.

Our re-evaluation of subdivisions and representations, along with their materials and *matériel*, need not be confined to the specialized disciplines we have been discussing. On the contrary, it should extend to philosophy, which after all does propose representations of space and time. Nor should a critique of philosophical ideologies be assumed to release us from the need to examine political ideologies in so far as they relate to space. And in point of fact such ideologies relate to space in a most significant way, because they intervene in space in the form of *strategies*. Their effectiveness in this role – and especially a new development, the fact that worldwide strategies are now seeking to generate a global space, their *own* space, and to set it up as an absolute – is another reason, and by no means an insignificant one, for developing a new concept of space.

VI

Reduction is a scientific procedure designed to deal with the complexity and chaos of brute observations. This kind of simplification is necessary at first, but it must be quickly followed by the gradual restoration of what has thus been temporarily set aside for the sake of analysis. Otherwise a methodological necessity may become a servitude, and the legitimate operation of reduction may be transformed into the abuse of *reductionism*. This is a danger that ever lies in wait for scientific endeavour. No method can obviate it, for it is latent in every method. Though indispensable, all reductive procedures are also traps.

Reductionism thus infiltrates science under the flag of science itself. Reduced models are constructed – models of society, of the city, of institutions, of the family, and so forth – and things are left at that. This is how social space comes to be reduced to mental space by means of a 'scientific' procedure whose scientific status is really nothing but a veil for ideology. Reductionists are unstinting in their praise for basic scientific method, but they transform this method first into a mere posture and then, in the name of the 'science of science' (epistemology), into a supposed absolute knowledge. Eventually, critical thought (where it is not proscribed by the orthodox) wakes up to the fact that systematic reduction and reductionism are part and parcel of a political practice. The state and political power seek to become, and indeed succeed in becoming, reducers of contradictions. In this sense reduction and reductionism appear as tools in the service of the state and of power: not as ideologies but as established knowledge; and not in the service of any specific state or government, but rather in the service of the state and power in general. Indeed, how could the state and political power reduce contradictions (i.e. incipient and renewed intrasocial conflicts) other than via the mediation of knowledge, and this by means of a strategy based on an admixture of science and ideology?

It is now generally acknowledged that not too long ago a functionalism held sway which was reductionistic with respect to the reality and comprehension of societies; such functional reductionism is readily subjected to criticism from all sides. What is not similarly acknowledged, and indeed passed over in silence, is that structuralism and formalism propose, after their fashion, equally reductive schemata. They are reductionist in that they give a privileged status to one concept – because they extrapolate; conversely, their reductionism encourages them to extrapolate. And, when the need to correct this error, or to compensate for it, makes itself felt, ideology stands ready to step into the breach with its verbiage (its 'ideological discourse', to use the jargon) and with its abuse of all signs whether verbal or not.

Reduction can reach very far indeed in its implications. It can 'descend' to the level of practice, for instance. Many people, members of a variety of groups and classes, suffer (albeit unevenly) the effects of a multiplicity of reductions bearing on their capacities, ideas, 'values' and, ultimately, on their possibilities, their space and their bodies. *Reduced models* constructed by one particular specialist or other are not always abstract in the sense of being 'empty' abstractions. Far from it, in fact: designed with a *reductive practice* in mind, they manage, with a little luck, to impose an order, and to constitute the elements of that order. Urbanism and architecture provide good examples of this. The working class, in particular, suffers the effects of such 'reduced models', including models of space, of consumption and of so-called culture.

Reductionism presses an exclusively analytic and non-critical knowledge, along with its attendant subdivisions and interpretations, into the service of power. As an ideology that does not speak its name, it successfully passes itself off as 'scientific' – and this despite the fact that it rides roughshod over established knowledge on the one hand and denies the possibility of *knowing* on the other. This is the scientific ideology *par excellence*, for the reductionist attitude may be actualized merely by passing from method to dogma, and thence to a homogenizing practice camouflaged as science.

At the outset, as I pointed out above, every scientific undertaking must proceed reductively. One of the misfortunes of the specialist is that he makes this methodological moment into a permanent niche for himself where he can curl up happily in the warm. Any specialist who clearly stakes out his 'field' may be sure that as long as he is prepared to work it a little he will be able to grow something there. The field he selects, and what he 'cultivates', are determined by the local conditions in his speciality and by that speciality's position in the knowledge market. But these are precisely the things that the specialist does not want to know about. As for the reduction upon which his procedures are founded, he adopts a posture that serves in its own way to justify it: a posture of denial.

Now, it is hard to think of any specialized discipline that is not involved, immediately or mediately, with space.

In the first place, as we have already learnt, each specialization stakes out its own particular mental and social space, defining it in a somewhat arbitrary manner, carving it out from the whole constituted by 'nature/ society', and at the same time concealing a portion of the activity of segmentation and rearrangement involved in this procedure (the sectioning-off of a 'field', the assembling of statements and reduced models relating to that field, and the shift from mental to social). All of which necessarily calls in addition for the adduction of propositions justifying – and hence interpreting – that activity.

Secondly, all specialists must work within the confines of systems for naming and classifying things found in space. The verification, description and classification of objects in space may be viewed as the 'positive' activity of a particular specialization – of geography, say, or anthropology, or sociology. At best (or at worst) a given discipline – as for example political science or 'systems analysis' – may concern itself with statements *about* space.

Lastly, specialists may be counted on to oppose a reduced model of the knowledge of space (based either on the mere noting of objects in space or else on propositions concerning – and segmenting – space) to any overall theory of (social) space. For them this stance has the added advantage of eliminating time by reducing it to a mere 'variable'.

We should not, therefore, be particularly surprised if the concept of the production of space, and the theory associated with it, were challenged by specialists who view social space through the optic of their methodology and their reductionistic schemata. This is all the more likely in view of the fact that both concept and theory threaten interdisciplinary boundaries themselves: they threaten, in other words, to alter, if not to erase, the specialists' carefully drawn property lines.

Perhaps I may be permitted at this point to imagine a dialogue with an interlocutor at once fictitious (because indeed imaginary) and real (because his objections are real enough).

'I am not convinced by your arguments. You talk of "producing space". What an absolutely unintelligible phrase! Even to speak of a *concept* in this connection would be to grant you far too much. No, there are only two possibilities here. Either space is part of nature or it is a concept. If it is part of nature, human – or "social" – activity marks it, invests it and modifies its geographical and ecological characteristics; the role of knowledge, on this reading, would be limited to the description of these changes. If space is a concept, it is as such already a part of knowledge and of mental activity, as in mathematics for example, and the job of scientific thought is to explore, elaborate upon and develop it. In neither case is there such a thing as the production of space.'

'Just a moment. The separations you are taking for granted between nature and knowledge and nature and culture are simply not valid. They are no more valid than the widely accepted "mind–matter" split. These distinctions are simply no improvement on their equally unacceptable opposite – namely, confusion. The fact is that technological activity and the scientific approach are not satisfied with simply modifying nature. They seek to master it, and in the process they tend to destroy it; and, before destroying it, they misinterpret it. This process began with the invention of tools.'

'So now you are going back to the Stone Age! Isn't that a little early?'

'Not at all. The beginning was the first premeditated act of murder; the first tool and the first weapon – both of which went hand in hand with the advent of language.'

'What you seem to be saying is that humankind emerges from nature. It can thus only understand nature from without – and it only gets to understand it by destroying it.'

'Well, if one accepts the generalization "humankind" for the sake of the argument, then, yes, humankind is born in nature, emerges from nature and then turns against nature with the unfortunate results that we are now witnessing.'

'Would you say that this ravaging of nature is attributable to capitalism?'

'To a large degree, yes. But I would add the rider that capitalism and the bourgeoisie have a broad back. It is easy to attribute a multitude of misdeeds to them without addressing the question of how they themselves came into being.'

'Surely the answer is to be found in mankind itself, in human nature?'

'No. In the nature of *Western* man perhaps.'

'You mean to say that you would blame the whole history of the West, its rationalism, its Logos, its very language?'

'It is the West that is responsible for the transgression of nature. It would certainly be interesting to know how and why this has come about, but those questions are strictly secondary. The simple fact is that the West has broken the bounds. "O felix culpa!" a theologian might say. And, indeed, the West is thus responsible for what Hegel calls the power of the negative, for violence, terror and permanent aggression directed against life. It has generalized and globalized violence – and forged the global level itself through that violence. Space as locus of production, as itself product and production, is both the weapon and the sign of this struggle. If it is to be carried through to the end – there is in any case no way of turning back – this gigantic task now calls for the immediate production or creation of something other than nature: a second, different or new nature, so to speak. This means the production of space, urban space, both as a product and as a work, in the sense in which art created works. If this project fails, the failure will be total, and the consequences of that are impossible to foresee.'

VII

Every social space is the outcome of a process with many aspects and many contributing currents, signifying and non-signifying, perceived and directly experienced, practical and theoretical. In short, every social space has a history, one invariably grounded in nature, in natural conditions that are at once primordial and unique in the sense that they are always and everywhere endowed with specific characteristics (site, climate, etc.).

When the history of a particular space is treated as such, the relationship of that space to the time which gave rise to it takes on an aspect that differs sharply from the picture generally accepted by historians. Traditional historiography assumes that thought can perform cross-sections upon time, arresting its flow without too much difficulty; its analyses thus tend to fragment and segment temporality. In the history of space as such, on the other hand, the historical and diachronic realms and the generative past are forever leaving their inscriptions upon the writing-tablet, so to speak, of space. The uncertain traces left by events are not the only marks on (or in) space: society in its actuality also deposits its script, the result and product of social activities. Time has more than one writing-system. The space engendered by time is always actual and synchronic, and it always presents itself as of a piece; its component parts are bound together by internal links and connections themselves produced by time.

Let us consider a primary aspect, the simplest perhaps, of the history of space as it proceeds from nature to abstraction. Imagine a time when each people that had managed to measure space had its own units of measurement, usually borrowed from the parts of the body: thumb's breadths, cubits, feet, palms, and so on. The spaces of one group, like their measures of duration, must have been unfathomable to all others. A mutual interference occurs here between natural peculiarities of space and the peculiar nature of a given human group. But how extraordinary to think that the body should have been part and parcel of so idiosyncratically gauged a space. The body's relationship to space, a social relationship of an importance quite misapprehended in later times, still retained in those early days an immediacy which would subsequently degenerate and be lost: space, along with the way it was measured and spoken of, still held up to all the members of a society an image and a living reflection of their own bodies.

The adoption of another people's gods always entails the adoption of their space and system of measurement. Thus the erection of the Pantheon in Rome pointed not only to a comprehension of conquered gods but also to a comprehension of spaces now subordinate to the master space, as it were, of the Empire and the world.

The status of space and its measurement has changed only very slowly; indeed the process is still far from complete. Even in France, cradle of the metric system, odd customary measures are still used when it comes, for example, to garment or shoe sizes. As every French schoolchild knows, a revolution occurred with the imposition of the abstract generality of the decimal system, yet we continue to make use of the duodecimal system in dealing with time, cycles, graphs, circumferences, spheres, and so on. Fluctuations in the use of measures, and thus in representations of space, parallel general history and indicate the direction it has taken – to wit, its trend towards the quantitative, towards homogeneity and towards the elimination of the body, which has had to seek refuge in art.

VIII

As a way of approaching the history of space in a more concrete fashion, let us now for a moment examine the ideas of the nation and of nationalism. How is the nation to be defined? Some people – most, in fact – define it as a sort of substance which has sprung up from nature (or more specifically from a territory with 'natural' borders) and grown to maturity within historical time. The nation is thus endowed with a consistent 'reality' which is perhaps more definitive than well defined. This thesis, because it justifies both the bourgeoisie's national state and its general attitude, certainly suits that class's purposes when it promotes patriotism and even absolute nationalism as 'natural' and hence eternal truths. Under the influence of Stalinism, Marxist thought has been known to endorse the same or a very similar position (with a dose of historicism thrown in for good measure). There are other theorists, however, who maintain that the nation and nationalism are merely ideological constructs. Rather than a 'substantial reality' or a body corporate, the nation is on this view scarcely more than a fiction projected by the bourgeoisie onto its own historical conditions and origins, to begin with as a way of magnifying these in imaginary fashion, and later on as a way of masking class contradictions and seducing the working class into an illusory national solidarity. It is easy, on the basis of this hypothesis, to reduce national and regional questions to linguistic and cultural ones – that is to say, to matters of secondary importance. We are thus led to a kind of abstract internationalism.

Both of these approaches to the question of the nation, the argument from nature and the argument from ideology, leave space out of the picture. The concepts used in both cases are developed in a mental space which thought eventually identifies with real space, with the space of social and political practice, even though the latter is really no more than a representation of the former, a representation itself subordinate to a specific representation of historical time.

When considered in relationship to space, the nation may be seen to have two moments or conditions. First, nationhood implies the existence of a *market* gradually built up over a historical period of varying length. Such a market is a complex ensemble of commercial relations and communication networks. It subordinates local or regional markets to the national one, and thus must have a hierarchy of levels. The social, economic and political development of a national market has been somewhat different in character in places where the towns came very early on to dominate the country, as compared with places where the towns grew up on a pre-existing peasant, rural and feudal foundation. The outcome, however, is much the same everywhere: a focused space embodying a hierarchy of centres (commercial centres for the most part, but also religious ones, 'cultural' ones, and so on) and a main centre – i.e. the national capital.

Secondly, nationhood implies *violence* – the violence of a military state, be it feudal, bourgeois, imperialist, or some other variety. It implies, in other words, a political power controlling and exploiting the resources of the market or the growth of the productive forces in order to maintain and further its rule.

We have yet to ascertain the exact relationship between 'spontaneous' economic growth on the one hand and violence on the other, as well as their precise respective effects, but our hypothesis does affirm that these two 'moments' indeed combine forces and *produce a space*: the space of the nation state. Such a state cannot therefore be defined in terms of a substantive 'legal person' or in terms of a pure ideological fiction or 'specular centre'. Yet to be evaluated, too, are the connections between national spaces of this kind and the world market, imperialism and its strategies, and the operational spheres of multinational corporations.

Let us now turn to a very general view of our subject. Producing an object invariably involves the modification of a raw material by the application to it of an appropriate knowledge, a technical procedure, an effort and a repeated gesture (labour). The raw material comes, whether directly or indirectly, from nature: wood, wool, cotton, silk, stone, metal. Over the centuries, more and more sophisticated – and hence less and less 'natural' – materials have replaced substances obtained directly from nature. The importance of technical and scientific mediation has increased constantly. One only has to think of concrete, of man-made fibres, or of plastics. It is true, none the less, that many of the earliest materials, such as wool, cotton, brick and stone, are still with us.

The object produced often bears traces of the *matériel* and time that have gone into its production – clues to the operations that have modified the raw material used. This makes it possible for us to reconstruct those operations. The fact remains, however, that productive operations tend in the main to cover their tracks; some even have this as their prime goal: polishing, staining, facing, plastering, and so on. When construction is completed, the scaffolding is taken down; likewise, the fate of an author's rough draft is to be torn up and tossed away, while for a painter the distinction between a study and a painting is a very clear one. It is for reasons such as these that products, and even works, are further characterized by their tendency to detach themselves from productive labour. So much so, in fact, that productive labour is sometimes forgotten altogether, and it is this 'forgetfulness' – or, as a philosopher might say, this mystification – that makes possible the fetishism of commodities: the fact that commodities imply certain social relationships whose misapprehension they also ensure.

It is never easy to get back from the object (product or work) to the activity that produced and/or created it. It is the only way, however, to illuminate the object's nature, or, if you will, the object's relationship to nature, and reconstitute the process of its genesis and the development of its meaning. All other ways of proceeding can succeed only in constructing an abstract object – a model. It is not sufficient, in any case, merely to bring out an object's structure and to understand that structure: we need to generate an object in its entirety – that is, to reproduce, by and in thought, that object's forms, structures and functions.

How does one (where 'one' designates any 'subject') perceive a picture, a landscape or a monument? Perception naturally depends on the 'subject': a peasant does not perceive 'his' landscape in the same way as a town-dweller strolling through it. Take the case of a cultured art-lover looking at a painting. His eye is neither that of a professional nor that of an uncultivated person. He considers first one and then another of the objects depicted in the painting; he starts out by apprehending the relationships between these objects, and allows himself to experience the effect or effects intended by the painter. From this he derives a certain pleasure – assuming that the painting in question is of the type supposed to give pleasure to eye or mind. But our amateur is also aware that the picture is framed, and that the internal relations between colours and forms are governed by the work as a whole. He thus moves from consideration of the objects in the painting to consideration of the picture as an object, from what he has perceived in the pictural space to what he can comprehend about that space. He thus comes to sense or understand various 'effects', including some which have not been intentionally sought by the painter. He deciphers the picture and finds surprises in it, but always within the limits of its formal framework, and in the ratios or proportions dictated by that framework. His discoveries occur on the plane of (pictural) *space*. At this point in his aesthetic inquiry, the 'subject' asks a number of questions: he seeks to solve one problem in particular, that of the relationship between effects of meaning that have been sought by means of technique and those which have come about independently of the artist's intentions (some of which depend on him, the 'looker'). In this way he begins to trace a path back from the effects he has experienced to the meaning-producing activity that gave rise to them; his aim is to rediscover that activity and to try and identify (perhaps illusorily) with it. His 'aesthetic' perception thus operates, as one would expect, on several levels.

It is not hard to see that this paradigm case is paralleled by a trend in the history of philosophy that was taken up and advanced by Marx and by Marxist thought. The post-Socratic Greek philosophers analysed knowledge as social practice; reflecting the state of understanding itself, they inventoried the ways in which known *objects* were apprehended. The high-point of this theoretical work was Aristotelian teaching on *discourse* (*Logos*), and on the *categories* as at once elements of discourse and means for apprehending (or classifying) objects. Much later, in Europe, Cartesian philosophy refined and modified the definition of 'Logos'. Philosophers were now supposed to question the Logos – and put it into question: to demand its credentials, its pedigree, its certificate of origin, its citizenship papers. With Descartes, therefore, philosophy shifted the position of both questions and answers. It changed its focus, moving from 'thought thought' to 'thinking thought', from the objects

of thought to the act of thinking, from a discourse upon the known to
the operation of knowing. The result was a new 'problematic' – and
new difficulties.

Marx recommenced this Cartesian revolution, perfecting and broadening it in the process. His concern was no longer merely with works
generated by knowledge, but now also with *things* in industrial practice.
Following Hegel and the British economists, he worked his way back
from the results of productive activity to productive activity itself. Marx
concluded that any reality presenting itself in space can be expounded
and explained in terms of its genesis in time. But any activity developed
over (historical) time engenders (produces) a space, and can only attain
practical 'reality' or concrete existence within that space. This view of
matters emerged in Marx's thinking only in an ill-defined form; it was
in fact inherited by him in that form from Hegel. It applies to any
landscape, to any monument, and to any spatial ensemble (so long as
it is not 'given' in nature), as it does to any picture, work or product.
Once deciphered, a landscape or a monument refers us back to a creative
capacity and to a signifying process. This capacity may in principle be
dated, for it is a historical fact. Not, however, in the sense that an event
can be dated: we are not referring to the exact date of a monument's
inauguration, for example, or to the day that the *command* that it be
erected was issued by some notability. Nor is it a matter of a date in
the institutional sense of the word: the moment when a particular social
organization acceded to a pressing *demand* that it embody itself in a
particular edifice – the judiciary in a courthouse, for instance, or the
Church in a cathedral. Rather, the creative capacity in question here is
invariably that of a community or collectivity, of a group, of a fraction
of a class in action, or of an 'agent' (i.e. 'one who acts'). Even though
'commanding' and 'demanding' may be the functions of distinct groups,
no individual or entity may be considered ultimately responsible for
production itself: such responsibility may be attributed only to a social
reality capable of investing a space – capable, given the resources
(productive forces, technology and knowledge, means of labour, etc.),
of producing that space. Manifestly, if a countryside exists, there must
have been peasants to give it form, and hence too communities (villages),
whether autonomous or subject to a higher (political) power. Similarly,
the existence of a monument implies its construction by an urban group
which may also be either free or subordinate to a (political) authority.
It is certainly necessary to describe such states of affairs, but it is hardly
sufficient. It would be utterly inadequate from the standpoint of an
understanding of space merely to describe first rural landscapes, then

industrial landscapes, and finally urban spatiality, for this would simply leave all transitions out of the picture. Inasmuch as the quest for the relevant productive capacity or creative process leads us in many cases to political power, there arises the question of how such power is exercised. Does it merely command, or does it 'demand' also? What is the nature of its relationship to the groups subordinate to it, which are themselves 'demanders', sometimes also 'commanders', and invariably 'participants'? This is a historical problem – that of all cities, all monuments, all landscapes. The analysis of any space brings us up against the dialectical relationship between demand and command, along with its attendant questions: 'Who?', 'For whom?', 'By whose agency?', 'Why and how?' If and when this dialectical (and hence conflictual) relationship ceases to obtain – if demand were to outlive command, or vice versa – the history of space must come to an end. The same goes for the capacity to create, without a doubt. The production of space might proceed, but solely according to the dictates of Power: production without creation – mere reproduction. But is it really possible for us to envision an end to demand? Suffice it to say that silence is not the same thing as quietus.

What we are concerned with, then, is the long *history of space*, even though space is neither a 'subject' nor an 'object' but rather a social reality – that is to say, a set of relations and forms. This history is to be distinguished from an inventory of things *in space* (or what has recently been called material culture or civilization), as also from ideas and discourse *about space*. It must account for both representational spaces and representations of space, but above all for their interrelationships and their links with social practice. The history of space thus has its place between anthropology and political economy. The nomenclature, description and classification of objects certainly has a contribution to make to traditional history, especially when the historian is concerned with the ordinary objects of daily life, with types of food, kitchen utensils and the preparation and presentation of meals, with clothing, or with the building of houses and the materials and *matériel* it calls for. But everyday life also figures in representational spaces – or perhaps it would be more accurate to say that it forms such spaces. As for representations of space (and of time), they are part of the history of ideologies, provided that the concept of ideology is not restricted, as it too often is, to the ideologies of the philosophers and of the ruling classes – or, in other words, to the 'noble' ideas of philosophy, religion and ethics. A history of space would explain the development, and hence the temporal conditions, of those realities which some geographers call 'networks' and which are subordinated to the frameworks of politics. The history of space does not have to choose between 'processes' and 'structures', change and invariability, events and institutions. Its periodizations, moreover, will differ from generally accepted ones. Naturally, the history of space should not be distanced in any way from the history of time (a history clearly distinct from all philosophical theories of time in general). The departure point for this history of space is not to be found in geographical descriptions of natural space, but rather in the study of natural rhythms, and of the modification of those rhythms and their inscription in space by means of human actions, especially work-related actions. It begins, then, with the spatio-temporal rhythms of nature as transformed by a social practice.

The first determinants to consider will be anthropological ones, necessarily bound up with the elementary forms of the appropriation of nature: numbers, oppositions and symmetries, images of the world, myths.[6] In dealing with these elaborated forms, it is often hard to separate knowledge from symbolism, practice from theory, or denotation from connotation (in the rhetorical sense); the same goes for the distinctions between spatial arrangements (subdivision, spacing) and spatial interpretations (representations of space), and between the activities of partial groups (family, tribe, etc.) and those of global societies. At the most primitive level, behind or beneath these elaborate forms, lie the very earliest demarcations and orienting markers of hunters, herders and nomads, which would eventually be memorized, designated and invested with symbolism.

Thus mental and social activity impose their own meshwork upon nature's space, upon the Heraclitean flux of spontaneous phenomena, upon that chaos which precedes the advent of the body; they set up an order which, as we shall see, coincides, but *only up to a point*, with the order of words.

Traversed now by pathways and patterned by networks, natural space changes: one might say that practical activity writes upon nature, albeit in a scrawling hand, and that this writing implies a particular represen-

---

[6] As representative examples of a vast literature, see Viviana Pâques, *L'arbre cosmique dans la pensée populaire et dans la vie quotidienne du Nord-Ouest africain* (Paris: Institut d'Ethnologie du Muséum National d'Histoire Naturelle, 1964); Leo Frobenius, *Mythologie de l'Atlantide*, tr. from the German (Paris: Payot, 1949); Georges Balandier, *La vie quotidienne au royaume de Kongo du XVIᶜ au XVIIIᶜ siècle* (Paris: Hachette, 1965); Luc de Heusch, 'Structure et praxis sociales chez les Lele du Kasai', *L'homme: revue française d'anthropologie*, 4, no. 3 (Sep.–Dec. 1964), pp. 87–109. See also A. P. Logopoulos et al., 'Semeiological Analysis of the Traditional African Settlement', *Ekistics*, Feb. 1972.

tation of space. Places are marked, noted, named. Between them, within the 'holes in the net', are blank or marginal spaces. Besides *Holzwege* or woodland paths, there are paths through fields and pastures. Paths are more important than the traffic they bear, because they are what endures in the form of the reticular patterns left by animals, both wild and domestic, and by people (in and around the houses of village or small town, as in the town's immediate environs). Always distinct and clearly indicated, such traces embody the 'values' assigned to particular routes: danger, safety, waiting, promise. This graphic aspect, which was obviously not apparent to the original 'actors' but which becomes quite clear with the aid of modern-day cartography, has more in common with a spider's web than with a drawing or plan. Could it be called a text, or a message? Possibly, but the analogy would serve no particularly useful purpose, and it would make more sense to speak of texture rather than of texts in this connection. Similarly, it is helpful to think of architectures as 'archi-textures', to treat each monument or building, viewed in its surroundings and context, in the populated area and associated networks in which it is set down, as part of a particular production of space. Whether this approach can help clarify spatial practice is a question to which we shall be returning.

Time and space are not separable within a texture so conceived: space implies time, and vice versa. These networks are not closed, but open on all sides to the strange and the foreign, to the threatening and the propitious, to friend and foe. As a matter of fact, the abstract distinction between open and closed does not really apply here.

What modes of existence do these paths assume at those times when they are not being actualized through practice, when they enter into representational spaces? Are they perceived as lying within nature or as outside it? The answer is neither, for at such times people animate these paths and roads, networks and itineraries, through accounts of mythical 'presences', genies and good or evil spirits, which are conceived of as having a concrete existence. There is doubtless no such thing as a myth or symbol unassociated with a mythical or symbolic space which is *also* determined by practice.

It is certainly not impossible, moreover, that such anthropological determinants, carried down through the centuries by a particular group, perhaps abandoned only to be taken up once more, displaced or trans-ferred, should have survived into the present. On the other hand, careful investigation is called for before any conclusions can possibly be drawn about structural invariability or patterns of repetition and reproduction.

Let us turn with this in mind to the case of Florence.[7] In 1172 the commune of Florence reorganized its urban space in response to the growth of the town, its traffic and its jurisdiction. This was an undertaking of global intent, not a matter of separate architectural projects each having its own repercussions on the city; it included a town square, wharves, bridges and roads. The historian can fairly easily trace the interplay of command and demand in this instance. The 'demanders' were those people who wished to benefit from the protections and advantages, including an improved enceinte, that the city could vouchsafe them. The command aspect stemmed from an ambitious authority, with the wherewithal to back up its ambitions. The Roman walls were abandoned, and the four existing city gates were replaced by six main gates and four secondary ones on the right bank of the Arno, and three more in the Oltrarno, which was now incorporated into the city. The urban space thus produced had the form of a symbolic flower, the *rose des vents* or compass-card. Its configuration was thus in accord with an *imago mundi*, but the historian of space ought not to attribute the same degree of importance to this representational space, which originated in a far distant and far different place, as he does to the upheavals which were simultaneously transforming the *contado* or Tuscan countryside and its relationship to its centre, namely Florence, giving rise in the process to a new representation of space. The fact is that what was anthropologically essential in ancient times can become purely tangential in the course of history. Anthropological factors enter history as *material*, apt to be treated variously according to the circumstances, conjunctures, available resources and *matériel* used.[8] The process of historical change, which entails all kinds of displacements, substitutions and transfers, subordinates both materials and *matériel*. In Tuscany we have a period of transition from a representational space (an image of the world) to a representation of space, namely perspective. This allows us to date an important event in the history under consideration.

The history of space will begin at the point where anthropological

[7] Cf. J. Renouard, 'Les villes d'Italie' (duplicated course notes), fascicle 8, pp. 20ff.

[8] See above, pp. 77 ff., my remarks on the space of Tuscany and its repercussions for the art and science of the Quattrocento. We shall return to these issues later (see below, pp. 257 ff.) in connection with Erwin Panofsky's *Gothic Architecture and Scholasticism* and Pierre Francastel's *Art et technique au XIXe et XXe siècles*. So long as the focus is on architecture, the best discussion is still E. E. Viollet-le-Duc, *Entretiens sur l'architecture*, 4 vols (Paris: A. Morel, 1863–72); Eng. tr. by Benjamin Bucknall: *Lectures on Architecture*, 2 vols (Boston, Mass.: Ticknor, 1889).

factors lose their supremacy and end with the advent of a production of space which is expressly industrial in nature – a space in which reproducibility, repetition and the reproduction of social relationships are deliberately given precedence over works, over natural reproduction, over nature itself and over natural time. This area of study overlaps with no other. It is clearly circumscribed, for this history has a beginning and an end – a prehistory and a 'post-history'. In prehistory, nature dominates social space; in post-history, a localized nature recedes. Thus demarcated, the history of space is indispensable. Neither its beginning nor its end can be dated in the sense in which traditional historiography dates events. The beginning alone took up a period traces of which remain even now in our houses, villages and towns. In the course of this process, which may be properly referred to as historical, certain abstract relations were established: exchange value became general, first thanks to silver and gold (i.e. their functions), then thanks to capital. These abstractions, which are social relations implying forms, become tangible in two ways. In the first place, the instrument and general equivalent of exchange value, namely money, takes on concrete form in coins, in 'pieces' of money. Secondly, the commercial relations which the use of money presupposes and induces attain social existence only once they are projected onto the terrain in the shape of relational networks (communications, markets) and of hierarchically organized centres (towns). It must be presumed that in each period a certain balance is established between the centres (i.e. the functioning of each one) and the whole. One might therefore quite reasonably speak here of 'systems' (urban, commercial, etc.), but this is really only a minor aspect, an implication and consequence of that fundamental activity which is the production of space.

With the twentieth century, we are generally supposed to have entered the modern era. Despite – and because of – their familiarity, however, such crude terms as 'century', 'modern' and 'modernity' serve to conceal more than one paradox; these notions are in fact in urgent need of analysis and refinement. So far as space is concerned, decisive changes occurred at this juncture which are effectively obscured by invariant, surviving or stagnant elements, especially on the plane of represen-tational space. Consider the house, the dwelling. In the cities – and even more so in the 'urban fabric' which proliferates around the cities pre-cisely because of their disintegration – the House has a merely historico-poetic reality rooted in folklore, or (to put the best face on it) in ethnology. This *memory*, however, has an obsessive quality: it persists in art, poetry, drama and philosophy. What is more, it runs through the terrible urban reality which the twentieth century has instituted, embellishing it with a nostalgic aura while also suffusing the work of its critics. Thus both Heidegger's and Bachelard's writings – the importance and influence of which are beyond question – deal with this idea in a most emotional and indeed moving way. The dwelling passes everywhere for a special, still sacred, quasi-religious and in fact almost absolute space. With his 'poetics of space' and 'topophilia', Bachelard links representational spaces, which he travels through as he dreams (and which he distinguishes from representations of space, as developed by science), with this intimate and absolute space.[9] The contents of the House have an almost ontological dignity in Bachelard: drawers, chests and cabinets are not far removed from their natural analogues, as perceived by the philosopher–poet, namely the basic figures of nest, shell, corner, roundness, and so on. In the background, so to speak, stands Nature – maternal if not uterine. The House is as much cosmic as it is human. From cellar to attic, from foundations to roof, it has a density at once dreamy and rational, earthly and celestial. The relationship between Home and Ego, meanwhile, borders on identity. The shell, a secret and directly experienced space, for Bachelard epitomizes the virtues of human 'space'.

As for Heidegger's ontology – his notion of building as close to thinking, and his scheme according to which the dwelling stands opposed to a wandering existence but is perhaps destined one day to ally with it in order to welcome in Being – this ontology refers to things and non-things which are also far from us now precisely inasmuch as they are close to nature: the jug,[10] the peasant house of the Black Forest,[11] the Greek temple.[12] And yet space – the woods, the track – is nothing more and nothing other than 'being-there', than beings, than *Dasein*. And, even if Heidegger asks questions about its origin, even if he poses 'historical' questions in this connection, there can be no doubt about the main thrust of his thinking here: time counts for more than space; Being has a history, and history is nothing but the History of Being.

[9] See Gaston Bachelard, *La poétique de l'espace* (Paris: Presses Universitaires de France, 1957), p. 19. Eng. tr. by Maris Jolas: *The Poetics of Space* (Boston, Mass.: Beacon Press, 1969), p. xxxiv.

[10] See Martin Heidegger, 'The Thing', in *Poetry, Language, Thought*, tr. Albert Hofstadter (New York: Harper and Row, 1971), pp. 166ff. [Original: 'Das Ding', in *Vorträge und Aufsätze* (Pfullingen: Neske, 1954).]

[11] See Martin Heidegger, 'Building Dwelling Thinking', in *Poetry, Language, Thought*, p. 160. [Original: 'Bauen Wohnen Denken', in *Vorträge und Aufsätze*.]

[12] See the discussion in Martin Heidegger, *Holzwege* (Frankfurt a.M.: Klostermann, 1950).

This leads him to a restricted and restrictive conception of production, which he envisages as a causing-to-appear, a process of emergence which brings a thing forth as a thing now present amidst other already-present things. Such quasi-tautological propositions add little to Heidegger's admirable if enigmatic formulation according to which 'Dwelling is the basic character of Being in keeping with which mortals exist.'[13] Language for Heidegger, meantime, is simply the dwelling of Being.

This obsession with absolute space presents obstacles on every side to the kind of history that we have been discussing (the history of space / the space of history; representations of space / representational space). It pushes us back towards a purely descriptive understanding, for it stands opposed to any analytic approach and even more to any global account of the generative process in which we are interested. More than one specific and partial discipline has sought to defend this stance, notably anthropology (whose aims may readily be gauged from the qualifiers so often assigned to it: cultural, structural, etc.). It is from motives of this sort that anthropology lays hold of notions derived from the study of village life (usually the Bororo or Dogon village, but occasionally the Provençal or Alsatian one), or from the consideration of traditional dwellings, and, by transposing and/or extrapolating them, applies these notions to the modern world.

How is it that such notions can be transferred in this way and still retain any meaning at all? There are a number of reasons, but the principal one is nostalgia. Consider the number of people, particularly young people, who flee the modern world, the difficult life of the cities, and seek refuge in the country, in folk traditions, in arts and crafts or in anachronistic small-scale farming. Or the number of tourists who escape into an elitist (or would-be elitist) existence in underdeveloped countries, including those bordering the Mediterranean. Mass migrations of tourist hordes into rustic or urban areas which their descent only helps to destroy (woe unto Venice and Florence!) are a manifestation of a major spatial contradiction of modernity: here we see space being consumed in both the economic and the literal senses of the word.

The modern world's brutal liquidation of history and of the past proceeds in a very uneven manner. In some cases entire countries – certain Islamic countries, for example – are seeking to slow down industrialization so as to preserve their traditional homes, customs and representational spaces from the buffeting of industrial space and industrial representations of space. There are other – very modern – nations

<hr/>

[13] Heidegger, 'Building Dwelling Thinking', in *Poetry, Language, Thought*, p. 160.

which also try to maintain their living-arrangements and spaces unchanged, along with the customs and representations which go along with them. In Japan, for instance, which is a hyper-industrialized and hyper-urbanized nation, traditional living-quarters, daily life, and representational spaces survive intact – and this not in any merely folkloric sense, not as relics, not as stage management for tourists, not as consumption of the cultural past, but indeed as immediate practical 'reality'. This intrigues visitors, frustrates Japanese modernizers and technocrats, and delights humanists. There is an echo here, albeit a distant one, of the West's infatuation with village life and rustic homesteads.

This kind of perseveration is what makes Amos Rapoport's book on the 'anthropology of the home' so interesting.[14] The traditional peasant house of the Périgord is indeed just as worthy of study as those anthropological *loci classici*, the Eskimo's igloo and the Kenyan's hut. The limitations of anthropology are nonetheless on display here, and indeed they leap off the page when the author seeks to establish the general validity of reductionistic schemata based on a binary opposition – i.e. does the dwelling strengthen or does it reduce domesticity? – and goes so far as to assert that French people always (!) entertain in cafés rather than at home.[15]

Much as they might like to, anthropologists cannot hide the fact that the space and tendencies of modernity (i.e. of modern capitalism) will never be discovered either in Kenya or among French or any other peasants. To put studies such as these forward as of great importance in this connection is to avoid reality, to sabotage the search for knowledge, and to turn one's back on the actual 'problematic' of space. If we are to come to grips with this 'problematic', instead of turning to ethnology, ethnography or anthropology we must address our attention to the 'modern' world itself, with its dual aspect – capitalism, modernity – which makes it so hard to discern clearly.

The raw material of the production of space is not, as in the case of particular objects, a particular material: it is rather nature itself, nature transformed into a product, rudely manipulated, now threatened in its very existence, probably ruined and certainly – and most paradoxically – *localized*.

It might be asked at this juncture if there is any way of dating what might be called the moment of emergence of an awareness of space and its production: when and where, why and how, did a neglected knowl-

---

[14] *House Form and Culture* (Englewood Cliffs, N.J.: Prentice-Hall, 1969).
[15] Ibid., p. 69.

edge and a misconstrued reality begin to be recognized? It so happens that this emergence can indeed be fixed: it is to be found in the 'historic' role of the Bauhaus. Our critical analysis will touch on this movement at several points. For the Bauhaus did more than locate space in its real context or supply a new perspective on it: it developed a new conception, a global concept, of space. At that time, around 1920, just after the First World War, a link was discovered in the advanced countries (France, Germany, Russia, the United States), a link which had already been dealt with on the practical plane but which had not yet been rationally articulated: that between industrialization and urbanization, between workplaces and dwelling-places. No sooner had this link been incorporated into theoretical thought than it turned into a project, even into a programme. The curious thing is that this 'programmatic' stance was looked upon at the time as both rational and revolutionary, although in reality it was tailor-made for the state – whether of the state-capitalist or the state-socialist variety. Later, of course, this would become obvious – a truism. For Gropius or for Le Corbusier, the programme boiled down to the production of space. As Paul Klee put it, artists – painters, sculptors or architects – do not show space, they create it. The Bauhaus people understood that things could not be created independently of each other in space, whether movable (furniture) or fixed (buildings), without taking into account their interrelationships and their relationship to the whole. It was impossible simply to accumulate them as a mass, aggregate or collection of items. In the context of the productive forces, the technological means and the specific problems of the modern world, things and objects could now be produced in their relationships, along with their relationships. Formerly, artistic ensembles – monuments, towns, furnishings – had been created by a variety of artists according to subjective criteria: the taste of princes, the intelligence of rich patrons or the genius of the artists themselves. Architects had thus built palaces designed to house specific objects ('furniture') associated with an aristocratic mode of life, and, alongside them, squares for the people and monuments for social institutions. The resulting whole might constitute a space with a particular style, often even a dazzling style – but it was still a space never rationally defined which came into being and disappeared for no clear reason. As he considered the past and viewed it in the light of the present, Gropius sensed that henceforward social practice was destined to change. The production of spatial ensembles as such corresponded to the capacity of the productive forces, and hence to a specific rationality. It was thus no longer a question of introducing forms, functions or structures in isolation, but rather one of mastering global space by bringing forms, functions and structures together in accordance with a unitary conception. This insight confirmed after its fashion an idea of Marx's, the idea that industry has the power to open before our eyes the book of the creative capacities of 'man' (i.e. of social being).

The Bauhaus group, as artists associated in order to advance the total project of a total art, discovered, along with Klee,[16] that an observer could move around any object in social space – including such objects as houses, public buildings and palaces – and in so doing go beyond scrutinizing or studying it under a single or special aspect. Space opened up to perception, to conceptualization, just as it did to practical action. And the artist passed from objects in space to the concept of space itself. Avant-garde painters of the same period reached very similar conclusions: all aspects of an object could be considered simultaneously, and this simultaneity preserved and summarized a temporal sequence. This had several consequences.

1 A *new consciousness of space* emerged whereby space (an object in its surroundings) was explored, sometimes by deliberately reducing it to its outline or plan and to the flat surface of the canvas, and sometimes, by contrast, by breaking up and rotating planes, so as to reconstitute depth of space in the picture plane. This gave rise to a very specific dialectic.

2 *The façade* – as face directed towards the observer and as privileged side or aspect of a work of art or a monument – *disappeared*. (Fascism, however, placed an increased emphasis on façades, thus opting for total 'spectacularization' as early as the 1920s.)

3 *Global space* established itself in the abstract as a void waiting to be filled, as a medium waiting to be colonized. How this could be done was a problem solved only later by the social practice of capitalism: eventually, however, this space would come to be filled by commercial images, signs and objects. This development would in turn result in the advent of the pseudo-concept of the environment (which begs the question: the environment of whom or of what?).

The historian of space who is concerned with modernity may quite confidently affirm the historic role of the Bauhaus. By the 1920s the

---

[16] In 1920 Klee had this to say: 'Art does not reflect the visible; it renders visible.'

great philosophical systems had been left behind, and, aside from the investigations of mathematics and physics, all thinking about space and time was bound up with social practice – more precisely, with industrial practice, and with architectural and urbanistic research. This transition from philosophical abstraction to the analysis of social practice is worth stressing. While it was going on, those responsible for it, the Bauhaus group and others, believed that they were more than innovators, that they were in fact revolutionaries. With the benefit of fifty years of hindsight, it is clear that such a claim cannot legitimately be made for anyone in that period except for the Dadaists (and, with a number of reservations, a few surrealists).

It is easy enough to establish the historic role of the Bauhaus, but not so easy to assess the breadth and limits of this role. Did it cause or justify a change of aesthetic perspective, or was it merely a symptom of a change in social practice? More likely the latter, *pace* most historians of art and architecture. When it comes to the question of what the Bauhaus's audacity produced in the long run, one is obliged to answer: the worldwide, homogeneous and monotonous architecture of the state, whether capitalist or socialist.

How and why did this happen? If there is such a thing as the history of space, if space may indeed be said to be specified on the basis of historical periods, societies, modes of production and relations of production, then there is such a thing as a space characteristic of capitalism – that is, characteristic of that society which is run and dominated by the bourgeoisie. It is certainly arguable that the writings and works of the Bauhaus, of Mies van der Rohe among others, outlined, formulated and helped realize that particular space – the fact that the Bauhaus sought to be and proclaimed itself to be revolutionary notwithstanding. We shall have occasion to discuss this irony of 'History' at some length later on.[17]

The first initiative taken towards the development of a history of space was Siegfried Giedeon's.[18] Giedeon kept his distance from practice but worked out the theoretical object of any such history in some detail; he put space, and not some creative genius, not the 'spirit of the times', and not even technological progress, at the centre of history as he conceived it. According to Giedeon there have been three successive

---

[17] See Michel Ragon, *Histoire mondiale de l'architecture et de l'urbanisme modernes*, 3 vols (Tournai: Casterman, 1971–8), esp. vol. II, pp. 147ff.

[18] Siegfried Giedeon, *Space, Time, and Architecture* (Cambridge, Mass.: Harvard University Press, 1941).

periods. During the first of these (ancient Egypt and Greece), architectural volumes were conceived and realized in the context of their social relationships – and hence from *without*. The Roman Pantheon illustrates a second conception, under which the *interior* space of the monument became paramount. Our own period, by contrast, supposedly seeks to surmount the exterior–interior dichotomy by grasping an interaction or unity between these two spatial aspects. Actually, Giedeon succeeds here only in *inverting* the reality of social space. The fact is that the Pantheon, as an image of the world or *mundus*, is an opening to the light; the *imago mundi*, the interior hemisphere or dome, symbolizes this exterior. As for the Greek temple, it encloses a sacred and consecrated space, the space of a localized divinity and of a divine locality, and the political centre of the city.[19] The source of such confusion is to be found in an initial error of Giedeon's, echoes of which occur throughout his work: he posits a pre-existing space – Euclidean space – in which all human emotions and expectations proceed to invest themselves and make themselves tangible. The spiritualism latent in this philosophy of space emerges clearly in Giedeon's later work *The Eternal Present*.[20] Giedeon was indeed never able to free himself from a naïve oscillation between the geometrical and the spiritualistic. A further problem was that he failed to separate the history he was developing from the history of art and architecture, although the two are certainly quite different.

The idea that space is essentially empty but comes to be occupied by visual messages also limits the thinking of Bruno Zevi.[21] Zevi holds that a geometrical space is animated by the gestures and actions of those who inhabit it. He reminds us, in a most timely manner, of the basic fact that every building has an interior as well as an exterior. This means that there is an architectural space defined by the inside–outside relationship, a space which is a tool for the architect in his social action. The remarkable thing here, surely, is that it should be necessary to recall this duality several decades after the Bauhaus, and in Italy to boot, supposedly the 'birthplace' of architecture. We are obliged to conclude that the critical analysis of the façade mentioned above has simply never taken hold, and that space has remained *strictly visual*, entirely subordinate to a 'logic of visualization'. Zevi considers that the visual conception of space rests upon a bodily (gestural) component which the

---

[19] Cf. Heidegger's discussion of the Greek temple in *Holzwege*.
[20] Siegfried Giedeon, *The Eternal Present*, 2 vols (New York: Bollingen Foundation/Pantheon, 1962–4).
[21] See Bruno Zevi, *Architecture as Space: How to Look at Architecture*, tr. Milton Gendel, rev. edn (New York: Horizon Press, 1974).

trained eye of the expert observer must take into account. Zevi's book
brings this 'lived' aspect of spatial experience, which thanks to its
corporal nature has the capacity to 'incarnate', into the realm of knowledge, and hence of 'consciousness', without ever entertaining the idea
that such a bodily component of optical (geometrico-visual) space might
put the priority of consciousness itself into question. He does not appear
to understand the implications of his findings beyond the pedagogical
sphere, beyond the training of architects and the education of con-
noisseurs, and he certainly does not pursue the matter on a theoretical
level. In the absence of a viewer with an acquired mastery of space,
how could any space be adjudged 'beautiful' or 'ugly', asks Zevi, and
how could this aesthetic yardstick attain its primordial value? To answer
one question with another, how could a constructed space subjugate or
repel otherwise than through *use*?[22]

Contributions such as those of Giedeon and Zevi undoubtedly have
a place in the development of a history of space, but they herald that
history without helping to institute it. They serve to point up its problems, and they blaze the trail. They do not tackle the tasks that still
await the history of space proper: to show up the growing ascendancy
of the abstract and the visual, as well as the internal connection between
them; and to expose the genesis and meaning of the 'logic of the visual'
– that is, to expose the *strategy* implied in such a 'logic' in light of the
fact that any particular 'logic' of this kind is always merely a deceptive
name for a strategy.

IX

Historical materialism will be so far extended and borne out by a history
so conceived that it will undergo a serious transformation. Its objectivity
will be deepened inasmuch as it will come to bear no longer solely upon
the production of things and works, and upon the (dual) history of that
production, but will reach out to take in space and time and, using
nature as its 'raw material', broaden the concept of production so as to
include the production of space as a process whose product – space –
itself embraces both things (goods, objects) and works.

The outline of history, its 'compendium' and 'index', is not to be
found merely in philosophies, but also beyond philosophy, in that

---

[22] Ibid., pp. 23ff. See also Philippe Boudon's comments in his *L'espace architectural*
(Paris: Dunod, 1971), pp. 27ff.

production which embraces concrete and abstract, historicizing both instead of leaving them in the sphere of philosophical absolutes. Likewise history is thus thoroughly relativized instead of being made into a substitute metaphysics or 'ontology of becoming'. This gives real meaning to the distinctions between prehistorical, historical and post-historical. Thus the properly historical period of the history of space corresponds to the accumulation of capital, beginning with its primitive stage and ending with the world market under the reign of abstraction.

As for dialectical materialism, it also is amplified, verified – and transformed. New dialectics make their appearance: work *versus* product, repetition *versus* difference, and so on. The dialectical movement immanent to the division of labour becomes more complex when viewed in the light of an exposition of the relationship between productive activity (both global labour – i.e. social labour – and divided or parcelled-out labour) and a specific product, unique in that it is also itself a tool – namely, space. The alleged 'reality' of space as natural substance and its alleged 'unreality' as transparency are simultaneously exploded by this advance in our thinking. Space still appears as 'reality' inasmuch as it is the milieu of accumulation, of growth, of commodities, of money, of capital; but this 'reality' loses its substantial and autonomous aspect once its development – i.e. its production – is traced.

There is one question which has remained open in the past because it has never been asked: what exactly is the mode of existence of social relationships? Are they substantial? natural? or formally abstract? The study of space offers an answer according to which the social relations of production have a social existence to the extent that they have a spatial existence; they project themselves into a space, becoming inscribed there, and in the process producing that space itself. Failing this, these relations would remain in the realm of 'pure' abstraction – that is to say, in the realm of representations and hence of ideology: the realm of verbalism, verbiage and empty words.

Space itself, at once a product of the capitalist mode of production and an economico-political instrument of the bourgeoisie, will now be seen to embody its own contradictions. The dialectic thus emerges from time and actualizes itself, operating now, in an unforeseen manner, in space. The contradictions of space, without abolishing the contradictions which arise from historical time, leave history behind and transport those old contradictions, in a worldwide simultaneity, onto a higher level; there some of them are blunted, others exacerbated, as this contradictory whole takes on a new meaning and comes to designate 'something else' – another mode of production.

X

Not everything has been said – far from it – about the inscription of time in space: that is, about the temporal process which gives rise to, which produces, the spatial dimension – whether we are concerned with bodies, with society, with the universe or with the world.

Philosophy has left us but the poorest of indications here. The world is described as a sequence of ill-defined events occurring in the shadows. The Cosmos amounts to a luminous simultaneity. Heraclitus and his followers propose an ever-new universal flux which carries 'beings' along and in which all stability is merely appearance. For the Eleatics, on the other hand, only stability constitutes the 'real' world and renders it intelligible, so that any *change* is merely appearance. Hence the absolute primacy of now difference (always and continually – and tragically – the new), now repetition (always and everywhere – and comically – the same thing over and over again). For some, then, space means decline, ruin – a slipping out of time as time itself slips out of (eternal) Being. As a conglomeration of things, space separates, disperses, and shatters unity, enveloping the finite and concealing its finiteness. For others, by contrast, space is the cradle, birthplace and medium of nature's communications and commerce with society; thus it is always fertile – always full of antagonisms and/or harmonies.

It is surely a little-explored view of time and space which proposes that time's self-actualization in space develops from a kernel (i.e. from a relative and not an absolute origin), that this actualizing process is liable to run into difficulties, to halt for rest and recuperation, that it may even at such moments turn in upon itself, upon its own inner uniqueness as both recourse and resource, before starting up again and continuing until it reaches its point of exhaustion. 'Feedback', to the extent that it played any part at all in such a view of things, would not set in motion a system appropriate to the moment; rather, it would establish synchrony with that diachronic unity which never disappears from any living 'being'. As for time's aforementioned inner resources, and fundamental availability, these stem from the real origins.

XI

I have already ventured a few statements concerning the relations between language and space. It is not certain that systems of non-verbal signs answer to the same concepts and categories as verbal systems, or even that they are properly systems at all, since their elements and moments are related more by contiguity and similarity than by any coherent systematization. The question, however, is still an open one. It is true that parts of space, like parts of discourse, are articulated in terms of reciprocal inclusions and exclusions. In language as in space, there is a before and an after, while the present dominates both past and future.

The following, therefore, are perfectly legitimate questions.

1 Do the spaces formed by practico-social activity, whether land-scapes, monuments or buildings, have meaning?
2 Can the space occupied by a social group or several such groups be treated as a message?
3 Ought we to look upon architectural or urbanistic works as a type of mass medium, albeit an unusual one?
4 May a social space viably be conceived of as a language or discourse, dependent upon a determinate practice (reading/writing)?

The answer to the first question must, obviously, be yes. The second calls for a more ambiguous 'yes and no': spaces contain messages – but can they be reduced to messages? It is tempting to reply that they imply more than that, that they embody functions, forms and structures quite unconnected with discourse. This is an issue that calls for careful scrutiny. As for the third and fourth questions, our replies will have to include the most serious reservations, and we shall be returning to them later.

We can be sure, at any rate, that an understanding of language and of verbal and non-verbal systems of signs will be of great utility in any attempt to understand space. There was once a tendency to study each fragment or element of space separately, seeking to relate it to its own particular past – a tendency to proceed, as it were, etymologically. Today, on the other hand, the preferred objects of study are ensembles, configurations or textures. The result is an extreme formalism, a fetishization of consistency in knowledge and of coherence in practice: a cult, in short, of *words*.

This trend has even generated the claim that discourse and thought have nothing to express but themselves, a position which leaves us with no truth, but merely with 'meaning'; with room for 'textual' work, and such work only. Here, however, the theory of space has something to contribute. Every language is located in a space. Every discourse says something about a space (places or sets of places); and every discourse is emitted from a space. Distinctions must be drawn between discourse *in* space, discourse *about* space and the discourse *of* space. There are thus relationships between language and space which are to a greater or lesser extent misconstrued or disregarded. There is doubtless no such thing as a 'true space', as once postulated by classical philosophy – and indeed still postulated by that philosophy's continuation, namely epistemology and the 'scientific criteria' it promotes. But there is certainly such a thing as a 'truth of space' which embodies the movement of critical theory without being reducible to it. Human beings – why do we persist in saying 'man'? – are in space; they cannot absent themselves from it, nor do they allow themselves to be excluded from it.

Apart from what it 're-marks' in relation to space, discourse is nothing more than a lethal void – mere verbiage. The analogy between the theory of space (and of its production) and the theory of language (and of its production) can only be carried so far. The theory of space describes and analyses textures. As we shall see, the straight line, the curve (or curved line), the check or draughtboard pattern and the radial–concentric (centre *versus* periphery) are forms and structures rather than textures. The production of space lays hold of such structures and integrates them into a great variety of wholes (textures). A texture implies a meaning – but a meaning for whom? For some 'reader'? No: rather, for someone who lives and acts in the space under consideration, a 'subject' with a body – or, sometimes, a 'collective subject'. From the point of view of such a 'subject' the deployment of forms and structures corresponds to functions of the whole. Blanks (i.e. the contrast between absence and presence) and margins, hence networks and webs, have a *lived* sense which has to be raised intact to the *conceptual* level.

Let us now try to pursue this discussion to its logical conclusion. At present, in France and elsewhere, there are two philosophies or theories of language. These two orientations transcend squabbles between different schools of thought and, though they often overlap, they are basically distinct.

1 According to the first view, no sign can exist in isolation. The links between signs and their articulation are of major importance, for it is only through such concatenation that signs can have meaning, can signify. The sign thus becomes the focal point of a system of knowledge, and even of theoretical knowledge in general (semiology, semiotics).

Language, the vehicle of understanding, gives rise to an understanding of itself which is an absolute knowledge. The (unknown or misconstrued) 'subject' of language can only attain self-certitude to the extent that it becomes the subject of knowledge via an understanding of language as such.

The methodical study of chains of signifiers is thus placed at the forefront of the search for knowledge (*connaissance*). This search is assumed to begin with linguistic signs and then to extend to anything susceptible of carrying significance or meaning: images, sounds, and so on. In this way an absolute Knowledge (*Savoir*) can construct a mental space for itself, the connections between signs, words, things and concepts not differing from each other in any fundamental manner. Linguistics will thus have established a realm of certainty which can gradually extend its sovereignty to a good many other areas. The science of language embodies the essence of knowledge, the principle of absolute knowledge, and determines the order in which knowledge is acquired. It provides our understanding with a stable basis to which a series of extensions may be added – epistemology, for example, which indeed deals with acquired knowledge and the language of that knowledge; or semiology, which concerns itself with systems of non-verbal signs; and so on. Seen from this angle, everything – music, painting, architecture – is language. Space itself, reduced to signs and sets of signs, becomes part of knowledge so defined. As, little by little, do all objects in that space.

The theory of signs is connected to set theory, and hence to logic – that is, to 'pure' relationships such as those of commutativity, transitivity and distributivity (and their logical opposites). Every mental and social relationship may thus be reduced to a formal relation of the type: A is to B as B is to C. Pure formalism becomes an (albeit empty) hub for the totalization of knowledge, of discourse, of philosophy and science, of perceptibility and intelligibility, of time and space, of 'theoretical practice' and social practice.

It is scarcely necessary to evoke the great success that this approach has enjoyed recently in France. (In the English-speaking countries it is generally considered to be a substitute for logical empiricism.) But what are the reasons for this success? One is, certainly, that such an orientation helps ensconce knowledge, and hence the university, in a central position whence, it is thought, they may dominate social space in its entirety. Another reason is that in the last analysis this view of things attempts to save a Cartesian, Western, and Europe-centred Logos which is compromised, shaken, and assailed on all sides, from within as from without. The notion is, and everyone is surely familiar with it by now, that linguistics, along with its auxiliary disciplines, can be set up as a 'science of sciences' capable of rectifying the shortcomings, wherever they might occur, of other sciences such as political economy, history or sociology. The irony is that linguistics, in seeking to furnish knowledge with a solid core, has succeeded only in establishing a void, a dogmatically posited vacuum which, when not surrounded by silence, is buried in a mass of metalanguage, empty words and chit-chat about discourse. Caution – scientific caution – forbids any rash attempt to bridge the (epistemological) chasm between known and not-known; the forbidden fruit of lived experience flees or disappears under the assaults of reductionism; and silence reigns around the fortress of knowledge.

2 'Ich kann das Wort so hoch unmöglich schätzen': 'I cannot grant the word such sovereign merit.' Thus Goethe's Faust, Part I.[23] And indeed it is impossible to put such a high value upon language, on speech, on words. The Word has never saved the world and it never will.

For the second view of language alluded to above, an examination of signs reveals a terrible reality. Whether letters, words, images or sounds, signs are rigid, glacial, and abstract in a peculiarly menacing way. Furthermore, they are harbingers of death. A great portion of their importance lies in the fact that they demonstrate an intimate connection between words and death, between human consciousness and deadly acts: breaking, killing, suicide. In this perspective, all signs are bad signs, threats – and weapons. This accounts for their *cryptic* nature, and explains why they are liable to be hidden in the depths of grottoes or belong to sorcerers (Georges Bataille evokes Lascaux in this connection). Signs and figures of the invisible threaten the visible world. When associated with weapons, or found amidst weapons, they serve the purposes of the will to power. Written, they serve authority. What are they? They are the doubles of things. When they assume the properties of things, when they pass for things, they have the power to move us emotionally, to cause frustrations, to engender neuroses. As replicas capable of disassembling the 'beings' they replicate, they make possible the breaking and destruction of those beings, and hence also their reconstruction in different forms. The power of the sign is thus extended both by the power of knowledge over nature and by the sign's own hegemony over human beings; this capacity of the sign for action embodies what Hegel called the 'terrible power of negativity'. As compared with what is signified, whether a thing or a 'being', whether actual

[23] Goethe, Faust, Part I, l. 1226; tr. Walter Arndt (New York: Norton, 1976), p. 30.

or possible, a sign has a repetitive aspect in that it adds a corresponding representation. Between the signified and the sign there is a mesmerizing difference, a deceptive gap: the shift from one to the other seems simple enough, and it is easy for someone who has the words to feel that they possess the things those words refer to. And, indeed, they do possess them up to a certain point – a terrible point. As a vain yet also effective trace, the sign has the power of destruction because it has the power of abstraction – and thus also the power to construct a new world different from nature's initial one. Herein lies the secret of the Logos as foundation of all power and all authority; hence too the growth in Europe of knowledge and technology, industry and imperialism.

Space is also felt to have this deadly character: as the locus of communication by means of signs, as the locus of separations and the milieu of prohibitions, spatiality is characterized by a death instinct inherent to life – which only proliferates when it enters into conflict with itself and seeks its own destruction.

This pessimistic view of signs has a long pedigree. It is to be found in Hegel's notion of a negativity later compensated for by the positivity of knowledge.[24] It occurs, in a more acute and emphatic form, in Nietzsche the philologist–poet and philosopher (or metaphilosopher).[25] For Nietzsche, language has an anaphorical even more than a metaphorical character. It always leads beyond presentness, towards an elsewhere, and above all towards a hypervisualization which eventually destroys it. Prior to knowledge, and beyond it, are the body and the actions of the body: suffering, desire, pleasure. For Nietzsche the poet, poetry consists in a metamorphosis of signs. In the course of a struggle which overcomes the antagonism between work and play, the poet snatches words from the jaws of death. In the chain of signifiers, he substitutes life for death, and 'decodes' on this basis. The struggle is as terrible as the trap-ridden and shifting terrain upon which it is waged. Happily for the poet, he does not fight without succour: musicians, dancers, actors – all travel the same road; and, even if there is much anguish along the way, incomparable pleasures are the prize.

It is facile in this context – and simply too convenient – to draw a distinction between a poetry which intensifies life (Goethe's *Faust*, or Nietzsche's *Zarathustra*) and a poetry of death (Rilke, Mallarmé).[26]

[24] See my *Le langage et la société* (Paris: Gallimard, 1966), pp. 84ff.
[25] See Friedrich Nietzsche, *Das Philosophenbuch/Le Livre du philosophe* (Paris: Aubier-Flammarion, 1969), pp. 170ff.
[26] Cf. Maurice Blanchot, *L'espace littéraire* (Paris: Gallimard, 1955).

These two orientations in the theory (or philosophy) of language have rarely been presented separately – in their 'pure' forms, so to speak. French authors have for the most part sought a compromise of some kind, though Georges Bataille and Antonin Artaud are notable exceptions. This widespread eclecticism has been facilitated by psychoanalysis. A transition from discourse-as-knowledge to a 'science of discourse' is made suspiciously painlessly, as though there were no abyss between them. The science of discourse is next easily made to embrace the spoken, the unspoken and the forbidden, which are conceived of as the essence and meaning of lived experience. By which point the science of discourse is well on the way to bringing social discourse as a whole under its aegis. The death instinct, prohibitions (especially that against incest), castration and the objectification of the phallic, writing as the projection of the voice – these are just so many way-stations along this expansionist route. Semiotics, we are told, is concerned with the instincts of life and death, whereas the symbolic and semantic areas are the province of signs properly speaking.[27] As for space, it is supposedly given along with and in language, and is not formed separately from language. Filled with signs and meanings, an indistinct intersection point of discourses, a container homologous with whatever it contains, space so conceived is comprised merely of functions, articulations and connections – in which respect it closely resembles discourse. Signs are a necessity, of course, but they are sufficient unto themselves, because the system of verbal signs (whence written language derives) already embodies the essential links in the chain, spatial links included. Unfortunately, this proposed compromise, which sacrifices space by handing it on a platter to the philosophy of language, is quite unworkable. The fact is that signifying processes (a signifying practice) occur in a space which cannot be reduced either to an everyday discourse or to a literary language of texts. If indeed signs as deadly instruments transcend themselves through poetry, as Nietzsche claimed and sought to show in practice, they must of necessity accomplish this perpetual self-transcendence in space. There is no need to reconcile the two theses concerning signs by means of an eclecticism which is somehow respectful of both

27 See Julia Kristeva's doctoral thesis, 'Langage, sens, poésie' (1973), which puts much emphasis on this distinction between the semiotic realm (involving instincts) and the symbolic one (involving language as a system of communications). Indeed, Kristeva goes even further in this direction than Jacques Lacan in his *Ecrits* (Paris: Seuil, 1966). The author most adept at keeping both these balls in the air is Roland Barthes, as witness his entire work. The problem is forcefully posed by Hermann Hesse in his *Glass Bead Game* (see above, p. 24, note 30), but Hesse offers no solution.

'pure' knowledge and 'impure' poetry. The task confronting us is not to speculate on an ambiguity but rather to demonstrate a contradiction in order to resolve it, or, better, in order to show that space resolves it. The deployment of the energy of living bodies in space is forever going beyond the life and death instincts and harmonizing them. Pain and pleasure, which are poorly distinguished in nature, become clearly discernible in (and thanks to) social space. Products, and *a fortiori* works, are destined to be enjoyed (once labour, a mixture of painful effort and the joy of creation, has been completed). Although spaces exist which give expression to insurmountable separations – tombs being a case in point – there are also spaces devoted to encounter and gratification. And, if poets struggle against the iciness of words and refuse to fall into the traps set by signs, it is even more appropriate that architects should conduct a comparable campaign, for they have at their disposal both materials analogous to signs (bricks, wood, steel, concrete) and *matériel* analogous to those 'operations' which link signs together, articulating them and conferring meaning upon them (arches, vaults, pillars and columns; openings and enclosures; construction techniques; and the conjunction and disjunction of such elements). Thus it is that architectural genius has been able to realize spaces dedicated to voluptuousness (the Alhambra of Granada), to contemplation and wisdom (cloisters), to power (castles and châteaux) or to heightened perception (Japanese gardens). Such genius produces spaces full of meaning, spaces which first and foremost escape mortality: enduring, radiant, yet also inhabited by a specific local temporality. Architecture produces living bodies, each with its own distinctive traits. The animating principle of such a body, its presence, is neither visible nor legible as such, nor is it the object of any discourse, for it reproduces itself within those who *use* the space in question, within their lived experience. Of that experience the tourist, the passive spectator, can grasp but a pale shadow.

Once brought back into conjunction with a (spatial and signifying) *social practice*, the concept of space can take on its full meaning. Space thus rejoins material production: the production of goods, things, objects of exchange – clothing, furnishings, houses or homes – a production which is dictated by necessity. It also rejoins the productive process considered at a higher level, as the result of accumulated knowledge; at this level labour is penetrated by a materially creative experimental science. Lastly, it rejoins the freest creative process there is – the signifying process, which contains within itself the seeds of the 'reign of freedom', and which is destined in principle to deploy its possibilities under that reign as soon as labour dictated by blind and immediate necessity comes to an end – as soon, in other words, as the process of creating true works, meaning and pleasure begins. (It may be noted in passing that such creations are themselves very diverse: for example, contemplation may involve sensual pleasure, which, though it includes sexual gratification, is not limited to it.)

Let us now consider a seminal text of Nietzsche's on language, written in 1873. More of a philologist than a philosopher, and a lover of language because he approached it as a poet, Nietzsche here brought forward two concepts which were then already classic, and which have since been vulgarized: metaphor and metonymy. For the modern school of linguistics, which takes its inspiration from Saussure, these two figures of speech go beyond primary language; in other words, they transcend the first level of discourse. This is consistent with the meaning of the Greek prefix *meta*-: metaphor and metonymy are part of metalanguage – they belong to the second level of language.

In Nietzschean thought (which appears very different today from the way it appeared at the turn of the century), *meta*- is understood in a very radical manner. Metaphor and metonymy make their appearance here at the simplest level of language: words as such are already metaphoric and metonymic for Nietzsche – Kofmann, who seems to think that these terms apply only to concepts, notwithstanding.[28] Words themselves go beyond the immediate, beyond the perceptible – that is to say, beyond the chaos of sense impressions and stimuli. When this chaos is replaced by an image, by an audible representation, by a word and then by a concept, it undergoes a metamorphosis. The words of spoken language are simply metaphors for things.[29] The concept arises from an identification of things which are not identical – i.e. from metonymy. We take a language for an instrument of veracity and a structure of accumulated truths. In reality, according to Nietzsche, it is 'A mobile army of metaphors, metonyms, and anthropomorphisms – in short, a sum of human relations, which have been enhanced, transposed, and embellished poetically and rhetorically, and which after long use seem firm, canonical, and obligatory to a people'.[30] In more modern terms: language in action is more important than language in general or discourse in general; and speech is more creative than language as a system – and *a fortiori* than writing or reading. Language in action and

---

[28] See S. Kofmann, *La métaphore nietzschéenne* (Paris: Payot, 1972).
[29] See Nietzsche, *Philosophenbuch*, p. 179.
[30] Friedrich Nietzsche, 'On Truth and Lie in an Extra-Moral Sense' (1873), in Walter Kaufmann, ed. and tr., *The Portable Nietzsche* (New York: Viking, 1954), pp. 46–7.

the spoken word are inventive; they restore life to signs and concepts that are worn down like old coins. But just what is it that 'figures of speech', metaphors, metonyms and metamorphoses invent, call forth, translate or betray? Could it be that reality is grounded in the imagination? That the world was created by a god who was a poet or a dancer? The answer – at least so far as the social realm is concerned – must be no. The fact is that a 'pyramidal order', and hence a world of castes and classes, of laws and privileges, of hierarchies and constraints, stands opposed to the world of first impressions as 'that which is firmest, most general, best known, most human, and hence that which regulates and rules'.[31] A society is a space and an architecture of concepts, forms and laws whose abstract truth is imposed on the reality of the senses, of bodies, of wishes and desires.

At several points in his philosophical (or metaphilosophical) and poetic work, Nietzsche stresses the visual aspect predominant in the metaphors and metonyms that constitute abstract thought: idea, vision, clarity, enlightenment and obscurity, the veil, perspective, the mind's eye, mental scrutiny, the 'sun of intelligibility', and so on. This is one of Nietzsche's great discoveries (to use another visual metaphor). He points out how over the course of history the visual has increasingly taken precedence over elements of thought and action deriving from the other senses (the faculty of hearing and the act of listening, for instance, or the hand and the voluntary acts of 'grasping', 'holding', and so on). So far has this trend gone that the senses of smell, taste, and touch have been almost completely annexed and absorbed by sight. The same goes for sexuality, and for desire (which survives in travestied form as *Sehnsucht*). Here we see the emergence of the anaphorical aspect of language, which embraces both metaphor and metonymy.

The following conclusions may thus be drawn.

1 Metaphor and metonymy are not figures of speech – at least not at the outset. They *become* figures of speech. In principle, they are acts. What do such acts accomplish? To be exact, they decode, bringing forth from the depths not what is there but what is sayable, what is susceptible of figuration – in short, language. Here is the source of the activities of speech, of language in action, of discourse, activities which might more properly be named 'metaphorization' and 'metonymization'. What is the point of departure of these processes? The body metamorphosed. Do representations of space and representational spaces, to the degree that

---

[31] Nietzsche, *Philosophenbuch*, p. 185.

they make use of such 'figures', tend to 'naturalize' the spatial realm? No – or not merely – because they also tend to make it evaporate, to dissolve it in a luminous (optical and geometrical) transparency.

2 These procedures involve displacement, and hence also transposition and transfer. Beyond the body, beyond impressions and emotions, beyond life and the realm of the senses, beyond pleasure and pain, lies the sphere of distinct and articulated unities, of signs and words – in short, of abstractions. Metaphorization and metonymization are defining characteristics of signs. It is a 'beyond', but a nearby one, which creates the illusion of great remoteness. Although 'figures of speech' express much, they lose and overlook, set aside and place parentheses around even more.

3 It is perhaps legitimate to speak of a logic of the metaphorical and a logic of the metonymic, because these 'figures of speech' give birth to a form, that of coherent and articulate discourse, which is analogous to a logical form, and above all because they erect a mental and social architecture above spontaneous life. In discourse, as in the perception of society and space, there is a constant to-and-fro both between the component elements and between the parts and the whole.

4 This immense movement has myriad connections: on the one hand with rationality, with the Logos, with reasoning by analogy and by deduction; and on the other hand with social structures which are bound up in their turn with political structures – that is to say, with power. Hence the ever-growing hegemony of vision, of the visible and the legible (of the written, and of writing). All these elements – these forms, functions and structures – have complex spatial interrelationships which can be analysed and explained.

So, if there is *fetishism* (of a visual, intelligible and abstract space), and if there is *fascination* (with a natural space which has been lost and/or rediscovered, with absolute political or religious spaces, or with spaces given over to voluptuousness or death), then theory is well able to trace their genesis, which is to say their production.

XII

What is it that obscures the concept of production as it relates to space? Sufficient attention has already been paid to the proponents of absolute knowledge and to the new dogmatists, and there is no further need here to examine their talk of an epistemological field or base, of the space of the episteme, and so forth. We saw earlier how they reduce the social to the mental and the practical to the intellectual, at the same time underwriting the extension of the laws of private property to knowledge itself. I have not dealt, however, with the fact that a number of notions which tend to confuse the concept with which we are concerned derive from semiology, notably the thesis according to which social space is the result merely of a *marking* of natural space, a leaving of traces upon it. Though made use of by the semiologists, notions such as those of marks, marking and traces do not actually originate with them. Anthropologists, among others, used them earlier. The semiological use, however, places more emphasis on meaning: marks are supposed to signify, to be part of a system, and to be susceptible of coding and decoding. Space may be marked physically, as with animals' use of smells or human groups' use of visual or auditory indicators; alternatively, it may be marked abstractly, by means of discourse, by means of signs. Space thus acquires symbolic value. Symbols, on this view, always imply an emotional investment, an affective charge (fear, attraction, etc.), which is so to speak deposited at a particular place and thereafter 'represented' for the benefit of everyone elsewhere. In point of fact, early agricultural and pastoral societies knew no such split between the practical and the symbolic. Only very much later was this distinction detected by analytical thinking. To separate these two spheres is to render 'physical' symbols incomprehensible, and likewise practice, which is thus portrayed as the practice of a society without the capacity for abstraction. It is reasonable to ask, however, whether one may properly speak of a production of space so long as marking and symbolization of this kind are the only way of relating to space. And the answer to this question has to be: not as yet, even though living bodies, mobile and active, may already be said to be extending both their spatial perception and their occupation of space, like a spider spinning its web. If and to the extent that production occurs, it will be restricted for a long time to marks, signs and symbols, and these will not significantly affect the material reality upon which they are imprinted. For all that the earth may become Mother Earth, cradle of life, a symbolically sexual ploughed field, or a tomb, it will still be the earth.

It should be noted that the type of activity that consists in marking particular locations and indicating routes by means of markers or blazes is characteristic only of the very earliest stages of organized society. During these primitive phases, the itineraries of hunters and fishermen, along with those of flocks and herds, are marked out, and *topoi* (soon to become *lieux-dits*, or 'places called' such and such) are indicated by stones or cairns wherever no natural landmarks such as trees or shrubs are to hand. These are times during which natural spaces are merely *traversed*. Social labour scarcely affects them at all. Later on, marking and symbolization may become individualized or playful procedures, as for example when a child indicates her own corner because it amuses her to leave behind some trace of her presence.

This mistaken notion of the semiologists has given rise to the diametrically opposite but complementary idea that 'artificial' space is solely the result of a denaturing or denaturalization of some objective, authentically 'natural' space. What forces are said to be responsible for this? The obvious ones: science and technology, and hence abstraction. The problem with this view is that it studiously ignores the diversity of social spaces and of their historical origins, reducing all such spaces to the common trait of abstraction (which is of course inherent to all conceivable activity involving knowledge).

Semiology is also the source of the claim that space is susceptible of a 'reading', and hence the legitimate object of a practice (reading/ writing). The space of the city is said to embody a discourse, a language.[32]

Does it make sense to speak of a 'reading' of space? Yes and no. Yes, inasmuch as it is possible to envisage a 'reader' who deciphers or decodes and a 'speaker' who expresses himself by translating his progression into a discourse. But no, in that social space can in no way be compared to a blank page upon which a specific message has been inscribed (by whom?). Both natural and urban spaces are, if anything, 'over-inscribed': everything therein resembles a rough draft, jumbled and self-contradictory. Rather than signs, what one encounters here are directions – multifarious and overlapping instructions. If there is indeed text, inscription or writing to be found here, it is in a context of conventions, intentions and order (in the sense of social order *versus* social disorder). That space signifies is incontestable. But what it signifies is dos and don'ts – and this brings us back to power. Power's message is invariably confused – deliberately so; dissimulation is necessarily part of any message from power. Thus space indeed 'speaks' – but it does not tell all. Above all, it prohibits. Its mode of existence, its practical 'reality' (including its form) differs radically from the reality (or being-there) of something written, such as a book. Space is at once result and cause, product and producer; it is also a *stake*, the locus of projects and actions

---

[32] See Roland Barthes in *Architecture d'aujourd'hui*, nos 132 and 153.

deployed as part of specific strategies, and hence also the object of *wagers* on the future – wagers which are articulated, if never completely. As to whether there is a spatial code, there are actually several. This has not daunted the semiologists, who blithely propose to determine the hierarchy of levels of interpretation and then find a residue of elements capable of getting the decoding process going once more. Fair enough, but this is to mistake restrictions for signs in general. Activity in space is restricted by that space; space 'decides' what activity may occur, but even this 'decision' has limits placed upon it. Space lays down the law because it implies a certain order – and hence also a certain disorder (just as what may be seen defines what is obscene). Interpretation comes later, almost as an afterthought. Space commands bodies, prescribing or proscribing gestures, routes and distances to be covered. It is produced with this purpose in mind; this is its *raison d'être*. The 'reading' of space is thus merely a secondary and practically irrelevant upshot, a rather superfluous reward to the individual for blind, spontaneous and *lived* obedience.

So, even if the reading of space (always assuming there is such a thing) comes first from the standpoint of knowledge, it certainly comes last in the genesis of space itself. No 'reading of the space' of Romanesque churches and their surroundings (towns or monasteries), for example, can in any way help us predict the space of so-called Gothic churches or understand their preconditions and prerequisites: the growth of the towns, the revolution of the communes, the activity of the guilds, and so on. This space was *produced* before being *read*; nor was it produced in order to be read and grasped, but rather in order to be *lived* by people with bodies and lives in their own particular urban context. In short, 'reading' follows production in all cases except those in which space is produced especially in order to be read. This raises the question of what the virtue of readability actually is. It turns out on close examination that spaces made (produced) to be read are the most deceptive and tricked-up imaginable. The graphic impression of readability is a sort of *trompe-l'oeil* concealing strategic intentions and actions. Monumentality, for instance, always embodies and imposes a clearly intelligible message. It says what it wishes to say – yet it hides a good deal more: being political, military, and ultimately fascist in character, monumental buildings mask the will to power and the arbitrariness of power beneath signs and surfaces which claim to express collective will and collective thought. In the process, such signs and surfaces also manage to conjure away both possibility and time.

We have known since Vitruvius – and in modern times since Labrouste

(d. 1875), who was forever harping on it – that in architecture form must express function. Over the centuries the idea contained in the term 'express' here has grown narrower and more precise. Most recently, 'expressive' has come to mean merely 'readable'.[33] The architect is supposed to construct a signifying space wherein form is to function as signifier is to signified; the form, in other words, is supposed to enunciate or proclaim the function. According to this principle, which is espoused by most 'designers', the environment can be furnished with or animated by signs in such a way as to appropriate space, in such a way that space becomes readable (i.e. 'plausibly' linked) to society as a whole. The inherence of function to form, or in other words the application of the criterion of readability, makes for an instantaneousness of reading, act and gesture – hence the tedium which accompanies this quest for a formal–functional transparency. We are deprived of both internal and external distance: there is nothing to code and decode in an 'environment without environs'. What is more, the significant contrasts in a code of space designed specifically to signify and to 'be' readable are extremely commonplace and simple. They boil down to the contrast between horizontal and vertical lines – a contrast which among other things masks the vertical's implication of hauteur. Versions of this contrast are offered in visual terms which are supposed to express it with great intensity but which, to any detached observer, any ideal 'walker in the city', have no more than the appearance of intensity. Once again, the impression of intelligibility conceals far more than it reveals. It conceals, precisely, what the visible/readable 'is', and what traps it holds; it conceals what the vertical 'is' – namely, arrogance, the will to power, a display of military and police-like machismo, a reference to the phallus and a spatial analogue of masculine brutality. Nothing can be taken for granted in space, because what are involved are real or possible acts, and not mental states or more or less well-told stories. In produced space, acts reproduce 'meanings' even if no 'one' gives an account of them. Repressive space wreaks repression and terror even though it may be strewn with ostensible signs of the contrary (of contentment, amusement or delight).

This tendency has gone so far that some architects have even begun to call either for a return to ambiguity, in the sense of a confused and not immediately interpretable message, or else for a diversification of

[33] See Charles Jencks, *Architecture 2000: Predictions and Methods* (New York: Praeger, 1971), pp. 114–16.

space which would be consistent with a liberal and pluralistic society.[34] Robert Venturi, as an architect and a theorist of architecture, wants to make space dialectical. He sees space not as an empty and neutral milieu occupied by dead objects but rather as a field of force full of tensions and distortions. Whether this approach can find a way out of functionalism and formalism that goes beyond merely formal adjustments remains (in 1972) to be seen. Painting on buildings certainly seems like a rather feeble way of retrieving the richness of 'classical' architecture. Is it really possible to use mural surfaces to depict social contradictions while producing something more than graffiti? That would indeed be somewhat paradoxical if, as I have been suggesting, the notions of 'design', of reading/writing as practice, and of the 'signifier–signified' relationship projected onto things in the shape of the 'form–function' one are all directed, whether consciously or no, towards the dissolving of conflicts into a general transparency, into a one-dimensional present – and onto an as it were 'pure' *surface*.

I daresay many people will respond to such thinking somewhat as follows.

Your arguments are tendentious. You want to re-emphasize the signified as opposed to the signifier, the content as opposed to the form. But true innovators operate on forms; they invent new forms by working in the realm of signifiers. If they are writers, this is how they produce a discourse. The same goes for other types of creation. But as for architects who concern themselves primarily with content, as for 'users', as for the activity of dwelling itself – all these merely reproduce outdated forms. They are in no sense innovative forces.

To which my reply might be something like this:

I have no quarrel with the proposition that work on signifiers and the production of a language are creative activities; that is an incontestable fact. But I question whether this is the whole story – whether this proposition covers all circumstances and all fields. Surely there comes a moment when formalism is exhausted, when only a new injection of content into form can destroy it and so open up the way to innovation. The harmonists invented a great

---

[34] See Robert Venturi, *Complexity and Contradiction in Architecture* (New York: Museum of Modern Art/Doubleday, 1966).

musical form, for instance, yet the formal discoveries about harmony made by the natural philosophers and by theorists of music
such as Rameau did not take the exploration and exploitation of
the possibilities that far. Such progress occurred only with the
advent of a Mozart or a Beethoven. As for architecture, the builders
of palaces worked with and on signifiers (those of power). They
kept within the boundaries of a certain monumentality and made
no attempt to cross them. They worked, moreover, not upon texts
but upon (spatial) textures. Invention of a formal kind could not
occur without a change in practice, without, in other words, a
dialectical interaction between signifying and signified elements, as
some signifiers reached the exhaustion point of their formalism,
and some signified elements, with their own peculiar violence,
infiltrated the realm of signifiers. The combinatorial system of the
elements of a set – for our purposes a set of signs, and hence of
signifiers – has a shorter life than the individual combinations that
it embraces. For one thing, any such combinatorial system of signs
loses its interest and emotional force as soon as it is known and
recognized for what it is; a kind of saturation sets in, and even
changing the combinations that are included or excluded from the
system cannot remedy matters. Secondly, work on signifiers and
the production of a discourse facilitate the transmission of messages
only if the labour involved is not patent. If the 'object' bears traces
of that labour, the reader's attention will be diverted to the writing
itself and to the one who does the writing. The reader thus comes
to share in the fatigue of the producer, and is soon put off.

It is very important from the outset to stress the destructive (because
reductive) effects of the predominance of the readable and visible, of
the absolute priority accorded to the visual realm, which in turn implies
the priority of reading and writing. An emphasis on visual space has
accompanied the search for an impression of weightlessness in architecture. Some theorists of a supposed architectural revolution claim Le
Corbusier as a pioneer in this connection, but in fact it was Brunelleschi,
and more recently Baltard and then Eiffel, who blazed the trail. Once
the effect of weightiness or massiveness upon which architects once
depended has been abandoned, it becomes possible to break up and
reassemble volumes arbitrarily according to the dictates of an architectural neoplasticism. Modernity expressly reduces so-called 'iconological'
forms of expression (signs and symbols) to surface effects. Volumes or
masses are deprived of any physical consistency. The architect considers

himself responsible for laying down the social function (or use) of buildings, offices, or dwellings, yet interior walls which no longer have any spatial or bearing role, and interiors in general, are simultaneously losing all character or content. Even exterior walls no longer have any material substance: they have become mere membranes barely managing to concretize the division between inside and outside. This does not prevent 'users' from projecting the relationship between the internal or private and a threatening outside world into an invented absolute realm; when there is no alternative, they use the signs of this antagonism, relying especially on those which indicate property. For an architectural thought in thrall to the model of transparency, however, all partitions between inside and outside have collapsed. Space has been comminuted into 'iconological' figures and values, each such fragment being invested with individuality or worth simply by means of a particular colour or a particular material (brick, marble, etc.). Thus the sense of circumscribed spaces has gone the same way as the impression of mass. Within and without have melted into transparency, becoming indistinguishable or interchangeable. What makes this tendency even more paradoxical is the fact that it proceeds under the banner of structures, of significant distinctions, and of the inside–outside and signifier–signified relationships themselves.

We have seen that the visual space of transparency and readability has a content – a content that it is designed to conceal: namely, the phallic realm of (supposed) virility. It is at the same time a repressive space: nothing in it escapes the surveillance of power. Everything opaque, all kinds of partitions, even walls simplified to the point of mere drapery, are destined to disappear. This disposition of things is diametrically opposed to the real requirements of the present situation. The sphere of private life ought to be enclosed, and have a finite, or finished, aspect. Public space, by contrast, ought to be an opening outwards. What we see happening is just the opposite.

XIII

Like any reality, social space is related methodologically and theoretically to three general concepts: form, structure, function. In other words, any social space may be subjected to formal, structural or functional analysis. Each of these approaches provides a code and a method for deciphering what at first may seem impenetrable.

These terms may seem clear enough, but in fact, since they cannot avoid polysemy, they all carry burdens of ambiguity.

The term 'form' may be taken in a number of senses: aesthetic, plastic, abstract (logico-mathematical), and so on. In a general sense, it evokes the description of contours and the demarcation of boundaries, external limits, areas and volumes. Spatial analysis accepts this general use of the term, although doing so does not eliminate all problems. A formal description, for example, may aspire to exactitude but still turn out to be shot through with ideological elements, especially when implicit or explicit reductionistic goals are involved. The presence of such goals is indeed a defining characteristic of *formalism*. Any space may be reduced to its formal elements: to curved and straight lines or to such relations as internal-*versus*-external or volume-*versus*-area. Such formal aspects have given rise in architecture, painting and sculpture to genuine systems: the system of the golden number, for example, or that of the Doric, Ionic and Corinthian orders, or that of moduli (rhythms and proportions).

Consideration of aesthetic effects or 'effects of meaning' has no particular right of precedence in this context. What counts from the methodological and theoretical standpoint is the idea that none of these three terms can exist in isolation from the other two. Forms, functions and structures are generally given in and through a material realm which at once binds them together and preserves distinctions between them. When we consider an organism, for example, we can fairly easily discern the forms, functions and structures within this totality. Once this threefold analysis has been completed, however, a residue invariably remains which seems to call for deeper analysis. This is the *raison d'être* of the ancient philosophical categories of being, nature, substance and matter. In the case of a produced 'object', this constitutive relationship is different: the application to materials of a practical action (technology, labour) tends to blur, as a way of mastering them, the distinctions between form, function and structure, so that the three may even come to imply one another in an immediate manner. This tendency exists only implicitly in works of art and objects antedating the Industrial Revolution, including furniture, houses, palaces and monuments; under the conditions of modernity, on the other hand, it comes close to its limit. With the advent of 'design', materiality tends to give way to transparency – to perfect 'readability'. Form is now merely the sign of function, and the relation between the two, which could not be clearer – that is, easier to produce and reproduce – is what gives rise to structure. A case where this account does not apply is that not uncommon one where 'designer' and manufacturer find it amusing to confuse the issue, as it were, and give a form (often a 'classical' one) to a function completely unconnected with it: they disguise a bed as a cupboard, for example, or a refrigerator as bookshelves. The celebrated signifier–signified dichotomy is singularly appropriate when applied to such objects, but this special application is just that – and a good deal more limited than semantico-semiological orthodoxy would probably care to admit. As for social 'realities', here the opposite situation obtains: the distances between forms, functions and structures lengthen rather than diminish. The three tend to become completely detached from one another. Their relationship is obscured and they become indecipherable (or undecodable) as the 'hidden' takes over from the 'readable' in favour of the predominance of the latter in the realm of objects. Thus a particular institution may have a variety of functions which are different – and sometimes opposed – to its apparent forms and avowed structures. One merely has to think of the institutions of 'justice', of the military, or of the police. In other words, the space of objects and the space of institutions are radically divergent in 'modern' society. This is a society in which, to take an extreme example, the bureaucracy is supposed to be, aspires to be, loudly proclaims itself to be, and perhaps even believes itself to be 'readable' and transparent, whereas in fact it is the very epitome of opacity, indecipherability and 'unreadability'. The same goes for all other state and political apparatuses.

The relationship between these key terms and concepts (form, function, structure) becomes much more complex when one considers only those very abstract forms, such as the logical form, which do not depend on description and which are inseparable from a content. Among these, in addition to the logical form, must be numbered identity, reciprocity, recurrence, repetition (iteration), and difference. Marx, following Adam Smith and Ricardo, showed how and why the form of *exchange* has achieved predominance in social practice in association with specific functions and structures. The form of social space – i.e. the centre–periphery relationship – has only recently come to occupy a place in our thinking about forms. As for the urban form – i.e. assembly, encounter and simultaneity – it has been shown to belong among the classic forms, in company with centrality, difference, recurrence, reciprocity, and so on.

These forms, which are almost 'pure' (at the extreme limit of 'purity' the form disappears, as in the case of pure identity: A's identity with A) cannot be detached from a content. The interaction between form and content and the invariably concrete relationship between them are the object of analyses about which we may repeat what we said earlier:

each analytic stage deals with a residue left over from the previous stage, for an irreducible element – the substrate or foundation of the object's 'presence' – always subsists.

Between forms close to the point of purity at which they would disappear and their contents there exist mediations. In the case of spatial forms, for example, the form of the curve is mediated by the curved line, and the straight form by the straight line. All spatial arrangements use curved and/or straight forms; naturally, one or the other may predominate.

When formal elements become part of a *texture*, they diversify, introducing both repetition and difference. They articulate the whole, facilitating both movement from the parts to the whole and, conversely, the mustering by the whole of its component elements. For example, the capitals of a Romanesque cloister differ, but they do so within the limits permitted by a model. They break space up and give it rhythm. This illustrates the function of what has been called the 'signifying differential'.[35] The semicircular or ogival arch, with its supporting pillars and columns, has a different spatial meaning and value according to whether it occurs in Byzantine or in Oriental, in Gothic or in Renaissance architecture. Arches have both repetitive and differential functions within a whole whose 'style' they help determine. The same sort of thing goes in music for the theme and its treatment in fugal composition. Such 'diaeretic' effects, which the semiologists compare to metonymy, are to be met with in all treatments of space and time.

The peopling and investment (or occupation) of a space always happens in accordance with discernible and analysable forms: as dispersal or concentration, or as a function of a specific (or for that matter a nebulous) orientation. By contrast, assembly and concentration as spatial forms are always actualized by means of geometric forms: a town may have a circular (radial–concentric) or a quadrangular form.

The content of these forms metamorphoses them. The quadrangular form, for example, occurs in the ancient Roman military camp, in medieval *bastides*, in the Spanish colonial town and in the modern American city. The fact is, however, that these urban realities differ so radically that the abstract form in question is their only common feature.

The Spanish-American colonial town is of considerable interest in this regard. The foundation of these towns in a colonial empire went hand in hand with the production of a vast space, namely that of Latin

---

[35] See Julia Kristeva, *Semeiotike* (Paris: Seuil, 1969), pp. 298ff. The 'signifying differential' is to be distinguished from Osgood's 'semantic differential'.

America. Their urban space, which was instrumental in this larger production process, has continued to be produced despite the vicissitudes of imperialism, independence and industrialization. It is an urban space especially appropriate for study in that the colonial towns of Latin America were founded at the time of the Renaissance in Europe – that is to say, at a time when the study of the ancient world, and of the history, constitution, architecture and planning of its cities, was being resumed.

The Spanish-American town was typically built according to a plan laid down on the basis of standing orders, according to the veritable code of urban space constituted by the *Orders for Discovery and Settlement*, a collection, published in 1573, of official instructions issued to founders of towns from 1513 on. These instructions were arranged under the three heads of discovery, settlement and pacification. The very building of the towns thus embodied a plan which would determine the mode of occupation of the territory and define how it was to be reorganized under the administrative and political authority of urban power. The orders stipulate exactly how the chosen sites ought to be developed. The result is a strictly hierarchical organization of space, a gradual progression outwards from the town's centre, beginning with the *ciudad* and reaching out to the surrounding *pueblos*. The plan is followed with geometrical precision: from the inevitable Plaza Mayor a grid extends indefinitely in every direction. Each square or rectangular lot has its function assigned to it, while inversely each function is assigned its own place at a greater or lesser distance from the central square: church, administrative buildings, town gates, squares, streets, port installations, warehouses, town hall, and so on. Thus a high degree of segregation is superimposed upon a homogeneous space.[36] Some historians have described this colonial town as an artificial product, but they forget that this artificial product is also an instrument of production: a superstructure foreign to the original space serves as a political means of introducing a social and economic structure in such a way that it may gain a foothold and indeed establish its 'base' in a particular locality. Within this spatial framework, Spanish colonial architecture freely (so to speak) deployed the Baroque motifs which are especially evident in the decoration of façades. The relation between the 'micro' (architectural) plane and the 'macro' (spatial–strategic) one does exist here, but it cannot be reduced to a logical relationship or put into terms of formal implication. The main point to be noted, therefore, is the production of a social

[36] See Emma Scovazzi in *Espaces et société*, no. 3.

space by political power – that is, by violence in the service of economic goals. A social space of this kind is generated out of a rationalized and theorized form serving as an instrument for the violation of an existing space.

One is tempted to ask whether the various urban spaces with a grid pattern might not have comparable origins in constraints imposed by a central power. It turns out upon reflection, however, that there is no real justification for generalizing from the particular development of urban space in Latin America. Consider, for example, that transformation of space in New York City which began around 1810. Obviously it is to be explained in part by the existence and the influence of an already powerful urban nucleus, and by the actions of a duly empowered authority. On the other hand, developments in New York had absolutely nothing to do with the extraction of wealth by a metropolitan power, the colonial relationship with Britain having come to an end. Geometrical urban space in Latin America was intimately bound up with a process of extortion and plunder serving the accumulation of wealth in Western Europe; it is almost as though the riches produced were riddled out through the gaps in the grid. In English-speaking North America, by contrast, a formally homologous meshwork served only the production and accumulation of capital *on the spot*. Thus the same abstract form may have opposing functions and give rise to diverse structures. This is not to say that the form is indifferent to function and structure: in both these cases the pre-existing space was destroyed from top to bottom; in both the aim was homogeneity; and in both that aim was achieved.

What of the equally cross-ruled space of the Asian town and countryside? Here, apropos, is a résumé of the remarks of a Japanese philosopher of Buddhist background who was asked about the relationships between space, language and ideograms.

You will no doubt take a long time to understand the Chinese characters and the thinking behind these forms, which are not signs. You should know that for us perceptibility and intelligibility are not clearly distinct; the same goes for the signifier and what it signifies. It is hard for us to separate image and concept. So the meaning of an ideogram does not exist independently of its graphic representation. To put it in terms of your distinctions, sensation and intellect are merged for us into a single level of apprehension. Consider one of the simplest characters: a square and two strokes joining its centre to the middle points of each of its sides. I read this character, and I pronounce it *ta*. What you see, no doubt, is a dry geometrical figure. If I were to try and translate for you what I see and understand simultaneously when I look at this character, I would begin by saying that it was a bird's-eye view of a rice field. The boundary lines between rice fields are not stone walls or barbed-wire fences, but rather dykes which are an integral part of the fields themselves. When I contemplate this character, this rice field, I become the bird looking down from the optimum vantage point vertically above the centre of the field. What I perceive, however, is more than a rice field: it is also the order of the universe, the organizing-principle of space. This principle applies as well to the city as to the countryside. In fact everything in the universe is divided into squares. Each square has five parts. The centre designates He who thinks and sustains the order of the universe – formerly, the Emperor. An imaginary perpendicular line rises from the centre of the square. This is the ideal line going up to the bird overhead, to the perceiver of space. It is thus the dimension of thought, of knowledge, identified here with Wisdom and hence with the Power of the wise man to conceive and conserve the order of nature.

The Japanese notion of *shin-gyo-sho* elaborates further on this view of things. A basic principle rather than simply a procedure for ordering spatial and temporal elements, it governs the precincts of temples and palaces as well as the space of towns and houses; it informs the composition of spatial ensembles accommodating the broadest possible range of activity, from family life to major religious and political events. Under its aegis, public areas (the spaces of social relationships and actions) are connected up with private areas (spaces for contemplation, isolation and retreat) via 'mixed' areas (linking thoroughfares, etc.). The term *shin-gyo-sho* thus embraces three levels of spatial and temporal, mental and social organization, levels bound together by relationships of reciprocal implication. These relationships are not merely logical ones, though the logical relationship of implication certainly under-lies them. The 'public' realm, the realm of temple or palace, has private and 'mixed' aspects, while the 'private' house or dwelling has public (e.g. reception rooms) and 'mixed' ones. Much the same may be said of the town as a whole.

We thus have a global perception of space rather than represen-tations of isolated spots. Meeting-places, intersections in the chequ-erwork pattern, crossroads – these are more important to us than other places. Whence a number of social phenomena which may seem strange to your anthropologists, such as Edward T. Hall in his *Hidden Dimension*,[37] but which seem perfectly normal to us. It is indeed true, for example, that before the Americans came to Japan crossroads had names but the roads themselves did not, and that our houses bear numbers based on their age, not on their positions in the street. We have never had fixed routes for getting from one place to another, as you do, but that does not mean that we do not know where we are coming from or going to. We do not separate the ordering of space from its form, its genesis from its actuality, the abstract from the concrete, or nature from society. There is no house in Japan without a garden, no matter how tiny, as a place for contemplation and for contact with nature; even a handful of pebbles *is* nature for us – not just a detached symbol of it. We do not think right away of the distances that separate objects from one another. For space is never empty: it always embodies a meaning. The perception of gaps itself brings the whole body into play. Every group of places and objects has a centre, and this is therefore true of the house, the city or the whole world. The centre may be perceived from every side, and reached from every angle of approach; thus to occupy any vantage point is to perceive and discover everything that occurs. The centre so conceived can never become neutral or empty. It cannot be the 'locus of an absence', because it is occupied by Divinity, Wisdom and Power, which by manifesting themselves show any impression of void to be illusory. The accentuation of and infusion of metaphysical value into centres does not imply a corresponding devaluation of what surrounds those centres. Nature and divinity in the first place, then social life and relationships, and finally individual and private life – all these aspects of human reality have their assigned places, all implicatively linked in a concrete fashion. Nor is this assertion affected by the fact that the emphasis may shift upwards in order to express the transcendence of divinity, wisdom or power, whereas private life with its attendant gestures remains on a 'horizontal' plane, pitching its tent, so to speak, at ground level. A single order embraces all. Thus urban space is comprised, first, of wide avenues leading to the temples and palaces, secondly of medium-sized squares and streets which are the transitional and

[37] Edward T. Hall, *The Hidden Dimension* (Garden City, NY: Doubleday, 1966).

connecting spaces, and, thirdly and lastly, of the charming flower-filled alleys that afford access to our houses.

The important thing here is not to reconstruct a view which, though different from the Western one, is no less viable and up-to-date (and hence only indirectly the concern of anthropology in the broad sense, and even more distantly of ethnology), but rather to understand the grid that underlies it. Interestingly, this religious or political space has retained its relevance for thousands of years because it was rational from the outset. If we let the letter G (for 'global') represent the level of the system which has the broadest extension – namely the 'public' level of temples, palaces and political and administrative buildings; if we let P represent the level of residence and the places set aside for it – houses, apartments, and so on; and if M is allowed to stand for intermediate spaces – for arteries, transitional areas, and places of business – then we arrive at the following scheme.

$$
G \begin{cases} g \\ m \\ p \end{cases}
$$

$$
M \begin{cases} g \\ m \\ p \end{cases}
$$

$$
P \begin{cases} g \\ m \\ p \end{cases}
$$

In general descriptive terms, the 'private' realm P subsumes (though they are clearly distinct) entrances, thresholds, reception areas and family living-spaces, along with places set aside for retreat and sleep. Each individual dwelling likewise has an entrance, a focus, a place of retreat and so on. The level M takes in avenues and squares, medium-sized thoroughfares and the passageways leading to the houses. As for level G, it may be subdivided into interior spaces open to the public and the closed headquarters of institutions, into accessible itineraries and places reserved for notables, priests, princes and leaders. Similar considerations apply for each element of the system. Each location, at each level, has

its characteristic traits: open or closed, low or high, symmetrical or
asymmetrical.

Let us return now to the Japanese philosopher's remarks, the conclusion of which is something of a diatribe, something of an indictment
of Western civilization:

> Your streets, squares and boulevards have ridiculous names which
> have nothing to do with them, nor with the people and things
> around them – lots of names of generals and battles. Your cities
> have smashed any reasonable conception of space to pieces. The
> grid on which they are based, and the way you have elaborated
> upon it, are the best that the West can manage in this area, but it
> is a poor best. It is based merely on a set of transformations – on
> a structure. It took one of your greatest researchers to discover
> the fact that complex spaces in the form of trellises or semi-trellises
> are superior in practice to simplified spaces planned out in a
> branched or rectilinear manner. Our system, which I have been
> describing to you, shows why this is true: it has a concrete logic,
> a logic of the senses. Why don't you take it as a gift from us?
> Work on the hypothesis of a discourse at once theoretical and
> practical, a discourse of the everyday which also transcends every-
> day life, a discourse both mental and social, architectural and
> urbanistic. Something like the discourse of your forebears – and I
> am talking about the ancient Greeks, not the Gauls. Such a dis-
> course does not *signify* the city: it is the urban discourse itself.
> True, it partakes of the absolute. But why shouldn't it? It is a
> living discourse – unlike your lethal use of signs. You say you can
> 'decode' your system. Well, we do better than that: we create ours.

Here is the 'pro-Western' rejoinder:

> Not so fast, my friend. You say that the East has possessed a secret
> from time immemorial that the West has either lost or never had
> – namely, the key to the relationship between what people living
> in society do and what they say. In other words, the East is
> supposedly well acquainted with a vital connection which brings
> the religious, political and social realms into harmony with one
> another, whereas the West has destroyed all prospect of such
> harmony through its use of signs and its analytical proclivities.
> And you propose that your experience and thinking be made the

basis for the definition of a scheme closely akin to what Erwin
Panofsky calls, apropos of the Middle Ages, a *modus operandi* –
a scheme responsible at once for a specific way of life, a specific
space, specific monuments, specific ideas – in short, for a specific
civilization. You suggest that there is an underlying grid, or deep
structure, which explains the nature of places, the ways in which
they are put to use, the routes followed by their occupants, and
even the everyday gestures of those occupants. Permit me to point
out just how complicated such a scheme becomes as soon as one
tries to reconstruct it. Take a space Gg, closed, elevated and
symmetrical. It has to be distinguished from a space Gm, open,
elevated and symmetrical, as also from a space Gp, closed, located
at a lower level and asymmetrical – and so on and so forth. The
combinatory system involved is vast – and hard to work with even
with the help of a computer. Furthermore, can you be sure that it
accounts adequately for actual reality? Is it true, or sufficient, to
say that a temple in Kyoto has a public part, a part set aside for
rites, and a part reserved for priests and meditators? I grant that
your scheme explains something very important: difference within
a framework of repetition. Considered in its various contexts, for
example, the Japanese garden remains the same yet is never the
same: it may be an imperial park, an inaccessible holy place, the
accessible annex of a sanctuary, a site of public festivity, a place
of 'private' solitude and contemplation, or merely a way from one
place to another. This remarkable institution of the garden is
always a microcosm, a symbolic work of art, an object as well as
a place, and it has diverse 'functions' which are never merely
functions. It effectively eliminates from your space that antagonism
between 'nature' and 'culture' which takes such a devastating toll
in the West: the garden exemplifies the appropriation of nature,
for it is at once entirely natural – and thus a symbol of the
macrocosm – and entirely cultural – and thus the projection of a
way of life. Well and good. But let's not go overboard with
analogies. You say you are the possessors of a rationality. What
exactly is that rationality? Does it include conceiving of space as
a discourse, with rooms, houses (not forgetting gardens), streets,
and so on functioning as that discourse's component and signifying
elements? Your space, which is indeed both abstract and concrete,
has one drawback: it belongs to Power. It implies (and is implied
by) Divinity and Empire – knowledge and power combined and
conflated. Is that what you would have the West adopt? Well, we

find it hard to accept the idea that space and time should be produced by political power. Such ultra-Hegelianism (to use our terminology) is very fine, but it is unacceptable to us. The state is not (or is no longer) and can never for us be Wisdom united with Power. There is every reason to fear that your scheme could become a terrible weapon of oppression. You want to formalize this scheme scientifically in the Western manner. Westerners, on the other hand, might be more inclined to see it as an authoritarian definition of the space–time totality.

XIV

Formal and functional analyses do not eliminate the need to consider scale, proportion, dimension and level. That is the task of structural analysis, which is concerned with the relations between the whole and the parts, between 'micro' and 'macro' levels. Methodologically and theoretically, structural analysis is supposed to complement and complete the other kinds of analysis, not to transcend them. It is responsible for defining the whole (the global level) and for ascertaining whether it embodies a logic – that is, a strategy accompanied by a measure of symbolism (hence an 'imaginary' component). The relationship between the whole and the parts is bound up with general and well-known categories such as those of anaphora, metonymy and metaphor, but structural analysis introduces other, specific, categories into the discussion.

We have already encountered a case where structural analysis adduces such specific categories: the case of the production of monumental space. The ancient world worked with heavy masses. Greek theory and practice achieved the effect of unity by using both gravity and the struggle against weight; vertical forces, both ascending and descending, were neutralized and balanced without destroying the perception of volumes. Basing themselves on an identical principle, on the use of great volumes, the Romans exploited a complex arrangement of counterposed loads, supports and props, to obtain an effect of massiveness and strength unabashedly founded on weight. A less blatant structure, the outcome of an interplay between opposing forces, was typical of the Middle Ages; balance and the effect of balance were assured by lateral thrusts; lightness and *élan* were the order of the day. The modern period has seen the triumph of weightlessness, though in a way still consistent with the orientation of medieval architecture. Structural analysis is concerned, therefore, with clearly determined forces, as with the material relationships obtaining between those forces – relationships which give rise to equally clearly determined spatial structures: columns, vaults, arches, pillars, and so on.

Might it be said, then, that our analytic concepts correspond to certain classical terms, still often used, referring to the production of architectural space: that form and formal analysis correspond to 'composition', function to 'construction', and structure to proportion, scale, rhythm and the various 'orders'? The answer is yes – up to a certain point. The correspondence is sufficient, at any rate, to allow for the translation of 'classical' texts, from Vitruvius to Viollet-le-Duc, into modern terms. But this terminological parallelism cannot be taken too far, because that would be to forget the context, the materials and *matériel* – to forget that 'composition' is informed by ideologies, that 'construction' is a function of social relations, and that techniques, which have a great influence upon rhythm and upon the order of space, are liable to change.

As for the rather widely espoused view that the Greeks discovered a completely rational unity of form, function and structure, that this unity has been broken up in the course of history and that it needs to be restored, this hypothesis is a not unattractive one, but it takes no account of the new set of problems associated with the construction of ordinary buildings. The Greeks' celebrated unity applies almost exclusively to monumentality – to temple, stadium or agora.

The nexus of problems relating to space and its production extends beyond the field of classical architecture, beyond monuments and public buildings, to take in the 'private' sphere, the sphere of 'residence' and 'housing'. Indeed the relationship between private and public is now fundamental: today the global picture includes both these aspects, along with their relationship, and partial analyses, whether formal, functional or structural, must take this into account. The West's 'classical' terminology and perceptions must therefore be modified. The East may have something to teach the West in this regard, for the 'Asiatic mode of production' was always more apt to take 'private' residence into consideration. At all events, the categories of private and public and the contrast between monuments and buildings must henceforth be integral to our paradigm.

The tripartite approach founded on formal, functional and structural analyses cannot therefore be unreservedly endorsed as *the* method for deciphering social spaces, for what is truly essential gets through the 'grid'. By all means let us adopt this approach, and make the best use of it we can, but caution is very much in order.

I attempted earlier to show that semantic and semiological categories such as message, code and reading/writing could be applied only to spaces already produced, and hence could not help us understand the actual production of space. Relationships basic to semantic or semiological discussion which may refer to space in one way or another include: with respect to signs, the relationship between signifier and signified, and that between symbol and meaning; with respect to value, that between the value-imparting element and the element invested with value, likewise that between the devaluing factor and the factor divested of value; and, lastly, the relationship between what has a referent and what does not. Of the fact that spaces may 'signify' there can be no doubt. Is what is signified invariably contained by the signifier? Here, as elsewhere, the relationship of signifier to signified is susceptible to disjunction, distortion, instability, disparity and substitutions. Consider the presence of Greek columns on the façade of a stock exchange or bank, for example, or that of a pseudo-agora in a suburban 'new town'. What do such cases signify? Certainly something other than what they appear or seek to signify: specifically, the inability of capitalism to produce a space other than capitalist space and its efforts to conceal that production as such, to erase any sign of the maximization of profit. Are there spaces which fail to signify anything? Yes – some because they are neutral or empty, others because they are overburdened with meaning. The former fall short of signification; the latter overshoot it. Some 'over-signifying' spaces serve to scramble all messages and make any decoding impossible. Thus certain spaces produced by capitalist promoters are so laden with signs – signs of well-being, happiness, style, art, riches, power, prosperity, and so on – that not only is their primary meaning (that of profitability) effaced but meaning disappears altogether.

It is possible, and indeed normal, to decipher or decode spaces. This presupposes coding, a message, a reading and readers. What codes are involved? I use the plural advisedly, for it is doubtless as correct apropos of space as it is in the cases of philosophical and literary 'readings'. The codes in question, however, still have to be named and enumerated – or else, should this prove impossible, the questions of how and why this is so should be answered, and the meaning of this state of affairs explained.

According to Roland Barthes, we all have five codes available to us when reading a text.[38] First and foremost, the code of knowledge: on arrival in St Mark's Square, 'Ego' knows a certain number of things about Venice – about the doges, the Campanile, and so on. Memory floods his mind with a multitude of facts. Before long, he elicits another kind of meaning as he begins reading this (materialized) text in a manner roughly corresponding to the use of concept of function, to the use of functional analysis. ('Roughly' is the operative word here, of course, because his comprehension does not extend much beyond some sense of the *raison d'être*, or former *raison d'être*, of the Doge's Palace, the Piombi or the Bridge of Sighs.) He will also inevitably latch onto a few symbols: the lion, the phallus (the Campanile), the challenge to the sea. Though he may have learnt to attach dates to these, he also perceives them as embodying 'values' that are still relevant – indeed eternal. The disentanglement of these impressions from knowledge allows another code or reading – the symbolic one – to come into play. Meanwhile, 'Ego' is bound to feel some emotion: he may have been here before, long ago, or always dreamt of coming; he may have read a book or seen a film – *Death in Venice* perhaps. Such feelings are the basis of the subjective and personal code which now emerges, giving the decoding activity the musical qualities of a fugue: the theme (i.e. this place – the Square, the Palace, and so on) mobilizes several voices in a counterpoint in which these are never either distinct or confused. Finally, the simple empirical evidence of the paving-stones, the marble, the café tables leads 'Ego' to ask himself quite unexpected questions – questions about truth *versus* illusion, about beauty *versus* the message, or about the meaning of a spectacle which cannot be 'pure' precisely because it arouses emotions.

This kind of semantico-semiological research has gradually become more diversified. At the outset its theoretical project, on the basis of a strictly interpreted distinction between signifier and signified, posited the existence of two codes and two codes only: a denotative code operating at a primary level (that of the literal, the signified) which was acceptable to all linguists, and a connotative code, operating at a secondary (rhetorical) level, which was rejected by the more scientifically minded linguists as too vague a conception. More recently, however, the theory's basic concepts (message, code, reading) have become more flexible; a pluralistic approach has replaced the earlier strict insistence on an integral unity, and the former emphasis on consistency has given way to an emphasis on differences. The question is: how far can this

[38] See Roland Barthes, *S/Z* (Paris: Seuil, 1970), pp. 25ff. Eng. tr. by Richard Miller: *S/Z* (New York: Hill and Wang, 1974), pp. 18ff.

emphasis be carried, and how is difference to be defined in this context?

Barthes, for example, as we have seen, proposes five codes of equal importance and interest, worked out analytically *a posteriori*. Why five, rather than four or six, or some other number? By what mechanism is the choice made between one and another of these codes? And how are transitions made between them? Is there nothing to which they do not apply? Do they permit a truly exhaustive decoding of a given text, whether it is made up of verbal or non-verbal signs? If, to the contrary, residual elements remain, are we to conclude that infinite analysis is possible? Or are we being referred implicitly to a 'non-code' realm?

In point of fact this approach leaves two areas untouched, one on the near side and the other on the far side, so to speak, of the readable/visible. On the near side, what is overlooked is the body. When 'Ego' arrives in an unknown country or city, he first experiences it through every part of his body – through his senses of smell and taste, as (provided he does not limit this by remaining in his car) through his legs and feet. His hearing picks up the noises and the quality of the voices; his eyes are assailed by new impressions. For it is by means of the body that space is perceived, lived – and produced. On the far side of the readable/visible, and equally absent from Barthes's perspective, is power. Whether or not it is constitutional, whether or not it is disseminated through institutions and bureaucracies, power can in no wise be decoded. For power has no code. The state has control of all existing codes. It may on occasion invent new codes and impose them, but it is not itself bound by them, and can shift from one to another at will. The state manipulates codes. Power never allows itself to be confined within a single logic. Power has only strategies – and their complexity is in proportion to power's resources. Similarly, in the case of power, signifier and signified coincide in the shape of violence – and hence death. Whether this violence is enacted in the name of God, Prince, Father, Boss or Patrimony is a strictly secondary issue.

It is pure illusion to suppose that thought can reach, grasp or define what is *in* space on the basis of propositions *about* space and general concepts such as message, code and readability. This illusion, which reduces both matter and space to a representation, is in fact simply a version of spiritualism or idealism – a version which is surely common to all who put political power, and hence state power, in brackets, and so see nothing but things. Cataloguing, classifying, decoding – none of these procedures gets beyond mere description. Empiricism, however, whether of the subtle or the crude variety, whether based on logic or on the facts themselves, presupposes a conception of space which contradicts the premises of empiricism itself in that it is incompatible as much with finite enumerations (including a restricted muster of codes) as with the indeterminacy of unlimited analysis. There is a proper role for the decoding of space: it helps us understand the transition from representational spaces to representations of space, showing up correspondences, analogies and a certain unity in spatial practice and in the theory of space. The limitations of the decoding-operation appear even greater, however, as soon as it is set in motion, for it then immediately becomes apparent just how many spaces exist, each of them susceptible of multiple decodings.

Beginning with space-as-matter, paradigmatic contrasts proliferated: abundance *versus* barrenness, congeniality *versus* hostility, and so on. It was upon this primary stratum of space, so to speak, that agricultural and pastoral activity laid down the earliest networks: *ur*-places and their natural indicators; blazes or way-markers with their initial duality of meaning (direction/orientation, symmetry/asymmetry). Later, absolute space – the space of religion – introduced the highly pertinent distinctions between speech and writing, between the prescribed and the forbidden, between accessible and reserved spaces, and between full and empty. Thus certain spaces were carved out of nature and made complete by being filled to saturation point with beings and symbols, while other spaces were withdrawn from nature only to be kept empty as a way of symbolizing a transcendent reality at once absent and present. The paradigm became more complex as new contrasts came into play: within/without, open/closed, movable/fixed. With the advent of historical space, places became much more diverse, contrasting much more sharply with one another as they developed individual characteristics. City walls were the mark of a material and brutal separation far more potent than the formal polarities they embodied, such as curved-*versus*-straight or open-*versus*-closed. This separation had more than one signification – and indeed implied more than any mere signification, in that the fortified towns held administrative sway over the surrounding countryside, which they protected and exploited at the same time (a common enough phenomenon, after all).

Once diversified, places opposed, sometimes complemented, and sometimes resembled one another. They can thus be categorized or subjected to a grid on the basis of 'topias' (isotopias, heterotopias, utopias, or in other words analogous places, contrasting places, and the places of what has no place, or no longer has a place – the absolute, the divine, or the possible). More importantly, such places can also be viewed in terms of the highly significant distinction between *dominated* spaces and *appropriated* spaces.

XV

Before considering the distinction between domination and appropriation, however, a word must be said about the relationship between the basic axes of diachronic and synchronic. No space ever vanishes utterly, leaving no trace. Even the sites of Troy, Susa or Leptis Magna still enshrine the superimposed spaces of the succession of cities that have occupied them. Were it otherwise, there would be no 'interpenetration', whether of spaces, rhythms or polarities. It is also true that each new addition inherits and reorganizes what has gone before; each period or stratum carries its own preconditions beyond their limits. Is this a case of metaphorization? Yes, but it is one which includes a measure of metonymization in that the superimposed spaces do constitute an ensemble or whole. These notions may not explain the process in question, but they do serve a real expository function: they help describe how it is that natural (and hence physical and physiological) space does not get completely absorbed into religious and political space, or these last into historical space, or any of the foregoing into that practico-sensory space where bodies and objects, sense organs and products all cohabit in 'objectality'. What are being described in this way are metamorphoses, transfers and substitutions. Thus natural objects – a particular mound of earth, tree or hill – continue to be perceived as part of their contexts in nature even as the surrounding social space fills up with objects and comes also to be apprehended in accordance with the 'objectality' shared by natural objects on the one hand and by products on the other.

Now let us consider *dominated* (and dominant) space, which is to say a space transformed – and mediated – by technology, by practice. In the modern world, instances of such spaces are legion, and immediately intelligible as such: one only has to think of a slab of concrete or a motorway. Thanks to technology, the domination of space is becoming, as it were, completely dominant. The 'dominance' whose acme we are thus fast approaching has very deep roots in history and in the historical sphere, for its origins coincide with those of political power itself. Military architecture, fortifications and ramparts, dams and irrigation systems – all offer many fine examples of dominated space. Such spaces are works of construction rather than 'works' in the sense in which we have been using the term, and they are not yet 'products' in its narrow, modern and industrial meaning; dominant space is invariably the realization of a master's project. This may seem simple enough, but in fact the concept of dominated space calls for some elucidation. In order to dominate space, technology introduces a new form into a pre-existing space – generally a rectilinear or rectangular form such as a meshwork or chequerwork. A motorway brutalizes the countryside and the land, slicing through space like a great knife. Dominated space is usually closed, sterilized, emptied out. The concept attains its full meaning only when it is contrasted with the opposite and inseparable concept of *appropriation*.

In Marx, the concept of appropriation is sharply opposed to that of property, but it is not thoroughly clarified – far from it, in fact. For one thing, it is not clearly distinguished from the anthropological and philosophical notion of human nature (i.e. what is 'proper' to human beings); Marx had not entirely abandoned the search for a specific human nature, but he rejected any idea that it might be constituted by laughter, by play, by the awareness of death, or by 'residence'; rather, it lay in (social) labour and – inseparably – in language. Nor did Marx discriminate between appropriation and domination. For him labour and technology, by dominating material nature, thereby immediately transformed it according to the needs of (social) man. Thus nature was converted directly from an enemy, an indifferent mother, into 'goods'.

Only by means of the critical study of space, in fact, can the concept of appropriation be clarified. It may be said of a natural space modified in order to serve the needs and possibilities of a group that it has been appropriated by that group. Property in the sense of possession is at best a necessary precondition, and most often merely an epiphenomenon, of 'appropriative' activity, the highest expression of which is the work of art. An appropriated space *resembles* a work of art, which is not to say that it is in any sense an *imitation* work of art. Often such a space is a structure – a monument or building – but this is not always the case: a site, a square or a street may also be legitimately described as an appropriated space. Examples of appropriated spaces abound, but it is not always easy to decide in what respect, how, by whom and for whom they have been appropriated.

Peasant houses and villages speak: they recount, though in a mumbled and somewhat confused way, the lives of those who built and inhabited them. An igloo, an Oriental straw hut or a Japanese house is every bit as expressive as a Norman or Provençal dwelling.[39] Dwelling-space may be that of a group (of a family, often a very large one) or that of a community (albeit one divided into castes or classes which tend to break it up). Private space is distinct from, but always connected with, public space. In the best of circumstances, the outside space of the community is dominated, while the indoor space of family life is appropriated.[40] A situation of this kind exemplifies a spatial practice which, though still immediate, is close, in concrete terms, to the work of art. Whence the charm, the enduring ability to enchant us, of houses of this kind. It should be noted that appropriation is not effected by an immobile group, be it a family, a village or a town; time plays a part in the process, and indeed appropriation cannot be understood apart from the rhythms of time and of life.

Dominated space and appropriated space may in principle be combined – and, ideally at least, they ought to be combined. But history – which is to say the history of accumulation – is also the history of their separation and mutual antagonism. The winner in this contest, moreover, has been domination. There was once such a thing as appropriation without domination – witness the aforementioned hut, igloo or peasant house. Domination has grown *pari passu* with the part played by armies, war, the state and political power. The dichotomy between dominated and appropriated is thus not limited to the level of discourse or signification, for it gives rise to a contradiction or conflictual tendency which holds sway until one of the terms in play (domination) wins a crushing victory and the other (appropriation) is utterly subjugated. Not that appropriation *disappears*, for it cannot: both practice and theory continue to proclaim its importance and demand its restitution.

Similar considerations apply to the body and to sexuality. Dominated by overpowering forces, including a variety of brutal techniques and an extreme emphasis on visualization, the body fragments, abdicates responsibility for itself – in a word, disappropriates itself. Body cultures and body techniques have been developed, in antiquity and since, which truly appropriate the body. Sports and gymnastics as we know them, however, to say nothing of the passive exposure of the body to the sun, are little more than parodies or simulations of a genuine 'physical culture'. Any revolutionary 'project' today, whether utopian or realistic, must, if it is to avoid hopeless banality, make the reappropriation of

[39] See Rapoport, *House Form and Culture*. Like Hall, Rapoport inflates the significance of socio-cultural factors and 'actors'.
[40] Cf. Bachelard, *La poétique de l'espace* (see above, p. 121, n. 9).

the body, in association with the reappropriation of space, into a non-negotiable part of its agenda.

As for sex and sexuality, things here are more complicated. It may reasonably be asked whether an appropriation of sexuality has ever occurred except perhaps under certain transitory sets of circumstances and for a very limited number of people (one thinks, for example, of Arab civilization in Andalusia). Any true appropriation of sex demands that a separation be made between the reproductive function and sexual pleasure. This is a delicate distinction which, for reasons that are still mysterious, and despite great scientific advances in the sphere of contraception, can only be made in practice with great difficulty and attendant anxiety. We do not really know how and why this occurs, but it seems that detaching the biological sexual function from the 'human' one – which cannot properly be defined in terms of functionality – results only in the latter being compromised by the elimination of the former. It is almost as though 'nature' were itself incapable of distinguishing between pleasure and pain, so that when human beings are encouraged by their analytical tendencies to seek the one in isolation from the other they expose themselves to the risk of neutralizing both. Alternatively, they may be obliged to limit all orgiastic pleasure to predictable states reached by codified routes (drugs, eroticism, reading/writing of ready-made texts, etc.).

The true space of pleasure, which would be an appropriated space par excellence, does not yet exist. Even if a few instances in the past suggest that this goal is in principle attainable, the results to date fall far short of human desires.

Appropriation should not be confused with a practice which is closely related to it but still distinct, namely 'diversion' (*détournement*). An existing space may outlive its original purpose and the *raison d'être* which determines its forms, functions, and structures; it may thus in a sense become vacant, and susceptible of being diverted, reappropriated and put to a use quite different from its initial one. A recent and well-known case of this was the reappropriation of the Halles Centrales, Paris's former wholesale produce market, in 1969–71. For a brief period, this urban centre, designed to facilitate the distribution of food, was transformed into a gathering-place and a scene of permanent festival – in short, into a centre of play rather than of work – for the youth of Paris.

The diversion and reappropriation of space are of great significance, for they teach us much about the production of new spaces. During a period as difficult as the present one is for a (capitalist) mode of production which is threatened with extinction yet struggling to win a new lease on life (through the reproduction of the means of production), it may even be that such techniques of diversion have greater import than attempts at creation (production). Be that as it may, one upshot of such tactics is that groups take up residence in spaces whose pre-existing form, having been designed for some other purpose, is inappropriate to the needs of their would-be communal life. One wonders whether this morphological maladaptation might not play a part in the high incidence of failure among communitarian experiments of this kind.

From a purely theoretical standpoint, diversion and production cannot be meaningfully separated. The goal and meaning of theoretical thinking is production rather than diversion. Diversion is in itself merely appropriation, not creation – a reappropriation which can call but a temporary halt to domination.

3

Spatial Architectonics

I

Having assigned ontological status by speculative diktat to the most extreme degree of formal abstraction, classical philosophical (or metaphysical) thought posits a substantial space, a space 'in itself'. From the beginning of the *Ethics*, Spinoza treats this absolute space as an attribute or mode of absolute being – that is, of God.[1] Now space 'in itself', defined as infinite, has no shape in that it has no content. It may be assigned neither form, nor orientation, nor direction. Is it then the unknowable? No: rather, it is what Leibniz called the 'indiscernible'.

In the matter of Leibniz's criticism of Spinoza and Descartes, as in that of Newton's and Kant's criticism of Leibniz, modern mathematics tends to find in favour of Leibniz.[2] For the most part, philosophers have taken the existence of an absolute space as a given, along with whatever it might contain: figures, relations and proportions, numbers, and so on. Against this posture, Leibniz maintains that space 'in itself', space as such, is neither 'nothing' nor 'something' – and even less the totality of things or the form of their sum; for Leibniz space was, indeed, the indiscernible. In order to discern 'something' therein, axes and an origin must be introduced, and a right and a left, i.e. the direction or orientation of those axes. This does not mean, however, that Leibniz espouses the 'subjectivist' thesis according to which the observer and the measure together constitute the real. To the contrary, what Leibniz means to say is that it is necessary for space to be *occupied*. What, then, occupies

---

[1] Baruch Spinoza, *Ethics*, I, proposition xiv, corollary 2, and proposition xv, Scholium.
[2] See Hermann Weyl, *Symmetry* (Princeton, N.J.: Princeton University Press, 1952), and my discussion of Weyl's work below.

space? A body – not bodies in general, nor corporeality, but a specific body, a body capable of indicating direction by a gesture, of defining rotation by turning round, of demarcating and orienting space. Thus for Leibniz space is *absolutely relative* – that is, endowed both with a perfectly *abstract* quality which leads mathematical thought to treat it as primordial (and hence readily to invest it with transcendence), and with a *concrete* character (in that it is in space that bodies exist, that they manifest their material existence). How does a body 'occupy' space? The metaphorical term 'occupy' is borrowed from an everyday experience of space as already specific, already 'occupied'. The connection between space as 'available' and space as 'occupied', however, has nothing simple or obvious about it. Unfortunately, a metaphor cannot do duty for thought. We know that space is not a pre-existing void, endowed with formal properties alone. To criticize and reject absolute space is simply to refuse a particular *representation*, that of a container waiting to be filled by a content – i.e. matter, or bodies. According to this picture of things, (formal) content and (material) container are *indifferent* to each other and so offer no graspable difference. Any thing may go in any 'set' of places in the container. Any part of the container can receive anything. This indifference becomes separation, in that contents and container do not impinge upon one another in any way. An empty container accepts any collection of separable and separate items; separateness thus extends even to the contents' component elements; fragmentation replaces thought, and thought, reflective thinking, becomes hazy and may eventually be swallowed up in the empirical activity of simply counting things. The constitution of such a 'logic of separation' entails and justifies a *strategy* of separation.

We are thus obliged to consider a contrary hypothesis. Can the body, with its capacity for action, and its various energies, be said to create space? Assuredly, but not in the sense that occupation might be said to 'manufacture' spatiality; rather, there is an immediate relationship between the body and its space, between the body's deployment in space and its occupation of space. Before *producing* effects in the material realm (tools and objects), before *producing itself* by drawing nourishment from that realm, and before *reproducing itself* by generating other bodies, each living body *is* space and *has* its space: it produces itself in space and it also produces that space. This is a truly remarkable relationship: the body with the energies at its disposal, the living body, creates or produces its own space; conversely, the laws of space, which is to say the laws of discrimination in space, also govern the living body and the deployment of its energies. Hermann Weyl demonstrates this very clearly in his work on symmetry.[3] In nature, whether organic or in-organic, symmetries (in a plane or about an axis) exist wherever there is bilaterality or duality, left and right, 'reflection', or rotation (in space); these symmetries are not properties external to bodies, however. Though definable in 'purely' mathematical terms – as applications, operations, transformations or functions – they are not imposed upon material bodies, as many philosophers suppose, by prior thought. Bodies – deployments of energy – produce space and produce themselves, along with their motions, according to the laws of space. And this remains true, Weyl argues, whether we are concerned with corpuscles or planets, crystals,[4] electromagnetic fields,[5] cell division,[6] shells, or architectural forms, to which last Weyl attributes great importance. Here then we have a route from abstract to concrete which has the great virtue of demonstrating their reciprocal inherence. This path leads also from mental to social, a fact which lends additional force to the concept of the production of space.

This thesis is so persuasive that there seems to be little reason for not extending its application – with all due precautions, naturally – to *social* space. This would give us the concept of a specific space produced by forces (i.e. productive forces) deployed within a (social and determined/determining) spatial practice. Such a space would embody 'properties' (dualities, symmetries, etc.) which could not be imputed either to the human mind or to any transcendent spirit, but only to the actual 'occupation' of space, an occupation which would need to be understood genetically – that is, according to the sequence of productive operations involved.

What does this mean for the ancient idea of nature? It means that it must undergo quite substantial transformation. Once the relationship of mutual inherence between space and what it contained was broken, reflective thought tended to bring occult qualities and forces into the picture. Everything which derives from biologico-spatial reality – every-thing which is, in a word, 'automorphic' or 'biomorphic' – was endowed in this way with goal-directedness: symmetries now seemed to have been

---

[3] Weyl, *Symmetry.*
[4] Ibid., pp. 28–9.
[5] In a discussion which starts out from the 'classical' theses of Leibniz, Newton and Kant (ibid., pp. 16ff.), Weyl is led to express some reservations about Ernst Mach's position. Does this mean that his own stance supports that taken by Lenin in *Materialism and Empirio-Criticism?* Not exactly; Weyl would probably feel that Lenin asked the right question – but took bad aim and missed the target.
[6] Ibid., pp. 33ff.

planned by a calculating God and realized on the material plane by order of a divine will or power. How did it come about that that flower which knew not that it was a flower, or that it was beautiful, possessed a symmetry of the nth order? The answer was that it had been *designed*, by Spinoza's *natura naturans* or by Leibniz's mathematician God.

Many, like Descartes and his followers, though they found it hard to believe in any such engineering, simply shifted agency to a 'spirit', whether human or not, without attending too closely to the matter of how that spirit's 'design' might be realized otherwise than through the providential or transcendent action of the Idea (in the Hegelian sense). How and in what sense nature as such can 'be' mathematical is a question which the philosophers, with their scientific-*cum*-ideological partitions, have rendered unintelligible. The observer stands perplexed before the beauty of a seashell, a village or a cathedral, even though what confronts him consists perhaps merely in the material modalities of an active 'occupation' – specifically, the occupation of space. Incidentally, one may well wonder whether the 'integrons' proposed by François Jacob as a way of accounting for organic unity are really anything more than a philosophical/ideological/scientific device standing in for divine providence.[7]

There is another way of approaching the question, however: development in nature may be conceived of as obeying laws of space which are also laws of nature. Space *as such* (as at once occupied and occupying, and as a set of places) may be understood in a materialist way. A space so understood implies differences by definition, which gets us out of a number of difficulties related to the genesis of variations: we are no longer obliged to appeal either to originality or to origins as the source of difference; nor need we risk falling under the axe of the materialist critique of empirio-criticism. From this perspective, the form of a seashell is the result neither of a 'design' nor of 'unconscious' thought, nor yet of a 'higher' plan. The poetry of shells – their metaphorical role[8] – has nothing to do with some mysterious creative force, but corresponds merely to the way in which energy, under specific conditions (on a specific scale, in a specific material environment, etc.), is deployed; the relationship between nature and space is *immediate* in the sense that it does not depend on the mediation of an external force, whether natural

---

[7] See François Jacob, *La logique du vivant: une histoire de l'hérédité* (Paris: Gallimard, 1976), pp. 320ff.

[8] See Gaston Bachelard, *La poétique de l'espace* (Paris: Presses Universitaires de France, 1958), pp. 125ff. Eng. tr. by Maria Jolas: *The Poetics of Space* (Boston, Mass.: Beacon Press, 1969), pp. 129ff.

or divine. The law of space resides within space itself, and cannot be resolved into a deceptively clear inside-*versus*-outside relationship, which is merely a *representation of space*. Marx wondered whether a spider could be said to *work*. Does a spider obey blind instinct? Or does it have (or perhaps better, *is* it) an intelligence? Is it aware in any sense of what it is doing? It produces, it secretes and it occupies a space which it engenders according to its own lights: the space of its web, of its stratagems, of its needs. Should we think of this space of the spider's as an abstract space occupied by such separate objects as its body, its secretory glands and legs, the things to which it attaches its web, the strands of silk making up that web, the flies that serve as its prey, and so on? No, for this would be to set the spider in the space of analytic intellection, the space of discourse, the space of this sheet of paper before me, thus preparing the ground too inevitably for a rejoinder of the type: 'Not at all! It is nature (or instinct, or providence) which governs the spider's activity and which is thus responsible for that admirable and totally marvellous creation, the spider's web with its amazing equilibrium, organization, and adaptability.' Would it be true to say that the spider spins the web as an extension of its body? As far as it goes, yes, but the formulation has its problems. As for the web's symmetrical and asymmetrical aspects and the spatial structures (anchorage points, networks, centre/periphery) that it embodies, is the spider's knowledge of these comparable to the human form of knowledge? Clearly not: the spider produces, which manifestly calls for 'thought', but it does not 'think' in the same way as we do. The spider's 'production' and the characteristics thereof have more in common with the seashell or with the flower evoked by the 'Angel of Silesia' than with verbal abstraction. Here the production of space, beginning with the production of the body, extends to the productive secretion of a 'residence' which also serves as a tool, a means. This construction is consistent with those laws classically described as 'admirable'. Whether any dissociation is conceivable in this connection between nature and design, organic and mathematical, producing and secreting, or internal and external, is a question which must be answered – resoundingly – in the negative. Thus the spider, for all its 'lowliness', is already capable, just like human groups, of demarcating space and orienting itself on the basis of angles. It can create networks and links, symmetries and asymmetries. It is able to project beyond its own body those *dualities* which help constitute that body as they do the animal's relationship to itself and its productive and reproductive acts. The spider has a sense of right and left, of high and low. Its 'here and now' (in Hegel's sense) transcends the realm of 'thingness', for it embraces relationships and movements. We may say, then, that for any living body, just as for spiders, shellfish and so on, the most basic places and spatial indicators are first of all *qualified* by that body. The 'other' is present, facing the ego: a body facing another body. The 'other' is impenetrable save through violence, or through love, as the object of expenditures of energy, of aggression or desire. Here external is also internal inasmuch as the 'other' is another body, a vulnerable flesh, an accessible symmetry. Only later on in the development of the human species were spatial indicators quantified. Right and left, high and low, central and peripheral (whether named or no) derived from the body in action. It seems that it is not so much *gestures* which do the qualifying as the body as a whole. To say that the qualification of space depends on the body implies that space is determined by something that at times threatens and at times benefits it. This determination appears to have three aspects: gestures, traces, marks. 'Gesture' should be taken here in a broad sense, so that turning around may be considered a gesture, one which modifies a person's orientation and points of reference. The word is preferable to 'behaviour', for a gestural action has a goal or aim (which is not, of course, to imply some immanent teleology). A spider moving around on its web or a shellfish emerging from its shell are performing gestures in this sense. As for traces and marks, these obviously do not exist as 'concepts' for the spider, and yet everything happens 'just as though' they did. Marks are made by living beings with the means readily available to them, notably excreta such as urine, saliva, and so on. Sexual marks must be very ancient (but to what – or to whom – were they first affixed?). As indicators merely of affect, however, marks would appear to be of much more recent origin, and limited to few species. Intentionality is a late development, accompanying that of brain and hands, but traces and marks play a part in animal life from a very early date. Places were already being marked (and 're-marked'). In the beginning was the Topos. Before – long before – the advent of the Logos, in the chiaroscuro realm of primitive life, lived experience already possessed its internal rationality; this experience was *producing* long before *thought* space, and spatial thought, began *reproducing* the projection, explosion, image and orientation of the body. Long before space, as perceived by and for the 'I', began to appear as split and divided, as a realm of merely virtual or deferred tensions and contacts. Long before space emerged as a medium of far-off possibilities, as the locus of potentiality. For, long before the analysing, separating intellect, long before formal knowledge, there was an intelligence of the body.

Time is distinguishable but not separable from space. The concentric rings of a tree trunk reveal the tree's age, just as a shell's spirals, with their 'marvellous' spatial concreteness, reveal the age of that shell's former occupant – this according to rules which only complicated mathematical operations can 'translate' into the language of abstraction. Times, of necessity, are local; and this goes too for the relations between places and their respective times. Phenomena which an analytical intelligence associates solely with 'temporality', such as growth, maturation and aging, cannot in fact be dissociated from 'spatiality' (itself an abstraction). Space and time thus appear and manifest themselves as different yet unseverable. Temporal cycles correspond to circular spatial forms of a symmetrical kind. It may even be that linear temporal processes of a repetitive and mechanical character are associated with the constitution of spatial axes (along which a repeated operation may be performed). At all events, the dissociation of spatial and temporal and the social actualization of that dissociation can only be a late development, a corollary of which has been the split between representations of space and representational spaces. It is by taking representational spaces as its starting-point that art seeks to preserve or restore this lost unity.[9]

All of which gives us some sense of how and to what degree duality is constitutive of the unity of the material living being. Such a being carries its 'other' within itself. It is symmetrical, hence dual – and doubly so, for its symmetry is both bilateral and rotational; and this state of affairs must in turn be viewed through the dual lens of space and time, of cyclical repetition and linear repetition.

Around the living being, and through its activity, which may legitimately be described as 'productive', is constituted the field which the behaviourists call 'behavioural'. This field comes into play as a network of relations, a network projected and simultaneously actualized by the living being as it acts within, in conjunction with, and upon, its spatial 'milieu'. The realm of 'behaviour' thus bears spatial characteristics determined by the projection in question: right–left symmetry, high versus low, and so on.

At the same time, the living being constitutes itself from the outset as an internal space. Very early on, in phylogenesis as in the genesis of the individual organism, an indentation forms in the cellular mass. A cavity

---

[9] See Claude Gaignebet's analysis of the spatio-temporal unity of the festivals of the Christian calendar, as evoked in Bruegel's Fight between Carnival and Lent: '"Le Combat de Carnaval et de Carême" de P. Bruegel', Annales: ESC, 27, no. 2 (1972), pp. 313–45.

gradually takes shape, simple at first, then more complex, which is filled by fluids. These fluids too are relatively simple to begin with, but diversify little by little. The cells adjacent to the cavity form a screen or membrane which serves as a boundary whose degree of permeability may vary. From now on external space will stand opposed to an internal space or milieu: here is the primary and most decisive differentiation in the history of biological being. The internal milieu will play an ever-greater role; and the space thus produced will eventually take on the most varied forms, structures and functions, beginning with an initial stage at which it has the form of what the embryologists call a 'gastrula'.

A closure thus comes to separate within from without, so establishing the living being as a 'distinct body'. It is a quite relative closure, however, and has nothing in common with a logical division or abstract split. The membranes in question generally remain permeable, punctured by pores and orifices. Traffic back and forth, so far from stopping, tends to increase and become more differentiated, embracing both energy exchange (alimentation, respiration, excretion) and information exchange (the sensory apparatus). The whole history of life has been characterized by an incessant diversification and intensification of the interaction between inside and outside.

Thus relativized and emancipated from extrapolations and systematizations, the notion of 'closure' has an operational utility: it helps to account for what happens in both natural and social life. In the social realm, closures tend to become absolute. A defining characteristic of (private) property, as of the position in space of a town, nation or nation state, is a closed frontier. This limiting case aside, however, we may say that every spatial envelope implies a barrier between inside and out, but that this barrier is always relative and, in the case of membranes, always permeable.

II

From a dynamic standpoint, the living organism may be defined as an apparatus which, by a variety of means, captures energies active in its vicinity. It absorbs heat, performs respiration, nourishes itself, and so on. It also, as a 'normal' thing, retains and stocks a surplus of available energy over and above what it needs for dealing with immediate demands and attacks. This allows the organism a measure of leeway for taking initiatives (these being neither determined nor arbitrary). This surplus or superfluity of energy is what distinguishes life from survival (the bare minimum needed to support life). Captive energy is not generally stored indefinitely or preserved in a stagnant state. When it is, the organism degenerates. It is in the nature of energy that it be expended – and expended *productively*, even when the 'production' involved is merely that of play or of gratuitous violence. The release of energy always gives rise to an effect, to damage, to a change in reality. It modifies space or generates a new space. Living or vital energy seems *active* only if there is an excess, an available surplus, superfluity and an actual expenditure thereof. In effect, energy must be *wasted*; and the explosive waste of energy is indistinguishable from its productive use: beginning on the plane of animal life, play, struggle, war and sex are coextensive. Production, destruction and reproduction overlap and intersect.

Energy accumulates: so much is obvious – a truism. It is difficult, however, to form a clear picture of the mechanisms of this accumulation, and even more so of its consequences. Even though the expenditure of energy always seems 'excessive', even 'abnormal', a living organism which does not have access to such a surplus, and hence to the possibilities which that surplus opens up, has quite different reactions to its immediate circumstances.

In other words, that 'principle of economy' which has so often been put forward by a particular kind of rationalism or crude functionalism is biologically and 'biomorphically' inadequate. It is a low-level principle applied only to situations where a short supply of energy calls for restrictions on expenditure. It applies, in other words, only at the level of survival.

In sharp contrast to the rationalism of the 'principle of economy' and its niggardly productivism (the minimum expenditure – and this only in order to satisfy 'needs') is the opposite thesis, espoused by a succession of philosophers, according to which waste, play, struggle, art, festival – in short, Eros – are themselves a necessity, and a necessity out of which the partisans of this view make a virtue. The pedigree of the philosophical endorsement of excess, of superfluity – and hence of transgression – in this connection goes back to Spinoza; it may be traced thence, via Schiller, Goethe, and Marx – who detested asceticism, even if he sometimes allowed himself to be seduced by the notion of a 'proletarian' version of it – to its culmination in Nietzsche. There is little trace of it, be it noted, in Freud, whose bio-energetic theories tend to collapse into mechanism. The psychoanalytic distinctions between Eros and Thanatos, pleasure principle and reality (or productivity) principle, and life and death instincts, too often lose all dialectical character and become little more than a mechanical interplay between pseudo-concepts – little more than metaphors for a supposed scarcity of energy. If the living organism indeed captures, expends and wastes a surplus of energy, it must do so in accordance with the laws of the universe. The Dionysian side of existence – excess, intoxication, risks (even mortal risks) – has its own peculiar freedom and value. The living organism and the total body contain within them the potential for play, violence, festival and love (which is not to say that this potential must necessarily be realized, nor even that any motivation to do so need be present).

The Nietzschean distinction between Apollonian and Dionysian echoes the dual aspect of the living being and its relationship to space – its own space and the other's: violence and stability, excess and equilibrium. Inadequate as this distinction may be, it is certainly meaningful.

It is not sufficient, however, to say of the living organism merely that it captures energy and uses it in an 'economic' manner: it does not capture just any energy, nor does it expend that energy in an arbitrary way. It has its own specific prey, surroundings and predators – in a word, its own *space*. It lives *in* that space, and it is a component part *of* it – a part, that is, of a fauna or flora, and of an ecology, a more or less stable ecological system. Within its space, the living being receives information. Originally, before the advent of the abstraction devised by human societies, information was no more distinct from material reality than the content of space was from its form: the cell receives information in material form. There is a systematic philosophical tendency among the investigators of such phenomena, however, to reduce the living being – whether at the level of the individual cell or at the level of the organism as a cellular whole – to terms of information reception; that is to say, to terms of minute quantities of energy.[10] They disregard or ignore the economy of the living body as recipient and reservoir of *massive* energies. Though they put all the emphasis on the organism's self-regulatory mechanisms, they no longer discern those mechanisms' dysfunctions, defects, errors, or excessive outlays of energy. The dual regulatory system based on organic substances and catalysts which biology proposes is apparently supposed to leave nothing out of account. It is true that energetic theories, for their part, have paid no attention to the informational, relational or situational realms, concentrating exclusively

---

[10] See for example Jacques Monod, *Le hasard et la nécessité, essai sur la philosophie naturelle de la biologie moderne* (Paris: Seuil, 1970). Eng. tr. by Austryn Wainhouse: *Chance and Necessity: An Essay on the Natural Philosophy of Modern Biology* (New York: Knopf, 1971).

on the grosser forms of energy – those which can, so to speak, be measured in calories. The truth is, however, that in its relationship to itself and its own space the living being uses both minimal and massive types of energy (which are in any case not strictly separable). The organism thus combines apparatuses storing enormous quantities of energy which are discharged explosively (musculature, sexual apparatus, members) with apparatuses designed to respond to very feeble stimuli – i.e. information – and to consume barely any energy (the sensory apparatus: the brain and sense organs).[11] What we find here, therefore – or, rather, what we come back to – is a constitutive dualism. The living being is not merely a data-processing machine, nor merely a desiring, killing or producing machine – it is both at once.

Around the living organism, both those energies which it captures and those which threaten it are *mobile*: they are 'currents' or 'flows'. By contrast, in order to capture available energies the organism must have at its disposal apparatuses which are *stable*. It must respond to aggression with defensive actions, setting up boundaries around the body that it can maintain and protect.

The fact that a surplus of energy is accumulated before being discharged is thus a defining aspect of the very concept of the 'living body' and its relationship with its space – i.e. with itself, its vicinity, its surroundings, and the world at large. A productive squandering of energy is not a contradiction in terms: an expenditure of energy may be deemed 'productive' so long as some change, no matter how small, is thereby effected in the world. The concept of production is thus sharpened and revived without becoming so broad as to lose all meaning: we see that a game may qualify as a piece of work, or as a work in the strong sense of the word, while a space designed for playful activity may legitimately be deemed a *product* in that it is the outcome of an activity which regulates itself (lays down rules for itself) as it unfolds. Furthermore, productive energy implies the living organism's relationship with itself, and in this connection takes the form of *reproductive* energy; as such it is characterized by repetition – repetition in the division and multiplication of cells, in actions, in reflexes. As for *sexual*

---

[11] This has been well brought out by Georges Bataille, elaborating on a Nietzschean theme in his *La part maudite, essai d'économie générale* (Paris: Editions de Minuit, 1949); Eng. tr. by Robert Hurley: *The Accursed Share: An Essay on General Economy*, vol. I (New York: Zone Books, 1988). It would be unjust not to give credit here to Wilhelm Reich for his contribution to the development of an energetic theory (and this in a much-disparaged period of his work). Cf. also a Yugoslavian film which comments not unhumorously on this issue: Dusan Makavejev's *WR: Mysteries of the Organism* (1972).

reproduction, it is merely one of many forms of reproduction essayed by nature, a form which owes its prominent status solely to its *success* down several lines of descent. In the case of sexual reproduction, the discontinuous or explosive aspect of productive energy has manifestly won out over continuous production, over sprouting and proliferation.

Surplus energy *qua* 'normal' energy relates on the one hand to itself, i.e. to the body which stores it, and on the other hand to its 'milieu', i.e. to space. In the life of every 'being' – species, individual or group – there are moments when the energy available is so abundant that it tends to be explosively discharged. It may be turned back against itself, or it may spread outwards, in gratuitousness or grace. The incidence of destruction, self-destruction, aimless violence and suicide is high in nature generally and even higher in the human species. Excesses of all kinds are the result of excess energy, as Bataille understood in the wake of Nietzsche (although he was perhaps somewhat excessive in his application of this rule itself).

It follows that Freud's celebrated 'death instinct' should be treated as a derivative phenomenon. The symptomatic study which psychoanalysts since Freud have made of morbid tendencies and drives has generated a great deal of accurate data in the 'fields' which fall under such rubrics as Eros and Thanatos, narcissism, sado-masochism, self-destructiveness, eroticism, anxiety, and neurosis and psychosis, but all this work has only made any appeal to a primordial tendency here even more dubious. There is a drastic difference between the notion of a death instinct or drive, a force seeking annihilation and running counter to a forever thwarted life-affirming tendency, and the thesis of a whiplash effect resulting from basically justified excesses in the expenditure of vital energy. Even though we are bound to assume that the 'negation' of energy exists in space – that is, in the milieu in which energy is expended, diffused and dissipated – this is not to say that death and self-destruction are causes or reasons rather than effects. Thus the 'death instinct' simply implies an unproductive use or misuse – a 'misemployment', so to speak – of basic energy. It is the dialectical outcome of a conflictual relationship internal to this energy, a relationship which cannot be reduced to mere mechanisms of defence or of equilibrium and their failures. There is sense in a joyful pessimism.

III

In the foregoing discussion space has been taken *partes extra partes*, as Spinoza would say. That we are dealing with finiteness, with parts and subdivisions, with component elements, and with each part's uniqueness and origins (its 'etymology') – of this there can be no doubt. The very concept of a form, with an internal self-'reflection' or duplicate of itself as its defining characteristic – the concept, in other words, of symmetry with its constitutive dualisms (reflectional symmetry and rotational symmetry, asymmetry as itself determined by symmetry, and so on) – implies a circumscribed space: a body with contours and boundaries. Obviously, however, it is not enough merely to evoke subdivisions of space and allotments of energy: currents flow and propagate themselves within an *infinite* space. 'Infinity is the original fact; what has to be explained is the source of the finite', writes Nietzsche. 'In infinite time and in infinite space there are no terminal points.' Here thought is overcome by a kind of vertigo. Yet, he adds, 'Though it has nothing to hold on to, humanity must somehow stand upright – therein lies the immense task of the artist.' But Nietzsche assigns no absolute, general or total priority to the imagination.[12]

Could the infinite and the finite be mere illusion, the one just as much as the other, and each, as it were, the illusion *of* the other? Are they mirages, reflections or refractions, or in some sense that which lies short of – and beyond – each part? Time *per se* is an absurdity; likewise space *per se*. The relative and the absolute are reflections of one another: each always refers back to the other, and the same is true of space and time. We are confronted by a double surface, a double appearance which is governed by a single law and a single reality, that of reflection/refraction. The maximum difference is contained in every difference, even a minimal one. 'Every form belongs to the subject. It is the apprehension of the surface by the mirror.'[13]

---

[12] 'Die *Unendlichkeit* ist die uranfängliche Tatsache: es wäre nur zu erklären, woher das *Endliche* stamme. . . . In der unendlichen Zeit und dem unendlichen Raume gibt es keine Ziele . . . Ohne jede derartige Anlehnung muss die Menscheit *stehn* können – ungeheure Aufgabe der Künstler!' – Friedrich Nietzsche, *Das Philosophenbuch/Le livre du philosophe* (Paris: Aubier-Flammarion, 1969), fragment 120, p. 118.
[13] 'Alle Gestalt ist dem Subjekt zugehörig. Es ist das Erfassen der Oberflächen durch Spiegel' (ibid., fragment 121, p. 118). '*Durch Spiegel*' – i.e. in, by, and through the mirror.

IV

By thus engendering surface, image[14] and mirror, reflection pierces the surface and penetrates the depths of the relationship between repetition and difference. Duplication (symmetry) implies repetition, yet it also gives rise to a difference constitutive of a space. It should not be conceived of on the model of numerical iteration (1 and 1 and 1, etc.), nor on that of serial recurrence. Rather the opposite: duplication and symmetry/asymmetry call for causal notions irreducible to classical (serial and linear) ideas. When the mirror is 'real', as is constantly the case in the realm of objects, the space in the mirror is imaginary – and (cf. Lewis Carroll) the locus of the imagination is the 'Ego'. In a living body, on the other hand, where the mirror of reflection is imaginary, the *effect* is real – so real, indeed, that it determines the very structure of the higher animals.[15] It is for all the world as though the left side of the bodies of these animals were the reflection in a plane mirror of the right, the result being a perfect reflectional symmetry; this is completed, moreover, by a rotational symmetry: the life of the spinal column.

From the social standpoint, space has a dual 'nature' and (in any given society) a dual general 'existence'. On the one hand, one (i.e. each member of the society under consideration) relates oneself to space, situates oneself in space. One confronts both an immediacy and an objectivity of one's own. One places oneself at the centre, designates oneself, measures oneself, and uses oneself as a measure. One is, in short, a 'subject'. A specific social status – assuming always a stable situation, and hence determination by and in a *state* – implies a role and a function: an individual and a public identity. It also implies a

---

[14] Symmetry in the sense of *bilateral symmetry* is a strictly mathematical and absolutely precise concept, according to Weyl: 'A body, a spatial configuration, is symmetric with respect to a given plane E if it is carried into itself by reflection in E. Take any line l perpendicular to E and any point p on l: there exists one and only one point p' on l which has the same distance from E but lies on the other side. The point p' coincides with p only if p is on E. Reflection is that mapping of space upon itself, $S : p \rightarrow p'$, that carries the arbitrary point p into this its mirror image p', with respect to E' (*Symmetry*, pp. 4–5).

The interest and importance of the mirror derives not, therefore, from the fact that it projects the 'subject's' (or Ego's) image back to the 'subject' (or Ego), but rather from the fact that it extends a repetition (symmetry) immanent to the body into space. The Same (Ego) and the Other thus confront each other, as alike as it is possible to imagine, all but identical, yet differing absolutely, for the image has no density, no weight. Right and left are there in the mirror, reversed, and the Ego perceives its double.

[15] See Weyl, *Symmetry*, p. 4.

location, a place in society, a position. On the other hand, space serves an intermediary or mediating role: beyond each plane surface, beyond each opaque form, 'one' seeks to apprehend something else. This tends to turn social space into a transparent medium occupied solely by light, by 'presences' and influences. On the one hand, therefore, space contains opacities, bodies and objects, centres of efferent actions and effervescent energies, hidden – even impenetrable – places, areas of viscosity, and black holes. On the other, it offers sequences, sets of objects, concatenations of bodies – so much so, in fact, that anyone can at any time discover new ones, forever slipping from the non-visible realm into the visible, from opacity into transparency.[16] Objects touch one another, feel, smell and hear one another. Then they contemplate one another with eye and gaze. One truly gets the impression that every shape in space, every spatial plane, constitutes a mirror and produces a mirage effect; that within each body the rest of the world is reflected, and referred back to, in an ever-renewed to-and-fro of reciprocal reflection, an interplay of shifting colours, lights and forms. A mere change of position, or a change in a place's surroundings, is enough to precipitate an object's passage into the light: what was covert becomes overt, what was cryptic becomes limpidly clear. A movement of the body may have a similar goal. Here is the point of intersection of the two sensory fields.

Were it not for this dual aspect and natural/social space, how could we understand language itself? 'Nature' can only be apprehended through objects and shapes, but this perception occurs within an overall context of illumination where bodies pass from their natural obscurity into the light, not in an arbitrary manner but according to a specific sequence, order or articulation. Where natural space exists, and even more so where social space exists, the movement from obscurity to enlightenment – the process of decipherment – is perpetual. It is in fact part and parcel of the way in which the existence of space is established. This incessant deciphering activity is objective as much as subjective – in which respect it indeed transcends the old philosophical distinction between objectivity and subjectivity. It becomes more acute as soon as concealed parts of space (the internal portions of things and things outside the field of perception) come to have associated with themselves symbols, or corresponding signs or indices, which are often tabooed, holy/evil, revelatory

[16] Apropos of this development and the dualism that underpins it, see the last writings of Maurice Merleau-Ponty, notably *L'oeil et l'esprit* (Paris: Gallimard, 1964), where he abandons a phenomenological account of perception in favour of a deeper analysis. Merleau-Ponty remained attached, however, to the philosophical categories of 'subject' and 'object', which have no relation to social practice.

or occult. It is in this sense that it cannot be properly described as either a subjective or an objective, a conscious or an unconscious, activity; rather, it is an activity which serves to *generate* consciousness: messages, by virtue of space and of the interplay of reflections and mirages within it, are intrinsic to lived experience itself.

Space – *my* space – is not the context of which I constitute the 'textuality': instead, it is first of all *my body*, and then it is my body's counterpart or 'other', its mirror-image or shadow: it is the shifting intersection between that which touches, penetrates, threatens or benefits my body on the one hand, and all other bodies on the other. Thus we are concerned, once again, with gaps and tensions, contacts and separations. Yet, through and beyond these various effects of meaning, space is actually experienced, *in its depths*, as duplications, echoes and reverberations, redundancies and doublings-up which engender – and are engendered by – the strangest of contrasts: face and arse, eye and flesh, viscera and excrement, lips and teeth, orifices and phallus, clenched fists and opened hands – as also clothed *versus* naked, open *versus* closed, obscenity *versus* familiarity, and so on.[17] None of these oppositions and conjunctions/disjunctions has anything to do with a logic or formal system.

Should we therefore conclude that mirror and mirage effects exist but that there is no such thing as an *anti-mirror* effect, a lived experience of blank opacity? Certainly not if we recall Tzara's description of mirrors as the 'fruits of dread', or Bataille's comparison of himself with a 'tarnished mirror'. Here, too, is Eluard: 'The reflection of the personality must be wiped away before inspiration can spring forth from the mirror.'[18] The mirror is a surface at once pure and impure, almost material

---

[17] See the works of Octavio Paz, especially *Conjunciones y disyunciones* (Mexico City: Joaquín Mortiz, 1969); Eng. tr. by Helen R. Lane: *Conjunctions and Disjunctions* (New York: Viking, 1974). Paz examines the body, the mirror, and a variety of dualisms and their dialectical interactions, in the light of poetry. He draws a distinction, and points up an antagonism, applicable to all societies, cultures or civilizations, between the signs of 'body' and the signs of 'non-body' (see pp. 51, 58ff; Eng. tr., pp. 45, 52ff).

[18] Oddly absent from Bachelard's *La poétique de l'espace*, mirrors held a special fascination for the surrealists. One, Pierre Mabille, devoted a whole book to the subject. Cocteau gave mirrors an important role in both his poetic and his cinematographic works; it was in this connection that he invented the superstition of the 'purely' visual. Consider too the immense part played by the mirror in every major tradition, whether popular or artistic. Cf. Jean-Louis Schefer, *Scénographie d'un tableau* (Paris: Seuil, 1969).

The psychoanalysts have made great play with the 'mirror effect' in their attempts to demolish the philosophical 'subject'. Indeed they have gone far too far in this direction, for they consider the mirror effect only out of its properly spatial context, as part of a space internalized in the form of mental 'topologies' and agencies. As for the generalization of the 'mirror effect' into a theory of ideologies, see Louis Althusser's article in *La Pensée*,

yet virtually unreal; it presents the Ego with its own material presence, calling up its counterpart, its absence from – and at the same time its inherence in – this 'other' space. Inasmuch as its symmetry is projected therein, the Ego is liable to 'recognize' itself in the 'other', but it does not in fact coincide with it: 'other' merely *represents* 'Ego' as an inverted image in which the left appears at the right, as a reflection which yet generates an extreme difference, as a repetition which transforms the Ego's body into an obsessing will-o'-the-wisp. Here what is identical is at the same time radically other, radically different – and transparency is equivalent to opacity.

V

If my body may be said to enshrine a generative principle, at once abstract and concrete, the mirror's surface makes this principle invisible, deciphers it. The mirror discloses the relationship between me and myself, my body and the consciousness of my body – not because the reflection constitutes my unity *qua* subject, as many psychoanalysts and psychologists apparently believe, but because it transforms what I am into the sign of what I am. This ice-smooth barrier, itself merely an inert sheen, reproduces and displays what I am – in a word, signifies what I am – within an imaginary sphere which is yet quite real. A process of abstraction then – but a *fascinating* abstraction. In order to know myself, I 'separate myself out from myself'.[19] The effect is dizzying. Should the 'Ego' fail to reassert hegemony over itself by defying its own image, it must become Narcissus – or Alice. It will then be in danger of never rediscovering itself, space *qua* figment will have swallowed it up, and the glacial surface of the mirror will hold it forever captive in its emptiness, in an absence devoid of all conceivable presence or bodily warmth. The mirror thus presents or offers the most unifying but also

---

June 1970, p. 35. This is the product of a fantasy, and of a half-conscious wish to preserve dogmatic Marxism.

[19] In his *Le système des objets* (Paris: Gallimard, 1968), Jean Baudrillard sees the mirror as nothing more, for the bourgeois, than an extension of 'his' drawing-room or bedroom. This is to limit the mirror's real significance, and in effect to abolish the (psychoanalytic) notion of narcissism. The ambiguity (or duality) of these phenomena, along with their inherent complexity, emerges clearly from the analyses of Jacques Lacan (cf. 'his account of the mirror stage in 'La Famille', *Encyclopédie française*, Vol. VIII: Henri Wallon, ed., *La vie mentale*, Paris, 1938), but Lacan does not provide much in the way of elucidation. For him the mirror helps to counteract the tendency of language to break up the body into pieces, but it freezes the Ego into a rigid form rather than leading it towards transcendence in and through a space which is at once practical and symbolic (imaginary).

the most disjunctive relationship between form and content: forms therein have a powerful reality yet remain unreal; they readily expel or contain their contents, yet these contents retain an irreducible force, an irreducible opacity, and this is as true for *my* body (the content of 'my consciousness') as for other bodies, for bodies in general. So many objects have this dual character: they are transitional inasmuch as they tend towards something else, yet they are also aims or 'objectives' in their own right. Among all such objects, the mirror undoubtedly has a special place. All the same, to argue (as some overzealous proponents of psychoanalysis do) that all property can be defined in terms of a kind of mirror effect, on the grounds that possession of an object by the 'Ego' makes that object the Ego's own is to overstep the bounds that 'culture' places on stupidity in general.

There is in fact little justification for any systematic generalization from the effects of this particular object, whose role is properly confined to a sphere within the immediate vicinity of the body.

The mirror is thus at once an object among others and an object different from all others, evanescent, fascinating. In and through the mirror, the traits of other objects in relationship to their spatial environment are brought together; the mirror is an object in space which informs us about space, which speaks of space. In some ways a kind of 'picture', the mirror too has a frame which specifies it, a frame that can be either empty or filled. Into that space which is produced first by natural and later by social life the mirror introduces a truly dual spatiality: a space which is imaginary with respect to origin and separation, but also concrete and practical with respect to coexistence and differentiation. Many philosophers – and non-philosophers, such as Lenin – have sought to define thought in terms of a mirror effect, in terms of *reflection*. But in so doing they confuse act and symbol. Prior to its practical realization, to its material manufacture, the mirror already existed in magical or mythic modes: the surface of water symbolizes the surface of consciousness and the material (concrete) process of decipherment which brings what is obscure forth into the light.

For our present purposes, we need to consider and elaborate upon a number of relationships usually treated as 'psychic' (i.e. relating to the psyche). We shall treat them, however, as *material*, because they arise in connection with the (material) body/subject and the (material) mirror/object; at the same time we shall look upon them as particular instances of a 'deeper' and more general relationship which we shall be coming back to later in our discussion – that between repetition and differentiation. The relationships in question are the following.

1 *Symmetry* (planes and axes): duplication, reflection – also asymmetry as correlated with symmetry.
2 *Mirages* and mirage effects: reflections, surface *versus* depth, the revealed *versus* the concealed, the opaque *versus* the transparent.
3 *Language* as 'reflection', with its familiar pairs of opposites: connoting *versus* connoted, or what confers value *versus* what has value conferred upon it; and refraction through discourse.
4 *Consciousness* of oneself and of the other, of the body and of the abstract realm of otherness and of becoming-other (alienation).
5 *Time*, the immediate (directly experienced, hence blind and 'unconscious') link between repetition and differentiation.
6 Lastly, *space*, with its double determinants: imaginary/real, produced/producing, material/social, immediate/mediated (milieu/transition), connection/separation, and so on.

Only late on was the realm of symbols and signs integrated into the larger realm of shadows. Bearers of a clarity at once auspicious and ill-starred, symbols and signs were at first cryptic in character (but in a material sense); concealed in grottoes or caves, they sometimes caused these places to be cursed, sometimes to become holy, as sanctuaries or temples. The truth of signs and the signs of truth are contained within the same enigma: the enigma of the Italiot and Roman *mundus* – the hole, the bottomless pit. The enigma, too, of the Christian reliquaries – those underground churches or chapels so aptly named 'crypts'. And the enigma, finally, of an opaque body – or opaque bodies – whence truth emerges in stunning clarity: the body that brings light into the darkness.

Of the relationship between the sexes (which is in no sense a special case), may not comparable things be said?

1 Here too we find symmetry (and asymmetry): male and female.
2 Here too we find displaced illusional effects (transparency *versus* opacity). The *other* emerges and turns out to be the *same*, albeit in an ambiguous and shadowy manner: the same desire as fails to recognize itself as such. A fragmentation ensues and, thanks to the oscillation between knowing and misapprehending, a will (to power) is able to intrude itself.
3 This fragmentation of desire, heralding the explosive fragmentation of pleasure, naturally leads to a separation, but this in no way eliminates 'reflection' (in the sense of that relationship between self and other in which each person seeks himself in

hopes of finding the other, while what he seeks in the other is a projection of himself).

4 Hence the great nostalgia we feel about an absolute love which always leads us back to a relative one, a 'pure' love which always disappoints, which is inconceivable apart from a flesh that reverses the original tendency and releases the original tension, and replaces them with a fulfilment that is no less disappointing for being more attainable. A nostalgia, then, that contains dissent – and resentment. The imaginary plane of the mirror is here too – the divider between two mirror images or doubles: to perceive oneself in this space is to meld one's features with those of the counterpart.

It goes without saying that no 'theory of doubles' would stop here, although this line of thought would certainly constitute the initial focus of any such theory of reflections and mirages. It would have to be extended, for one thing, to take in theatrical space, with its interplay between fictitious and real counterparts and its interaction between gazes and mirages in which actor, audience, 'characters', text, and author all come together but never become one. By means of such theatrical interplay bodies are able to pass from a 'real', immediately experienced space (the pit, the stage) to a perceived space – a third space which is no longer either scenic or public. At once fictitious and real, this third space is classical theatrical space.

To the question of whether such a space is a representation of space or a representational space, the answer must be neither – and both. Theatrical space certainly implies a *representation of space* – scenic space – corresponding to a particular *conception* of space (that of the classical drama, say – or the Elizabethan, or the Italian). The *representational space*, mediated yet directly experienced, which infuses the work and the moment, is established as such through the dramatic action itself.

VI

Identifying the foundations upon which the space of each particular society is built, the underpinnings of that space's gradual development, is only the beginning of any exploration of a reality that to begin with seems transparently clear. Thus representations of space, which confuse matters precisely because they offer an already clarified picture, must be dispelled.

The mirage effects whose preconditions I have tried to establish (though not to elaborate upon) in the foregoing discussion can be extraordinary – more specifically, they can introduce an extraordinary element into an ordinary context. They cannot be reduced solely to the surprise of the Ego contemplating itself in the glass, and either discovering itself or slipping into narcissism. The power of a landscape does not derive from the fact that it offers itself as a spectacle, but rather from the fact that, as mirror and mirage, it presents any susceptible viewer with an image at once true and false of a creative capacity which the subject (or Ego) is able, during a moment of marvellous self-deception, to claim as his own. A landscape also has the seductive power of all *pictures*, and this is especially true of an urban landscape – Venice, for example – that can impose itself immediately as a *work*. Whence the archetypal touristic delusion of being a participant in such a work, and of understanding it completely, even though the tourist merely passes through a country or countryside and absorbs its image in a quite passive way. The work in its concrete reality, its products, and the productive activity involved are all thus obscured and indeed consigned to oblivion.

Mirage effects have far-ranging consequences. Under the conditions of modernity, as absolute political space extends its sway, the impression of transparency becomes stronger and stronger, and the illusion of a new life is everywhere reinforced. Real life indeed appears quite close to us. We feel able, from within everyday life, to reach out and grasp it, as though nothing lay between us and the marvellous reality on the other side of the mirror. All the prerequisites for it exist – so what is missing? An utterance of some kind, spoken or written? A gesture? A successful attack on some particular aspect of things, or the removal of some particular obstacle – ideology perhaps, or established knowledge, or some repressive institution or other, or religion, or theatricality, or the educational system, or the spectacle? The list is endless.

The idea of a new life is at once realistic and illusory – and hence neither true nor false. What is true is that the preconditions for a different life have already been created, and that that other life is thus on the cards. What is false is the assumption that being on the cards and being imminent are the same thing, that what is immediately possible is necessarily a world away from what is only a distant possibility, or even an impossibility. The fact is that the space which contains the realized preconditions of another life is the same one as prohibits what those preconditions make possible. The seeming limpidity of that space is therefore a delusion: it appears to make elucidation unnecessary, but in reality it urgently requires elucidation. A total revolution – material, economic, social, political, psychic, cultural, erotic, etc. – seems to be in the offing, as though already immanent to the present. To change life, however, we must first change space. Absolute revolution is our self-image and our mirage – as seen through the mirror of absolute (political) space.

VII

A social space is not a *socialized* space.[20] The would-be general theory of the 'socialization' of whatever precedes society – i.e. nature, biology, physiology (needs, 'physical' life), and so on – is really just the basic tenet of an *ideology*. It is also a 'reactive' mirage effect. To hold, for example, that natural space, the space described by the geographer, existed as such and was then at some point socialized leads either to the ideological posture of nostalgic regret for a space that is no longer, or else to the equally ideological view that this space is of no consequence because it is disappearing. In reality, whenever a society undergoes a transformation, the materials used in the process derive from another, historically (or developmentally) anterior social practice. A purely natural or original state of affairs is nowhere to be found. Hence the notoriously difficult problems encountered by (philosophical) thinking on the subject of origins. The notion of a space which is at first empty, but is later filled by a social life and modified by it, also depends on this hypothetical initial 'purity', identified as 'nature' and as a sort of ground zero of human reality. Empty space in the sense of a mental and social void which facilitates the socialization of a not-yet-social realm is actually merely a *representation of space*. Space is conceived of as being transformed into 'lived experience' by a social 'subject', and is governed by determinants which may be practical (work, play) or bio-social (young people, children, women, active people) in character. This representation subtends the notion of a space in which the 'interested parties', individuals or groups, supposedly dwell and have their being. Of any actual historically generated space, however, it would be more

---

[20] *Pace* Georges Matoré, whose *L'espace humain* (Paris: La Colombe, 1962), though one of the best discussions of semantics and spatial metaphors, is limited in its significance because of the author's espousal of this erroneous thesis.

accurate to say that it played a socializing role (by means of a multiplicity of networks) than that it was itself socialized.

Can the space of work, for example (when indeed it is legitimate to speak of such a space), be envisaged as a void occupied by an entity called work? Clearly not: it is produced within the framework of a global society, and in accordance with that society's constitutive production relations. In capitalist society, the space of work consists of production units: businesses, farms, offices. The various networks which link these units are also part of the space of work. As for the agencies that govern these networks, they are not identical to those that govern work itself, but they are articulated with them in a relatively coherent manner which does not, however, exclude conflicts and contradictions. The space of work is thus the result, in the first place, of the (repetitive) gestures and (serial) actions of productive labour, but also – and increasingly – of the (technical and social) division of labour; the result therefore, too, of the operation of markets (local, national and worldwide) and, lastly, of property relationships (the ownership and management of the means of production). Which is to say that the space of work has contours and boundaries only for and through a thought which *abstracts*; as one network among others, as one space among many interpenetrating spaces, its existence is strictly relative.

Social space can never escape its basic duality, even though triadic determining factors may sometimes override and incorporate its binary or dual nature, for the way in which it presents itself and the way in which it is represented are different. Is not social space always, and simultaneously, both a *field of action* (offering its extension to the deployment of projects and practical intentions) and a *basis of action* (a set of places whence energies derive and whither energies are directed)? Is it not at once *actual* (given) and *potential* (locus of possibilities)? Is it not at once *quantitative* (measurable by means of units of measurement) and *qualitative* (as concrete extension where unreplenished energies run out, where distance is measured in terms of fatigue or in terms of time needed for activity)? And is it not at once a collection of *materials* (objects, things) and an ensemble of *matériel* (tools – and the procedures necessary to make efficient use of tools and of things in general)?

Space appears as a realm of objectivity, yet it exists in a social sense only for activity – for (and by virtue of) walking or riding on horseback, or travelling by car, boat, train, plane, or some other means. In one sense, then, space proposes homologous paths to choose from, while in another sense it invests particular paths with special value. The same goes for angles and turns: what is to the left may also be sinister; what is to the right may also be 'right' in the sense of rectitude. A would-be homogeneous space, open to whatever actions may be reasonable, authorized or ordered, can, under its other aspect, take responsibility for prohibitions, embody occult traits and bestow favour or disfavour upon individuals and the groups to which they belong. Localization is answered by divergence, and focus on a central point is answered by radiation, by influx and diffusion. Like energy in a material form such as a molecule or an atom, social energy is both directed and dispersed; it becomes concentrated in a certain place, yet continues to act upon the sphere outside. This means that social spaces have foundations that are at once material and formal, including concentricity and grids, straight lines and curves – all the modalities of demarcation and orientation. Social spaces cannot be defined, however, by reducing them to their basic dualism; rather, this dualism supplies the materials for the realization of a very great variety of projects. In natural or (later) 'geographical' space, routes were inscribed by means of simple linear markings. Ways and tracks were pores which, without colliding, gradually widened and lengthened, leading to the establishment of *places* (way-stations, localities made special for one reason or another) and boundaries. Through these pores, which accentuated local particularities by making use of them, flowed increasingly dense human streams: simple herding, the seasonal movement of flocks, migrations of masses of people, and so on.

These activities and spatio-temporal determinants may be said to belong to the *anthropological* stage of social reality. We have defined this stage as the stage of demarcation and orientation. Dominant in archaic and agricultural–pastoral societies, these later became recessive and subordinate activities. There is no stage, however, at which 'man' does not demarcate, beacon or sign his space, leaving traces that are both symbolic and practical; changes of direction and turns in this space always need to be represented, and 'he' meets this figurative need either by taking his own body as a centre or by reference to other bodies (celestial bodies, for example, the angle of incidence of whose light serves to refine the human perception of angles in general).

It should not be supposed that 'primitive' people – seasonally migrant herders, let us say – formed abstract representations of straight and curved lines, of obtuse and acute angles, or – even virtually – of measures. Their indicators remained purely qualitative in character, like those of animals. Different directions appeared as either benevolent or ill-omened. The indicators themselves were objects invested with affective significance – what would later be called 'symbolic' objects. Egregious aspects of the terrain were associated perhaps with a memory, perhaps with particular actions which they facilitated. The networks of paths and roads made up a space just as concrete as that of the body – of which they were in fact an extension. Directions in space and time were inhabited for such a herder – and how could it be otherwise? – by real and fictitious, dangerous or lucky 'creatures'. Thus qualified, symbolically or practically, this space bore along the myths and stories attached to it. The concrete space constituted by such networks and frontiers had more in common with a spider's web than with geometrical space. We have already noted that calculation has to reconstruct in a complicated way what 'nature' produces in the living body and its extensions. We also know that symbolism and praxis cannot be separated.

The relationships established by *boundaries* are certainly of the greatest importance here, along with the relationship between boundaries and named places; thus the most significant features of a shepherd's space might include the place (often enclosed) where he gathers his sheep, the spring where he waters them, the bounds of the pasture available to him, and his neighbours' land, which is off limits. Every social space, then, once duly demarcated and oriented, implies a superimposition of certain relations upon networks of named places, of *lieux-dits*. This results in various kinds of space.

1 Accessible space for normal use: routes followed by riders or flocks, ways leading to fields, and so on. Such use is governed prescriptively – by established rules and practical procedures.
2 Boundaries and forbidden territories – spaces to which access is prohibited either relatively (neighbours and friends) or absolutely (neighbours and enemies).
3 Places of abode, whether permanent or temporary.
4 Junction points: these are often places of passage and encounter; often, too, access to them is forbidden except on certain occasions of ritual import – declarations of war or peace, for example.

Boundaries and junction points (which are also, in the nature of things, points of friction) will naturally have different aspects according to the type of society, according to whether we are considering relatively settled peasants, plundering warriors, or true nomads or herders given to seasonal migrations.

Social space does incorporate one three-dimensional aspect, inherited from nature, namely the fact that between what is above (mountains, highlands, celestial beings) and what is below (in grottoes or caves) lie the surfaces of the sea and of the earth's flatlands, which thus constitute planes (or plains) that serve both to separate and to unite the heights and the depths. Here is the basis of representations of the Cosmos. Similarly, caves, grottoes, hidden and underground places provide the starting-point for representations of the world and myths of the earth-as-mother. As perceived by our shepherd, however, such oppositions as those between west and east, north and south, high and low, or before and behind have nothing to do with abstract ideas. Rather, they are at once relationships and qualities. Space thus *qualifies* in terms of time, in terms of ill-defined measures (paces, degree of fatigue), or in terms of parts of the body (cubits, inches, feet, palms, etc.). Through displacement outwards from the centre, the body of the thinking and acting subject is replaced by a social object such as a chief's hut, a pole or, later, a temple or church. The 'primitive' situates or speaks of space as a member of a collectivity which itself occupies a regulated space closely bound up with time. He does not envisage himself in space as one point among others in an abstract milieu. That is a type of perception belonging to a much later period, and is contemporaneous with the space of 'plans' and maps.

VIII

The body serves both as point of departure and as destination. We have already encountered this body – our body – many times in the present discussion. But what body, precisely, are we talking about?

Bodies resemble each other, but the differences between them are more striking than the similarities. What is there in common between the body of a peasant leading his working ox, shackled to the soil by his plough, and the body of a splendid knight on his charger or show horse? These two bodies are as different as those of the bullock and the entire horse in whose company we find them! In either case, the animal intervenes as *medium* (means, instrument or intermediary) between man and space. The difference between the 'media' implies an analogous difference between the two spaces in question. In short, a wheatfield is a world away from a battlefield.

But what conception of the body are we to adopt or readopt, discover or rediscover, as our point of departure? Plato's? Aquinas's? The body that sustains the *intellectus* or the body that sustains the *habitus*? The body as glorious or the body as wretched? Descartes's body-as-object, or the body-as-subject of phenomenology and existentialism? A fragmented body, represented by images, by words, and traded retail? Must we start out from a *discourse* on the body? If so, how are we to avoid the deadly tendency of discourse towards abstraction? And, if indeed we must begin from an abstraction, how can we limit its impact, or go beyond its limitations?

Should we perhaps rather take off from the 'social body' – a body battered and broken by a devastating practice, namely the division of labour, and by the weight of society's demands? But how can we expect to define a critical space if we start out by accepting a body inserted into this already 'social' space – and mutilated by it? On the other hand, what basis do we have – and indeed what means – for defining this body *in itself*, without ideology?

When the body came up earlier on in our analysis, it did not present itself either as subject or as object in the philosophical sense, nor as an internal milieu standing in opposition to an external one, nor as a neutral space, nor as a mechanism occupying space partially or fragmentarily. Rather, it appeared as a 'spatial body'. A body so conceived, as produced and as the production of a space, is immediately subject to the determinants of that space: symmetries, interactions and reciprocal actions, axes and planes, centres and peripheries, and concrete (spatio-temporal) oppositions. The materiality of this body is attributable neither to a consolidation of parts of space into an apparatus, nor to a nature unaffected by space which is yet somehow able to distribute itself through space and so occupy it. Rather, the spatial body's material character derives from space, from the energy that is deployed and put to use there.

Considered as a 'machine', the spatial body is two-sided: one side is run by massive supplies of energy (from alimentary and metabolic sources), the other side by refined and minute energies (sense data). The question arises whether such a 'two-sided machine' is a machine at all. To treat it as such must at the very least introduce a dialectical element into – and hence concretize – the Cartesian concept of 'machine', a concept which is not only highly abstract but also embedded in a very abstractly conceived representation of space. The notion of a two-sided machine naturally implies interaction within its bipartite structure. It embraces the possibility of unpredictable effects, and rejects all strict mechanism, all hard-and-fast and unilateral definition. This machine's devices for the emission and reception of small-scale energies lie in the sensory organs, the afferent and efferent nerve pathways, and the brain.

The organs of massive-energy use are the muscles, and above all the sexual organs, which are the pole where such energy accumulates explosively. The body's organic constitution is itself directly linked to the body's spatial constitution (or organization). How could the tendencies intrinsic to this whole – the tendency to capture, withhold and accumulate energy on the one hand, and the tendency to discharge it suddenly on the other – fail to have a conflictual relationship? The same goes for the coexisting tendencies to explore space and to invade it. The conflicts inherent in the spatio-temporal reality of this body – which is neither substance, nor entity, nor mechanism, nor flux, nor closed system – culminate in the antagonisms in human beings between knowledge and action, head and genitals, and desires and needs. As for which of these conflicts is the most or least significant, that is a value-based question which is meaningless unless one posits a hierarchy. There is no sense in doing so, however – or, rather, doing so is a way of losing the sense of the matter. For the notion of hierarchy can only lead us into the realm of the Western Logos, into the Judaeo-Christian tradition. The conflicts in question, though, do not depend solely on language, on fractured words, fractured images or fractured places. They flow also – and indeed primarily – from an opposition constitutive of the living organism as a dialectical totality: the fact that in this organism the pole of small-scale energy (brain, nerves, senses) does not necessarily concord, in fact rather the opposite, with the massive-energy pole (sexual apparatus). The living organism has neither meaning nor existence when considered in isolation from its extensions, from the space that it reaches and produces (i.e. its 'milieu' – to use a fashionable term that tends to reduce activity to the level of mere passive insertion into a natural material realm). Every such organism is reflected and refracted in the changes that it wreaks in its 'milieu' or 'environment' – in other words, *in its space*.

At times the body, which we have yet to explore, gets covered up, concealed from view, but then it re-emerges – then it is as it were resuscitated. Does this suggest a connection between the history of the body and the history of space?

With its warts plainly visible, but also its strengths and triumphs, the body as here conceived is not susceptible to the simple (and in fact crude and ideological) distinction between normal and abnormal states, between health and pathology. In what is conventionally referred to as 'nature', where the fundamental rule is fertilization, is any discrimination made between pleasure and pain? Not in any obvious way, certainly. One is tempted to say, rather, that such a distinction is in fact the work – the great work even – of humanity, a work often diverted and misdirected, but one which enlists the contributions of learning and art. A heavy price attaches, however, to the attainment of this goal, for, once effected, this disassociation entails the separation of things that cannot or must not be separated.

Let us return, however, to our inventory of what the body has to give. Tangible space possesses (although these words are not ideal here) a basis or foundation, a ground or background, in the olfactory realm. If sensual rapture and its antithesis exist anywhere, if there is any sphere where, as a philosopher might say, an intimacy occurs between 'subject' and 'object', it must surely be the world of smells and the places where they reside.

> Next step, in they're plunged into some rot, some stump of dwarf birch, bark rubbed ass of raw by tail of bear or moose of caribou antlers eight years ago! . . . Into the open mouth of that remaining stump came the years of snow, sun, little jewels of bird shit, cries of sap from the long dying roots, the monomaniacal yodeling of insects, and wood rotting into rotting wood, into gestures of wood, into powder and punk all wet and stinking with fracture between earth and sky, yeah, D. J. could smell the break, gangrene in the wood, electric rot cleaner than meat and shit sick smell and red-hot blood of your blood in putrefaction, but a confirmed wood gangrene nonetheless, Burbank, a chaos of odor on the banks of the wound, nothing smells worse than half-life, life which has no life but don't know it – thank you, Mr. Philosopher . . .![21]

Such overwhelming and villainous smells are made up for in nature by their counterparts, by aromas and fragrances of all kinds, by the miraculous scents of flowers and by the odours of the flesh. It may be asked whether there is any point in dwelling on this space, which is in any case fast disappearing under the current onslaught of hygiene and asepticism. Is Hall perhaps right to assert that these are strictly anthropological or 'culturally' determined phenomena? Should the distaste unquestionably felt by some 'modern' people for natural odours be dismissed as the cause, or perhaps the effect, of the detergent industry? The search for answers to such questions may as well be left to the cultural anthropologists. For our purposes, the pertinent fact is that everywhere in the modern world smells are being eliminated. What is shown by this

---

[21] Norman Mailer, *Why Are We in Vietnam?* (1967; New York: Holt, Rinehart and Winston, 1982), p. 139.

immense deodorizing campaign, which makes use of every available means to combat natural smells whether good or bad, is that the transposition of everything into the idiom of images, of spectacle, of verbal discourse, and of writing and reading is but one aspect of a much vaster enterprise. Anyone who is wont (and every child falls immediately into this category) to identify places, people and things by their smells is unlikely to be very susceptible to rhetoric. Transitional objects to which desire becomes attached in seeking to escape subjectivity and reach out to 'the other' are founded primarily on the olfactory sense; this is true also for the erotic object in general.

Smells are not decodable. Nor can they be inventoried, for no inventory of them can have either a beginning or an end. They 'inform' only about the most fundamental realities, about life and death, and they are part of no significant dichotomies except perhaps that between life beginning and life ending. There is no pathway here other than the direct one between the receiving centre and the perimeter of its range – no pathway other than the nose and the scent themselves. Somewhere between information and the direct stimulation of a brutal response, the sense of smell had its glory days when animality still predominated over 'culture', rationality and education – before these factors, combined with a thoroughly cleansed space, brought about the complete atrophy of smell. One can't help feeling, though, that to carry around an atrophied organ which still claims its due must be somewhat pathogenic.

The rose of Angelus Silesius, which does not know that it is a flower, nor that it is beautiful, is also ignorant of the fact that it exudes a delightful scent. Though already threatened with extinction by the fruit, it unhesitatingly proffers its transient splendour. This act of self-display corresponds, however, to an 'unconscious' nature, striving and intent – to the interplay of life and death. Odours, which bespeak nature's violence and largesse, do not signify; they *are*, and they say what they are in all its immediacy: the intense particularity of what occupies a certain space and spreads outwards from that space into the surroundings. Nature's smells, be they foul or fragrant, are expressive. Industrial production, which often smells bad, also produces 'perfumes'; the aim is that these should be 'signifiers', and to this end words – advertising copy – link 'signifieds' to them: woman, freshness, nature, glamour, and so forth. But a perfume either induces or fails to induce an erotic mood – it does not carry on a discourse about it. It either fills a place with enchantment or else has no effect upon it at all.

Tastes are hard to distinguish both from smells and from the tactile sensations of lips and tongue. They do differ from smells, however, in that they tend to form pairs of opposites: sweet *versus* bitter, salty *versus* sugary, and so on. They are thus susceptible of coding, and of being produced according to a particular code; witness the way a cookery book can lay down practical rules for their creation. At the same time, tastes cannot constitute messages and their subjection to coding adds a determination that they do not possess in themselves. Sweet does not contain a reference to bitter, the elusive charm of the bittersweet notwithstanding. Sweet is opposed to sour as well as to bitter, although sourness and bitterness are not the same thing. Here it is social practice that separates what in nature is given together; this is a practice which seeks to produce pleasure. Opposing tastes only come into their own when they occur in conjunction with other attributes: cold and hot, crispy and soft, smooth and rough – attributes related to the sense of touch. Thus from that social practice known as 'cookery', from the arts of heating, chilling, boiling, preserving and roasting, there emerges a reality invested with a meaning which may properly be called 'human' – even though humanism rarely alludes to it; traditional humanism, like its modern opposite, sets little store by pleasure, both being content to remain on the level of *words*. Meanwhile, at the body's centre is a kernel resistant to such efforts to reduce it, a 'something' which is not truly differential but which is nevertheless neither irrelevant nor completely undifferentiated: it is within this primitive space that the intimate link persists between smells and tastes.

A philosopher might speak eloquently in this connection of a coextensive presence of space and Ego thanks to the mediation of the body, but in fact a good deal more – and indeed something quite different – is involved here. For the spatial body, becoming social does not mean being inserted into some pre-existing 'world': this body produces and reproduces – and it perceives what it reproduces or produces. Its spatial properties and determinants are contained within it. In what sense, then, does it perceive them? In the practico-sensory realm, the perception of right and left must be projected and imprinted into or onto things. Pairs of determinants – axes *versus* points of a compass, direction *versus* orientation, symmetry *versus* asymmetry – must be introduced into space, which is to say, produced in space. The preconditions and principles of the lateralization of space lie within the body, yet this must still be effected in such a way that right and left or up and down are indicated or marked – and choices thus offered to gesture and action.

According to Tomatis,[22] the hearing plays a decisive role in the lateralization of perceived space. Space is listened for, in fact, as much

---

[22] Alfred Ange Tomatis is a well-known authority on hearing, the inventor of a mechanical (electronic) ear, and the author of many contributions to orthophonics.

as seen, and heard before it comes into view. The perceptions of one ear differ from those of the other. This difference puts the child on alert, and lends volume and physical density to the messages it receives. The hearing thus plays a mediating role between the spatial body and the localization of bodies outside it. The organic space of the ear, which is brought into being through the child's relationship with its mother, is thus extended to sounds from beyond the sphere of that relationship – to other people's voices, for example. Hearing-disturbances, likewise, are accompanied by disturbances in lateralization in the perception of both external and internal space (dyslexias, etc.).

A homogeneous and utterly simultaneous space would be strictly imperceptible. It would lack the conflictual component (always resolved, but always at least suggested) of the contrast between symmetry and asymmetry. It may as well be noted at this juncture that the architectural and urbanistic space of modernity tends precisely towards this homogeneous state of affairs, towards a place of confusion and fusion between geometrical and visual which inspires a kind of physical discomfort. Everything is alike. Localization – and lateralization – are no more. Signifier and signified, marks and markers, are added after the fact – as decorations, so to speak. This reinforces, if possible, the feeling of desertedness, and adds to the malaise.

This modern space has an analogical affinity with the space of the philosophical, and more specifically the Cartesian tradition. Unfortunately it is also the space of blank sheets of paper, drawing-boards, plans, sections, elevations, scale models, geometrical projections, and the like. Substituting a verbal, semantic or semiological space for such a space only aggravates its shortcomings. A narrow and desiccated rationality of this kind overlooks the core and foundation of space, the total body, the brain, gestures, and so forth. It forgets that space does not consist in the projection of an intellectual representation, does not arise from the visible–readable realm, but that it is first of all *heard* (listened to) and *enacted* (through physical gestures and movements).

A theory of information that assimilates the brain to an apparatus for receiving messages puts that organ's particular physiology, and its particular role in the body, in brackets. Taken in conjunction with the body, viewed *in its body*, the brain is much more than a recording-machine or a decoding-mechanism. (Not, be it said, that it is merely a 'desiring-machine' either.) The total body constitutes, and produces, the space in which messages, codes, the coded and the decoded – so many choices to be made – will subsequently emerge.

The way for physical space, for the practico-sensory realm, to restore or reconstitute itself is therefore by struggling against the *ex post facto* projections of an accomplished intellect, against the reductionism to which knowledge is prone. Successfully waged, this struggle would overturn the Absolute Truth and the Realm of Sovereign Transparency and rehabilitate underground, lateral, labyrinthine – even uterine or feminine – realities. An uprising of the body, in short, against the signs of non-body: 'The history of the body in the final phase of Western culture is that of its rebellions.'[23]

Indeed the fleshly (spatio-temporal) body is already in revolt. This revolt, however, must not be understood as a harking-back to the origins, to some archaic or anthropological past: it is firmly anchored in the here and now, and the body in question is 'ours' – our body, which is disdained, absorbed, and broken into pieces by images. Worse than disdained – ignored. This is not a political rebellion, a substitute for social revolution, nor is it a revolt of thought, a revolt of the individual, or a revolt for freedom: it is an elemental and worldwide revolt which does not seek a theoretical foundation, but rather seeks by theoretical means to rediscover – and recognize – its own foundations. Above all it asks theory to stop barring its way in this, to stop helping conceal the underpinnings that it is at pains to uncover. Its exploratory activity is not directed towards some kind of 'return to nature', nor is it conducted under the banner of an imagined 'spontaneity'. Its object is 'lived experience' – an experience that has been drained of all content by the mechanisms of diversion, reduction/extrapolation, figures of speech, analogy, tautology, and so on. There can be no question but that social space is the locus of prohibition, for it is shot through with both prohibitions and their counterparts, prescriptions. This fact, however, can most definitely not be made into the basis of an overall definition, for space is not only the space of 'no', it is also the space of the body, and hence the space of 'yes', of the affirmation of life. It is not simply a matter, therefore, of a theoretical critique, but also of a 'turning of the world upon its head' (Marx), of an inversion of meaning, and of a subversion which 'breaks the tablets of the Law' (Nietzsche).

The shift, which is so hard to grasp, from the space of the body to the body-in-space, from opacity (warm) to translucency (cold), somehow facilitates the spiriting-away or scotomization of the body. How did this magic ever become possible – and how does it continue to be possible? What is the foundation of a mechanism which thus abolishes the foundations? What forces have been able in the past – and continue to be

[23] Paz, *Conjunciones*, p. 119; Eng. tr., p. 115.

able – to take advantage of what happens 'normally' along the particular route which leads from the Ego to the Other, or, more precisely, from the Ego to itself via its double the Other?

For the Ego to appear, to manifest itself as being in 'my body', is it sufficient for it to have oriented itself in terms of left and right, to have marked out directions relative to itself? Once a particular Ego has formulated the words 'my body', can it now perforce designate and locate other bodies and objects? The answer to these questions must be negative. Furthermore, the uttering of the words 'my body' presupposes the Ego's access to language and to a specific use of discourse – in short, it presupposes a whole history. What are the preconditions of such a history, such a use of discourse, such an intervention of language? What makes the coding of Ego and Alter Ego – and of the gap between them – possible?

For the Ego to appear, it must appear to itself, and its body must appear to it, as *subtracted* – and hence also extracted and abstracted – from the world. Being prey to the world's vicissitudes, and the potential victim of countless dangers, the Ego withdraws. It erects defences to seal itself off, to prevent access to itself. It sets up barriers to nature, because it feels vulnerable. It aspires to invulnerability. A pipe-dream? Of course – for what we are concerned with here is indeed *magic*. But is this magic performed before or after the act of denomination?

Imaginary and real barriers set up against attacks from outside can be reinforced. As Wilhelm Reich showed, defensive reactions may even give rise to a tough armouring.[24] Some non-Western cultures, however, proceed otherwise, relying upon a sophisticated discipline which places the body constantly beyond the reach of variations in its 'environment', safe from the onslaughts of the spatial realm. Such is the Eastern response to the spatio-temporal and practico-sensory body's humble *demands* – as opposed to the Western body's *commands*, which promote verbalization and the development of a hard protective shell.

In some circumstances a split occurs, and an interstice or interval is created – a very specific space which is at once magical and real. Might the unconscious not, after all, consist in an obscure nature or substantiality which wishes and desires? Perhaps it is not a source of language, nor a language *per se*? Perhaps, rather, it is that very interstice, that 'in-between' itself – along with whatever occupies it, gains access to it, and occurs therein? But, if an interstice, an interstice between what

---

[24] See J.-M. Palmier, *Wilhelm Reich, essai sur la naissance du freudo-marxisme* (Paris: Union Générale d'Editions, 1969).

and what? Between self and self, between the body and its Ego – or, better, between the Ego-seeking-to-constitute-itself and its body. The context here is necessarily that of a long learning-process, the process of formation and deformation which the immature and premature human child must undergo on the way to familial and social maturity. But what is it exactly that slips into the interstice in question? The answer is: language, signs, abstraction – all necessary yet fateful, indispensable yet dangerous. This is a lethal zone thickly strewn with dusty, mouldering words. What slips into it is what allows meaning to escape the embrace of lived experience, to detach itself from the fleshly body. Words and signs facilitate (indeed provoke, call forth and – at least in the West – command) metaphorization – the transport, as it were, of the physical body outside of itself. This operation, inextricably magical and rational, sets up a strange interplay between (verbal) disembodiment and (empirical) re-embodiment, between uprooting and reimplantation, between spatialization in an abstract expanse and localization in a determinate expanse. This is the 'mixed' space – still natural yet already *produced* – of the first year of life, and, later, of poetry and art. The space, in a word, of representations: representational space.

IX

The body does not fall under the sway of analytic thought and its separation of the cyclical from the linear. The unity which that reflection is at such pains to decode finds refuge in the cryptic opacity which is the great secret of the body. For the body indeed unites cyclical and linear, combining the cycles of time, need and desire with the linearities of gesture, perambulation, prehension and the manipulation of things – the handling of both material and abstract tools. The body subsists precisely at the level of the reciprocal movement between these two realms; their difference – which is lived, not thought – is its habitat. Is it not the body, in fact, since it preserves difference within repetition, that is also responsible for the emergence of the new from the repetitive? Analytic thought, by contrast, because it evacuates difference, is unable to grasp how repetition is able to secrete innovation. Such thought, such conceptualizing knowledge (*connaissance*), cannot acknowledge that it underwrites the body's trials and tribulations. Yet, once it has ensconced itself in the gap between lived experience and established knowledge (*savoir*), the work it does there is in the service of death. An empty body, a body conceived of as a sieve, or as a bundle of organs analogous to a bundle of things, a body 'dismembered' or treated as members unrelated to one another, a body without organs – all such supposedly pathological symptomatology stems in reality from the ravages of representation and discourse, which are only exacerbated by modern society, with its ideologies and contradictions (including that between permissiveness and repressiveness in space).

Can the breaking-into-pieces or fragmentation of the body – or, better, a bad relationship of the Ego to its body – be laid at the door of language alone? Do the decomposition of the body into localized functions and its abandonment as a totality whether subjective or objective occur as a result of the assignment to body parts, from earliest childhood, of discrete names, so that the phallus, the eyes, and so on, become so many dissociated elements within a representational space that is subsequently experienced in a pathological manner?

The problem with this thesis is that it exonerates the Christian (or rather the Judaeo-Christian) tradition, which misapprehends and despises the body, relegating it to the charnel-house if not to the Devil. It also exonerates capitalism, which has extended the division of labour into the very bodies of workers and even non-workers. Taylorism, one of the first 'scientific' approaches to productivity, reduced the body as a whole to a small number of motions subjected to strictly controlled linear determinations. A division of labour so extreme, whereby specialization extends to individual gestures, has undoubtedly had as much influence as linguistic discourse on the breaking-down of the body into a mere collection of unconnected parts.

The Ego's relationship to the body, which is annexed little by little to the realm of theoretical thought, turns out to be both complex and diverse. Indeed, there are as many different relationships between the Ego and its own body – as many forms of appropriation of that body, or of failure to appropriate it – as there are societies, 'cultures', or even perhaps individuals.

Furthermore, the Ego's practical relationship to its own body determines its relationship to other bodies, to nature, and to space. And vice versa: the relationship to space is reflected in the relationship to the other, to the other's body and the other's consciousness. The analysis – and self-analysis – of the total body, the way in which that body locates itself and the way in which it becomes fragmented, all are determined by a practice which includes discourse but which cannot be reduced to it. The detachment of work from play, from the gestures of ritual and from the erotic realm only serves to make whatever interaction or interference does occur that much more significant. Under the conditions of modern industry and city life, abstraction holds sway over the relationship to the body. As nature fades into the background, there is nothing to restore the total body – nothing in the world of objects, nothing in the world of action. The Western tradition, with its misapprehension of the body, remanifests itself in increasingly strange ways; laying the blame for all the damage at the door of discourse alone is to exculpate not only that tradition but also 'real' abstract space.

X

The body's inventiveness needs no demonstration, for the body itself reveals it, and deploys it in space. Rhythms in all their multiplicity interpenetrate one another. In the body and around it, as on the surface of a body of water, or within the mass of a liquid, rhythms are forever crossing and recrossing, superimposing themselves upon each other, always bound to space. They exclude neither primal tendencies nor any other energetic forces, whether these invest the interior or the surface of the body, whether they are 'normal' or excessive, whether they are responses to external action or endogenous and explosive in character. Such rhythms have to do with needs, which may be dispersed as tendencies, or distilled into desire. If we attempt to specify them, we find that some rhythms are easy to identify: breathing, the heartbeat, thirst, hunger, and the need for sleep are cases in point. Others, however, such as those of sexuality, fertility, social life, or thought, are relatively obscure. Some operate on the surface, so to speak, whereas others spring from hidden depths.

It is possible to envision a sort of 'rhythm analysis' which would address itself to the concrete reality of rhythms, and perhaps even to their use (or appropriation). Such an approach would seek to discover those rhythms whose existence is signalled only through mediations, through indirect effects or manifestations. Rhythm analysis might eventually even displace psychoanalysis, as being more concrete, more effective, and closer to a pedagogy of appropriation (the appropriation of the body, as of spatial practice). It might be expected to apply the principles and laws of a general rhythmology to the living body and its internal and external relationships. Such a discipline's field of application *par excellence*, its preferred sphere of experiment, would be the sphere of music and dance, the sphere of 'rhythmic cells' and their effects. The repetitions and redundancies of rhythms, their symmetries and asymmetries, interact in ways that cannot be reduced to the discrete and fixed determinants of analytic thought. Only if this is clearly grasped can the polyrhythmic body be understood and appropriated. Rhythms differ from one another in their amplitude, in the energies they ferry and deploy, and in their frequency. Such differences, conveyed and reproduced by the rhythms which embody them, translate into intensity or strength of anticipation, tension and action. All these factors interact with one another within the body, which is traversed by rhythms rather as the 'ether' is traversed by waves.

The way in which rhythms may be said to embrace both cyclical and linear is illustrated by music, where the measure and the beat are linear in character, while motifs, melody and particularly harmony are cyclical (the division of octaves into twelve half-tones, and the reiteration of sounds and intervals within octaves). Much the same may be said of dance, a gestural system whose organization combines two codes, that of the dancer and that of the spectator (who keeps time by clapping or with other body movements): thus, as evocative (paradigmatic) gestures recur, they are integrated into a ritually linked gestural chain.

What do we know about rhythms, as sequential relationships in space, as objective relationships? The notion of *flows* (of energy, matter, etc.) is self-sufficient only in political economy. It is in any case always subordinate to the notion of space. As for 'drive', this idea is a transposition onto the psychic level of the fundamental, but at the same time dissociated, idea of rhythm. What we *live* are rhythms – rhythms experienced subjectively. Which means that, here at least, 'lived' and 'conceived' are close: the laws of nature and the laws governing our bodies tend to overlap with each other – as perhaps too with the laws of so-called social reality.

An organ has a rhythm, but the rhythm does not have, nor is it, an organ; rather, it is an interaction. A rhythm invests places, but is not itself a place; it is not a thing, nor an aggregation of things, nor yet a simple flow. It embodies its own law, its own regularity, which it derives from space – from its own space – and from a relationship between space and time. Every rhythm possesses and occupies a spatio-temporal reality which is known by our science and mastered so far as its physical aspect (wave motion) is concerned, but which is misapprehended from the point of view of living beings, organisms, bodies and social practice. Yet social practice is made up of rhythms – daily, monthly, yearly, and so on. That these rhythms have become more complicated than natural rhythms is highly probable. A powerful unsettling factor in this regard is the practico-social dominance of linear over cyclical repetition – that is to say, the dominance of one aspect of rhythms over another.

Through the mediation of rhythms (in all three senses of 'mediation': means, medium, intermediary), an animated space comes into being which is an extension of the space of bodies. How exactly the laws of space and of its dualities (symmetries/asymmetries, demarcation/orientation, etc.) chime with the laws of rhythmic movement (regularity, diffusion, interpenetration) is a question to which we do not as yet have the answer.

XI

What is the unconscious if not consciousness itself, if not consciousness and its double, which it contains and keeps within itself – namely, 'self-consciousness'? Consciousness, then, *qua* mirror image, *qua* repetition and mirage. What does this mean? In the first place, it means that any substantification or naturalization of the unconscious, locating it above or below consciousness, must sooner or later fall into ideological fatuity.[25] Consciousness is not unaware of itself; if it were, of whom and of what would it be the consciousness? In essence, and by definition, self-consciousness is a reduplication, a self-reproduction, as much as it is a 'reflection' of objects. But does it know itself? No. It is acquainted neither with the conditions of its own existence nor with the laws (if any) which govern it. In this sense it may justifiably be compared to language, not only because there is no consciousness without a language, but also because those who speak, and even those who write, are unacquainted with the conditions and laws of language, of their language, even as they practise it. What then is the 'status' of language? Between knowledge and ignorance here there is a mediation which sometimes functions effectively as an intermediary but which may also block the way. This mediation is *misapprehension*. Like the flower which does not know it is a flower, self-consciousness, so much vaunted in Western thought from Descartes to Hegel (and even more recently, at least in philosophy), misapprehends its own preconditions whether natural (physical) or practical, mental or social. We have long known that from early childhood the consciousness of 'conscious beings' apprehends itself as a reflection of what it has wrought in 'the object' or in the other by means of certain privileged products, namely instrumental objects and speech. Consciousness thus apprehends itself in and through

---

[25] See *L'inconscient*, proceedings of the sixth Colloque de Bonneval, 1960 (Paris: Desclée de Brouwer, 1966), pp. 347ff.

what it *produces*: by playing with a simple stick, for example – by disordering or breaking things – the child begins to 'be'. Conscious beings apprehend themselves in a mélange of violence, lack, desires, needs, and knowledge properly (or improperly) so called.

In this sense, then – but not exactly after the fashion of language as such – consciousness misapprehends itself. Consciousness, itself the locus of knowing, thus permits the emergence of a knowledge characterized by misunderstanding: on the one hand, the illusion of a perfect or transparent knowledge (the Idea, divine knowledge, absolute Knowledge); on the other hand, notions of a mystery, of an unknowable realm, or of an unconscious. To return to this last term: it is neither true nor false to speak of an unconscious. Hence it is both true and false at the same time. The unconscious resembles an illusion with a *raison d'être* – a sort of mirage effect. People, and more particularly psychologists, psychoanalysts and psychiatrists, use the unconscious as an appropriate receptacle for whatever they please to consign to it, including the preconditions of consciousness in the nervous system or brain; action and language; what is remembered and what is forgotten; and the body and its own history. The tendency to fetishize the unconscious is inherent in the image of unconsciousness itself. This is why this idea opens the door so wide to ontology, metaphysics, the death drive, and so on.

Still, the term is meaningful in that it designates that unique process whereby every human 'being' is formed, a process which involves reduplication, doubling, repetition at another level of the spatial body; language and imaginary/real spatiality; redundancy and surprise; learning through experience of the natural and social worlds; and the forever-compromised appropriation of a 'reality' which dominates nature by means of abstraction but which is itself dominated by the worst of abstractions, the abstraction of power. The 'unconscious' in this sense, as the imaginary and real locus of a struggle, as the obscure counterweight to that 'luminous' entity known as *culture*, has nothing in common with the ragbag concept of the psychologists and other experts.

What an enigma sleep presents for philosophy! How can the *cogito* ever slumber? Its duty is to keep vigil till the end of time, as Pascal understood and reiterated. Sleep reproduces life in the womb and foreshadows death; yet this kind of rest has its own fullness. In sleep the body gathers itself together, building up its energy reserves by imposing silence on its information receptors. It closes down, and passes through a moment with its own truth, its own beauty, its own worth. This is one moment among others, a poetic moment. It is now that the 'space of the dream' makes its paradoxical appearance. At once imaginary and real, this space is different from the space of language, though of the same order, and the faithful guardian of sleep rather than of social learning. Is this then the space of 'drives'? It would be better described as a space where dispersed and broken rhythms are reconstituted, a space for the poetic reconstruction of situations in which wishes are present – but wishes which are not so much fulfilled as simply proclaimed. It is a space of enjoyment, indeed it establishes a virtual reign of pleasure, though erotic dreams break up on the reefs of the dreamer's pleasure and disillusion. The space of the dream is strange and alien, yet at the same time as close to us as is possible. Rarely coloured, even more rarely animated by music, it still has a sensual-sensory character. It is a theatrical space even more than a quotidian or poetic one: a putting into images of oneself, for oneself.

Visual space in its specificity contains an immense crowd, veritable hordes of objects, things, bodies. These differ by virtue of their place and that place's local peculiarities, as also by virtue of their relationship with 'subjects'. Everywhere there are privileged objects which arouse a particular expectation or interest, while others are treated with indifference. Some objects are known, some unknown, and some misapprehended. Some serve as relays: transitory or transitional in nature, they refer to other objects. Mirrors, though privileged objects, nevertheless have a transitional function of this kind.

Consider a window. Is it simply a void traversed by a line of sight? No. In any case, the question would remain: what line of sight – and whose? The fact is that the window is a non-object which cannot fail to become an object. As a transitional object it has two senses, two orientations: from inside to outside, and from outside to inside. Each is marked in a specific way, and each bears the mark of the other. Thus windows are differently framed outside (for the outside) and inside (for the inside).

Consider a door. Is it simply an aperture in the wall? No. It is framed (in the broadest sense of the term). A door without a frame would fulfil one function and one function only, that of allowing passage. And it would fulfil that function poorly, for something would be missing. Function calls for something other, something more, something better than functionality alone. Its surround makes a door into an object. In conjunction with their frames, doors attain the status of works, works of a kind not far removed from pictures and mirrors. Transitional, symbolic and functional, the object 'door' serves to bring a space, the space of a 'room', say, or that of the street, to an end; and it heralds the reception to be expected in the neighbouring room, or in the house or interior that awaits. The threshold or sill of an entrance is another transitional object, one which has traditionally enjoyed an almost ritual significance (crossing a threshold as analogous to passing through a lock, or 'graduating'). So objects fall spontaneously into such classes as transitional objects, functional objects, and so on. These classes, however, are always provisional: the classes themselves are subject to change, while objects are liable to move from one class to another.

This brings us to the articulation between sensory and practico-perceptual space on the one hand and specific or practico-social space, the space of this or that particular society, on the other. Can social space be defined in terms of the projection of an ideology into a neutral space? No. Ideologies dictate the locations of particular activities, determining that such and such a place should be sacred, for example, while some other should not, or that a temple, a palace or a church must be here, and not there. But ideologies do not produce space: rather, they are in space, and of it. It is the forces of production and the relations of production that produce social space. In the process a global social practice is brought into being, comprising all the diverse activities which, at least up to now, have characterized any society: education, administration, politics, military organization, and so on. It follows that not all localization should be attributed to ideology. 'Place' in society, high society *versus* the lower depths, the political 'left' and 'right' – all these apparent forms of localization derive not only from ideology but also from the symbolic properties of space, properties inherent to that space's practical occupation.

In what does sensory space, within social space, consist? It consists in an 'unconsciously' dramatized interplay of relay points and obstacles, reflections, references, mirrors and echoes – an interplay implied, but not explicitly designated, by this discourse. Within it, specular and transitional objects exist side by side with tools ranging from simple sticks to the most sophisticated instruments designed for hand and body. Does the body, then, retrieve its unity, broken by language, from its own image coming towards it, as it were, from the outside? More than this, and better, is required before that can happen. In the first place, a welcoming space is called for – the space of nature, filled with non-fragmented 'beings', with plants and animals. (It is architecture's job to reproduce such a space where it is lacking.) And then effective, practical actions must be performed, making use of the basic materials and *matériel* available.

Splits reappear continually; they are bridged by metaphor and metonymy. Language possesses a practical function but it cannot harbour knowledge without masking it. The playful aspect of space escapes it, and only emerges in play itself (by definition), and in irony and humour. Objects serve as markers for rhythms, as reference points, as centres. Their fixedness, however, is only relative. Distances here may be abolished by look, word or gesture; they may equally well be exaggerated thereby. Distantiation alternates with convergence, absence with presence, concealment with revelation, reality with appearance – and all overlap in a theatre of reciprocal implication and explication where the action halts only during sleep. Relations in the perceptual realm do not reflect social relations as such – on the contrary, they disguise them. Social relations properly so called – i.e. the relations of production – are not visible in sensory–sensual (or practico-perceptual) space. They are circumvented. They need therefore to be decoded, but even in their decoded form it is difficult for them to be extracted from mental and located in social space. Sensory–sensual space tends to establish itself within the visible–readable sphere, and in so doing it promotes the misapprehension of aspects, indeed the dominant aspects, of social practice: labour, the division of labour, the organization of labour, and so on. This space, which does not recognize its own potential for playfulness (for it is readily taken over by play) does enshrine social relationships, which appear in it as relationships of opposition and contrast, as linked sequences. Long predominant among such relationships have been right and left, high and low, central and peripheral, demarcated and oriented space, near and far, symmetrical and asymmetrical, and auspicious and inauspicious. Nor should we forget paternity and maternity, male places and female places, and their attendant symbols. Now, it is true that the aim of our discussion is to establish the paradigm of a space. All the same, it is very important not to overlook, in the immediate vicinity of the body, and serving to extend it into the surrounding networks of relationships and pathways, the various types of objects. Among them are everyday utensils or tools – pot, cup, knife, hammer, or fork – which extend the body in accord with its rhythms; and those, such as the implements of peasant or artisan, which leave the body further behind, and establish their own spatial realms. Social space is defined (also) as the locus and medium of speech and writing, which sometimes disclose and sometimes dissimulate, sometimes express what is true and sometimes what is false (with the false serving the truth as relay, resource and foundation). It is in this world that the quest for enjoyment takes place, a quest whose object, once found, is destroyed by the act of taking pleasure itself.

Enjoyment in this sense forever evades the grasp. A game of mirrors, then: plenitude followed by disillusion. And it is a game that never ends, as the Ego recognizes itself, and misapprehends itself, in the Alter Ego. Misunderstanding also nourishes attitudes of listening and expectancy. Then the tide of the visual with its clarity overwhelms what is merely audible or touchable.

We have yet to consider the space of production, and the production of space. Sensory–sensual space is simply a sediment destined to survive as one layer or element in the stratification and interpenetration of social spaces.

We have already noted one overall characteristic of production: from products, be they objects or spaces, all traces of productive activity are so far as possible erased. What of the mark of the worker or workers who did the producing? It has no meaning or value unless the 'worker' is also a user and owner – as in the case of craftsmen or peasants. Objects are only perfected by being 'finished'.

There is nothing new about this, but it is appropriate to reiterate it, for it has important consequences. The fact is that this erasure facilitates the procedure whereby the worker is deprived of the product of his labour. It is tempting to generalize, and argue that such erasures of traces make possible an immense number of transfers and substitutions, and indeed that this kind of concealment is the basis not only of myths, mystifications and ideologies, but also of all domination and all power. An extrapolation of this kind, however, cannot be justified. In space, nothing ever disappears – no point, no place. Still, the concealment of the productive labour that goes into the product has one significant implication: social space is not coextensive with the space of social labour. Which is not to say that social space is a space of enjoyment, of non-labour, but merely that produced or worked objects pass from the space of labour to the enveloping social space only once the traces of labour have been effaced from them. Whence, of course, the commodity.

XII

At one level of social space, or in one region of it, concatenations of gestures are deployed. In its broadest sense, the category of the 'gestural' takes in the gestures of labour – the gestures of peasants, craftsmen or industrial workers. In a narrower and more restrictive sense, it does not cover technical gestures or productive acts; it does not extend beyond the gestures and acts of 'civil' life exclusive of all specialized activities and places (such as those associated with war, religion or justice); in short, all institutional gestures, coded and located as such, are barred. But, whether understood in the broad or the narrow sense, gestures as a whole mobilize and activate the total body.

Bodies (each body) and interbodily space may be pictured as possessed of specific assets: the *materials* (heredity, objects) which serve as their starting-point, and the *matériel* which they have available to them (behaviour patterns, conditioning – what are sometimes called stereotypes). For these bodies, the natural space and the abstract space which confront and surround them are in no way separable, as they may be from an analytic perspective. The individual situates his body in its own space and apprehends the space around the body. The energy available to each seeks employment in that space, and the other bodies which that energy encounters, be they inert or living, constitute obstacles, dangers, coagents, or prizes for it. The actions of each individual involve his multiple affiliations and basic constitution, with its dual aspect: first, the axes and planes of symmetry, which govern the movements of arms, legs, hands and limbs in general; secondly, the rotations and the gyrations which govern all sorts of movements of trunk or head – circular, spiral, 'figures of eight', and so on. The accomplishment of gestures, for which this *matériel* is the prerequisite, further implies the existence of affiliations, of groups (family, tribe, village, city, etc.) and of activity. It also calls for specific materials – for those objects which the activity in question requires; such objects are 'real', and therefore material in nature, but they are also symbolic, and hence freighted with affect.

What shall we say of the human hand? It certainly seems no less complex or 'rich' than the eye, or than language. The hand can feel, caress, grasp, brutalize, hit, kill. The sense of touch is the discoverer of matter. Thanks to tools – which are separate from nature and responsible for severing from nature whatever they impinge upon, but which are nevertheless extensions of the body and its rhythms (for instance, the hammer with its linear and repetitive action, or the potter's wheel with its circular and continuous one) – the hand modifies materials. Muscular effort can mobilize energies of a massive kind, and often in enormous quantities, to support repetitive gestures such as those associated with labour (but also those called for by games). By contrast, the search for information about things through skin contact, through feeling, through caresses, relies on the use of subtle energies.

The chief *matériel* employed by social gestures, then, consists of *articulated* movements. The articulation of human limbs is refined and complex; if one takes the fingers, the hand, the wrist and the arm into account, the total number of segments involved is very large.

More than one theorist has drawn a distinction between inarticulate and articulate as a way of distinguishing nature from culture: on the one hand, the inarticulate sphere of cries, tears, expressions of pain or pleasure, the sphere of spontaneous and animal life; on the other, the articulate sphere of words, of language and discourse, of thought and of the clear consciousness of self, things and acts. What is missing from this account is the mediation of bodily gestures. Are not such gestures, articulated and linked together as they are, more likely than drives to lie at the origin (so to speak) of language? Bound together outside the realm of work as well as within it, could they not have contributed to the development of that part of the brain which 'articulates' linguistic and gestural activity? In childhood, in the body of the child, there arguably exists a pre-verbal, gestural capacity – that is, a capacity which is concretely practical or 'operational', and which constitutes the basis of the child's first relationship as 'subject' to perceptible objects. Pre-verbal gestures of this kind might fall under several rubrics: destructive gestures (foreshadowing later productive ones), gestures of displacement, gestures of seriation, and gestures of grouping (groups being closed series).

The most sophisticated gestural systems – those of Asian dance, for example – bring into play all segments of the limbs, even the fingertips, and invest them with symbolic (cosmic) significance. But less complex systems, too, qualify fully as wholes invested with meaning: that is to say, as coded – and decodable – entities. It is legitimate to speak of 'codes' here because the ordering of gestures is laid down beforehand, and has ritual and ceremonial aspects. Such ensembles of gestures are made up, like language, of symbols, signs and signals. Symbols embody their own meaning; signs refer from a signifier to what is signified; signals elicit an immediate or deferred action which may be aggressive, affective, erotic, or whatever. Space is perceived as an interval, separating a deferred action from the gesture which heralds, proposes or signifies it. Gestures are linked on the basis of oppositions (for instance, rapid versus slow, stiff versus loose, peaceful versus violent) and on the basis of ritualized (and hence coded) rules. They may then be said to constitute a language in which expressiveness (that of the body) and signification (for others – other consciousnesses, other bodies) are no further apart than nature from culture, than the abstract from the practical. A highly dignified demeanour, for instance, demands that the axes and planes of symmetry govern the body in motion, so that they are preserved even as it moves around: the posture is straight, the gestures are of the kind we think of as harmonious. By contrast, attitudes of humility and humiliation flatten the body against the ground: the vanquished are supposed to prostrate themselves, worshippers to kneel, and the guilty to lower their heads and kiss the earth. And in the display of clemency or indulgence the inclining of the body parallels the bending of the will in compromise.

It goes without saying that such codes are specific to a particular society; indeed they stipulate an affiliation to that society. To belong to a given society is to know and use its codes for politeness, courtesy, affection, parley, negotiation, trading, and so on – as also for the declaration of hostilities (for codes having to do with social alliance are inevitably subtended by codes of insolence, insult and open aggression).

The importance of places and space in gestural systems needs emphasizing. High and low have great significance: on the one hand the ground, the feet and the lower members, and on the other the head and whatever surmounts or covers it – hair, wigs, plumes, headdresses, parasols, and so forth. Right and left are similarly rich in meaning (in the West, the left hand has of course acquired negative – 'sinister' – connotations). Variations in the use of the voice, as in singing, serve to accentuate such meanings: shrill/deep, high/low, loud/soft.

Gestural systems embody ideology and bind it to practice. Through gestures, ideology escapes from pure abstraction and performs actions (for example, the clenched-fist salute or the sign of the cross). Gestural systems connect representations of space with representational spaces – or, at least, they do so under certain privileged conditions. With their liturgical gestures, for instance, priests evoke the divine gestures which created the universe by mimicking them in a consecrated space. Gestures are also closely bound up with the objects which fill space – with furniture, clothing, instruments (kitchen utensils, work tools), games, and places of residence. All of which testifies to the complexity of the gestural realm.

May this realm then be said to embrace an essentially indefinite – and hence indefinable – variety of codes? We should by now be able to clear up this rather thorny problem. The fact is that the multiplicity of codes has determinants which are susceptible of categorization: everyday gestures differ from the gestures associated with feasts, the rites of friendship contrast with the rights of antagonism, and the everyday microgestural realm is clearly distinct from the macrogestural one, which is the realm of crowds in action. There are also, are there not, gestures – signs or signals – which allow passage from one code or subcode to another, interrupting the one so as to open the way to the other? Undoubtedly so.

We have every reason to speak of 'subcodes' and general codes in this connection. In the first place it makes it possible, if so desired, to classify codes by species and genus, as it were. And it allows us to avoid the 'unnecessary multiplication of entities' (in the event, of codes): why should Occam's razor not be applied to the relatively new concepts of coding and decoding, of message and decipherment? Above all, however, we must avoid conceiving of or imagining a spatial code which is merely a subcode of discourse, so that constructed space is seen as somehow dependent on discourse or on a modality of it. The study of gestures certainly invalidates any such view of things.

My aim in the foregoing discussion has not been to find a rationale for gestures but rather to clarify the relationship between gestural systems and space. Why do many Oriental peoples live close to the ground, using low furniture and sitting on their heels? Why does the Western world, by contrast, have rigid, right-angled furniture which obliges people to assume constricted postures? And why do the dividing-lines between such attitudes or (unformulated) codes correspond exactly to religious and political frontiers? Diversity in this sphere is still as incomprehensible as the diversity of languages. Perhaps the study of social spaces will throw some light on these questions.

Organized gestures, which is to say ritualized and codified gestures, are not simply performed in 'physical' space, in the space of bodies. Bodies themselves generate spaces, which are produced by and for their gestures. The linking of gestures corresponds to the articulation and linking of well-defined spatial segments, segments which repeat, but whose repetition gives rise to novelty. Consider, for example, the cloister, and the solemn pace of the monks who walk there. The spaces produced in the way we have been discussing are often multifunctional (the agora, for instance), although some strictly defined gestures, such as those associated with sport or war, produced their own specific spaces very early on – stadia, parade grounds, tiltyards, and so forth. Many such social spaces are given rhythm by the gestures which are produced within them, and which produce them (and they are accordingly often measured in paces, cubits, feet, palms or thumbs). The everyday microgestural realm generates its own spaces (for example, footways, corridors, places for eating), and so does the most highly formalized macrogestural realm (for instance, the ambulatories of Christian churches, or podia). When a gestural space comes into conjunction with a conception of the world possessed of its own symbolic system, a grand creation may result. Cloisters are a case in point. What has happened here is that, happily, a gestural space has succeeded in mooring a mental space – a space of contemplation and theological abstraction – to the earth, thus allowing it to express itself symbolically and to become part of a practice, the practice of a well-defined group within a well-defined society. Here, then, is a space in which a life balanced between the contemplation of the self in its finiteness and that of a transcendent infinity may experience a happiness composed of quietude and a fully accepted lack of fulfilment. As a space for contemplatives, a place of promenade and assembly, the cloister connects a finite and determinate locality – socially particularized but not unduly restricted as to use, albeit definitely controlled by an order or rule – to a theology of the infinite. Columns, capitals, sculptures – these are semantic differentials which mark off the route followed (and laid down) by the steps of the monks during their time of (contemplative) recreation.

If the gestures of 'spiritual' exchange – the exchange of symbols and signs, with their own peculiar delights, have produced spaces, the gestures of material exchange have been no less productive. Parley, negotiation and trade have always called for appropriate spaces. Over the ages merchants have been an active and original group, and productive after their fashion. Today the realm of commodities has extended its sway, along with that of capital, to the entire planet, and it has consequently assumed an oppressive role. The commodity system thus comes in for a good deal of denigration, and tends to be blamed for all ills. It should be remembered, however, that for centuries merchants and merchandise stood for freedom, hope and expanding horizons relative to the constraints imposed by ancient communities, whether agrarian societies or the more political cities. Merchants brought both riches and essential goods such as cereals, spices or fabrics. Commerce was synonymous with communication, and the exchange of goods went hand in hand with the exchange of ideas and pleasures. Today there are rather more remnants of that state of affairs in the East than in the West. The earliest commercial areas – porticoes, basilicas or market halls dating from a time when merchants and their gestures created their own spaces – are thus not without beauty. (It is worth asking ourselves *en passant* why spaces devoted to sensual pleasures seem so much rarer than places of power, knowledge or wisdom, and exchange.)

In attempting to account for these multifarious creations, the evocation of 'proxemics', whether in connection with children or adults, couples or families, groups or crowds, is inadequate. Hall's anthropological descriptive term 'proxemic', which is related to the idea of neighbourhood, is restrictive (and reductive) as compared with 'gestural'.[26]

XIII

Structural distinctions between binary operations, levels and dimensions must not be allowed to obscure the great dialectical movements that traverse the world-as-totality and help define it.

First moment: things (objects) in space. Production, still respectful of nature, proceeds by selecting portions of space and using them along with their contents. Agriculture predominates, and societies produce palaces, monuments, peasant dwellings, and works of art. Time is inseparable from space. Human labour directed at nature deconsecrates it, but distils the sacredness of elements of it into religious and political edifices. Form (of thought or of action) is inseparable from content.

Second moment: from this prehistory certain societies emerge and accede to the historical plane – that is, to the plane of *accumulation* (of riches, knowledge, and techniques) – and hence to the plane of production, first for exchange, then for money and capital. It is now that artifice, which at first has the appearance of art, prevails over nature, and that form and the formal separate from their content; abstraction and signs as such are elevated to the rank of basic and ultimate truths; and consequently philosophical and scientific thought comes to conceive of a space without things or objects, a space which is somehow of a higher order than its contents, a means for them to exist or a medium in which they exist. Once detached from things, space understood as a form emerges either as substance (Descartes) or else, on the contrary, as 'pure *a priori*' (Kant). Space and time are sundered, but space brings time under its sway in the praxis of accumulation.

Third moment: relative now, space and things are reunited; through thought, the contents of space, and in the first place time, are restored to it. The fact is that space 'in itself' is ungraspable, unthinkable, unknowable. Time 'in itself', absolute time, is no less unknowable. But

[26] See Edward T. Hall, *The Hidden Dimension* (Garden City, N.Y.: Doubleday, 1966), p. 1.

that is the whole point: time is known and actualized in space, becoming a social reality by virtue of a spatial practice. Similarly, space is known only in and through time. Unity in difference, the same in the other (and vice versa), are thus made concrete. But with the development of capitalism and its praxis a difficulty arises in the relations between space and time. The capitalist mode of production begins by producing things, and by 'investing' in places. Then the reproduction of social relations becomes problematic, as it plays a part in practice, modifying it in the process. And eventually it becomes necessary to reproduce nature also, and to master space by producing it – that is, the political space of capitalism – while at the same time reducing time in order to prevent the production of new social relations. But capitalism is surely approaching a threshold beyond which reproduction will no longer be able to prevent the production, not of things, but of new social relations. What would those relations consist in? Perhaps in the unity, at once familiar and new, of space and time, a unity long misapprehended, split up and superseded by the rash attribution of priority to space over time.

The movement I am describing may seem abstract. And indeed it is! For here, at the present juncture, as in Marx's work (or at least in part of it), a reflection upon the *virtual* is what guides our understanding of the real (or actual), while also retroactively affecting – and hence illuminating – the antecedents and the necessary preconditions of that reality. At the present 'moment', *modernity* with its contradictions has only just entered upon the stage. Marx took a similar tack to the one we are taking when (in a chapter of *Capital* that has only recently been published) he envisaged the implications and consequences of the extension of the 'world of commodities' and of the world market, developments which were at that time no more than virtualities embedded in history (the history, that is, of accumulation).

How should the charge that this procedure or method is mere extrapolation be answered? By pointing out the legitimacy of pushing an idea or hypothesis as far as it will go. The idea of *producing*, for example, today extends beyond the production of this or that thing or work to the *production of space*. And this has its retroactive effect on our understanding of antecedents – in this case an understanding of productive forces and forms. Our *modus operandi*, then, is a sort of forcing-house approach. Extreme hypotheses are permissible. The hypothesis, for instance, that the commodity (or the world market) will come to occupy all space; that exchange value will impose the law of value upon the whole planet; and that in some sense world history is nothing but the history of commodities. Pushing a hypothesis to its limit helps us discover what obstacles to its application exist and what objections to it should be raised. Proceeding in the same manner apropos of space, we may wonder whether the state will eventually produce its own space, an absolute political space. Or whether, alternatively, the nation states will one day see their absolute political space disappearing into (and thanks to) the world market. Will this last eventuality occur through self-destruction? Will the state be transcended or will it wither away? And must it be one or the other, and not, perhaps, both?

XIV

For millennia, *monumentality* took in all the aspects of *spatiality* that we have identified above: the perceived, the conceived, and the lived; representations of space and representational spaces; the spaces proper to each faculty, from the sense of smell to speech; the gestural and the symbolic. Monumental space offered each member of a society an image of that membership, an image of his or her social visage. It thus constituted a collective mirror more faithful than any personal one. Such a 'recognition effect' has far greater import than the 'mirror effect' of the psychoanalysts. Of this social space, which embraced all the above-mentioned aspects while still according each its proper place, everyone partook, and partook fully – albeit, naturally, under the conditions of a generally accepted Power and a generally accepted Wisdom. The monument thus effected a 'consensus', and this in the strongest sense of the term, rendering it practical and concrete. The element of repression in it and the element of exaltation could scarcely be disentangled; or perhaps it would be more accurate to say that the repressive element was metamorphosed into exaltation. The codifying approach of semiology, which seeks to classify representations, impressions and evocations (as terms in the code of knowledge, the code of personal feelings, the symbolic code, or the hermeneutic code),[27] is quite unable to cover all facets of the monumental. Indeed, it does not even come close, for it is the residual, the irreducible – whatever cannot be classified or codified according to categories devised subsequent to production – which is, here as always, the most precious and the most essential, the diamond at the bottom of the melting-pot. The use of the cathedral's monumental space necessarily entails its supplying answers to all the

---

[27] See Roland Barthes, *S/Z* (Paris: Seuil, 1970), pp. 25ff. Eng. tr. by Richard Miller: *S/Z* (New York: Hill and Wang, 1974), pp. 18ff.

questions that assail anyone who crosses the threshold. For visitors are bound to become aware of their own footsteps, and listen to the noises, the singing; they must breathe the incense-laden air, and plunge into a particular world, that of sin and redemption; they will partake of an ideology; they will contemplate and decipher the symbols around them; and they will thus, on the basis of their own bodies, experience a total being in a total space. Small wonder that from time immemorial conquerors and revolutionaries eager to destroy a society should so often have sought to do so by burning or razing that society's monuments. Sometimes, it is true, they contrive to redirect them to their own advantage. Here too, use goes further and deeper than the codes of exchange.

The most beautiful monuments are imposing in their durability. A cyclopean wall achieves monumental beauty because it seems eternal, because it seems to have escaped time. Monumentality transcends death, and hence also what is sometimes called the 'death instinct'. As both appearance and reality, this transcendence embeds itself in the monument as its irreducible foundation; the lineaments of atemporality overwhelm anxiety, even – and indeed above all – in funerary monuments. A *ne plus ultra* of art – form so thoroughly denying meaning that death itself is submerged. The Empress's Tomb in the Taj Mahal bathes in an atmosphere of gracefulness, whiteness and floral motifs. Every bit as much as a poem or a tragedy, a monument transmutes the fear of the passage of time, and anxiety about death, into splendour.

Monumental 'durability' is unable, however, to achieve a complete illusion. To put it in what pass for modern terms, its credibility is never total. It replaces a brutal reality with a materially realized appearance; reality is changed into appearance. What, after all, is the durable aside from the will to endure? Monumental imperishability bears the stamp of the will to power. Only Will, in its more elaborated forms – the wish for mastery, the will to will – can overcome, or believe it can overcome, death. Knowledge itself fails here, shrinking from the abyss. Only through the monument, through the intervention of the architect as demiurge, can the space of death be negated, transfigured into a living space which is an extension of the body; this is a transformation, however, which serves what religion, (political) power, and knowledge have in common.

In order to define monumental space properly,[28] semiological categorization (codifying) and symbolic explanations must be restrained. But

[28] Clearly we are not concerned here with architectural space understood as the preserve of a particular profession within the established social division of labour.

'restrained' should not be taken to mean refused or rejected. I am not saying that the monument is not the outcome of a signifying practice, or of a particular way of proposing a meaning, but merely that it can be reduced neither to a language or discourse nor to the categories and concepts developed for the study of language. A spatial work (monument or architectural project) attains a complexity fundamentally different from the complexity of a text, whether prose or poetry. As I pointed out earlier, what we are concerned with here is not texts but texture. We already know that a texture is made up of a usually rather large space covered by networks or webs; monuments constitute the strong points, nexuses or anchors of such webs. The actions of social practice are expressible but not explicable through discourse; they are, precisely, *acted* – and not *read*. A monumental work, like a musical one, does not have a 'signified' (or 'signifieds'); rather, it has a *horizon of meaning*: a specific or indefinite multiplicity of meanings, a shifting hierarchy in which now one, now another meaning comes momentarily to the fore, by means of – and for the sake of – a particular action. The social and political operation of a monumental work traverses the various 'systems' and 'subsystems', or codes and subcodes, which constitute and found the society concerned. But it also surpasses such codes and subcodes, and implies a 'supercoding', in that it tends towards the all-embracing presence of the totality. To the degree that there are traces of violence and death, negativity and aggressiveness in social practice, the monumental work erases them and replaces them with a tranquil power and certitude which can encompass violence and terror. Thus the mortal 'moment' (or component) of the sign is temporarily abolished in monumental space. In and through the work in space, social practice transcends the limitations by which other 'signifying practices', and hence the other arts, including those texts known as 'literary', are bound; in this way a consensus, a profound agreement, is achieved. A Greek theatre presupposes tragedy and comedy, and by extension the presence of the city's people and their allegiance to their heroes and gods. In theatrical space, music, choruses, masks, tiering – all such elements converge with language and actors. A spatial action overcomes conflicts, at least momentarily, even though it does not resolve them; it opens a way from everyday concerns to collective joy.

Turmoil is inevitable once a monument loses its prestige, or can only retain it by means of admitted oppression and repression. When the subject – a city or a people – suffers dispersal, the *building* and its functions come into their own; by the same token, *housing* comes to prevail over *residence* within that city or amidst that people. The building has its roots in warehouses, barracks, depots and rental housing. Buildings have functions, forms and structures, but they do not integrate the formal, functional and structural 'moments' of social practice. And, inasmuch as sites, forms and functions are no longer focused and appropriated by monuments, the city's contexture or fabric – its streets, its underground levels, its frontiers – unravel, and generate not concord but violence. Indeed space as a whole becomes prone to sudden eruptions of violence.

The balance of forces between monuments and buildings has shifted. Buildings are to monuments as everyday life is to festival, products to works, lived experience to the merely perceived, concrete to stone, and so on. What we are seeing here is a new dialectical process, but one just as vast as its predecessors. How could the contradiction between building and monument be overcome and surpassed? How might that tendency be accelerated which has destroyed monumentality but which could well reinstitute it, within the sphere of buildings itself, by restoring the old unity at a higher level? So long as no such dialectical transcendence occurs, we can only expect the stagnation of crude interactions and intermixtures between 'moments' – in short, a continuing spatial chaos. Under this dispensation, buildings and dwelling-places have been dressed up in monumental *signs*: first their façades, and later their interiors. The homes of the moneyed classes have undergone a superficial 'socialization' with the introduction of reception areas, bars, nooks and furniture (divans, for instance) which bespeak some kind of erotic life. Pale echoes, in short, of the aristocratic palace or town house. The town, meanwhile, now effectively blown apart, has been 'privatized' – no less superficially – thanks to urban 'decor' and 'design', and the development of fake environments. Instead, then, of a dialectical process with three stages which resolves a contradiction and 'creatively' transcends a conflictual situation, we have a stagnant opposition whose poles at first confront one another 'face to face', then relapse into muddle and confusion.

XV

There is still a good deal to be said about the notion of the monument. It is especially worth emphasizing what a monument is *not*, because this will help avoid a number of misconceptions. Monuments should not be looked upon as collections of symbols (even though every monument embodies symbols – sometimes archaic and incomprehensible ones), nor as chains of signs (even though every monumental whole is made up of signs). A monument is neither an object nor an aggregation of diverse objects, even though its 'objectality', its position as a social object, is recalled at every moment, perhaps by the brutality of the materials or masses involved, perhaps, on the contrary, by their gentle qualities. It is neither a sculpture, nor a figure, nor simply the result of material procedures. The indispensable opposition between inside and outside, as indicated by thresholds, doors and frames, though often underestimated, simply does not suffice when it comes to defining monumental space. Such a space is determined by what may take place there, and consequently by what may not take place there (prescribed/proscribed, scene/obscene). What appears empty may turn out to be full – as is the case with sanctuaries, or with the 'ships' or naves of cathedrals. Alternatively, full space may be inverted over an almost heterotopic void at the same location (for instance, vaults, cupolas). The Taj Mahal, for instance, makes much play with the fullness of swelling curves suspended in a dramatic emptiness. Acoustic, gestural and ritual movements, elements grouped into vast ceremonial unities, breaches opening onto limitless perspectives, chains of meanings – all are organized into a monumental whole.

The affective level – which is to say, the level of the body, bound to symmetries and rhythms – is transformed into a 'property' of monumental space, into symbols which are generally intrinsic parts of a politico-religious whole, into co-ordinated symbols. The component elements of such wholes are disposed according to a strict order for the purposes of the use of space: some at a first level, the level of affective, bodily, lived experience, the level of the spoken word; some at a second level, that of the perceived, of socio-political signification; and some at a third level, the level of the conceived, where the dissemination of the written word and of knowledge welds the members of society into a 'consensus', and in doing so confers upon them the status of 'subjects'. Monumental space permits a continual back-and-forth between the private speech of ordinary conversations and the public speech of discourses, lectures, sermons, rallying-cries, and all theatrical forms of utterance.

Inasmuch as the poet through a poem gives voice to a way of living (loving, feeling, thinking, taking pleasure, or suffering), the experience of monumental space may be said to have some similarity to entering and sojourning in the poetic world. It is more easily understood, however, when compared with texts written for the theatre, which are composed of dialogues, rather than with poetry or other literary texts, which are monologues.

Monumental qualities are not solely plastic, not to be apprehended solely through looking. Monuments are also liable to possess acoustic properties, and when they do not this detracts from their monumentality. Silence itself, in a place of worship, has its music. In cloister or cathedral, space is measured by the ear: the sounds, voices and singing reverberate in an interplay analogous to that between the most basic sounds and tones; analogous also to the interplay set up when a reading voice breathes new life into a written text. Architectural volumes ensure a correlation between the rhythms that they entertain (gaits, ritual gestures, processions, parades, etc.) and their musical resonance. It is in this way, and at this level, in the *non-visible*, that bodies find one another. Should there be no echo to provide a reflection or acoustic mirror of presence, it falls to an object to supply this mediation between the inert and the living: bells tinkling at the slightest breeze, the play of fountains and running water, perhaps birds and caged animals.

Two 'primary processes', as described by certain psychoanalysts and linguists, might reasonably be expected to operate in monumental space: (1) displacement, implying metonymy, the shift from part to whole, and contiguity; and (2) condensation, involving substitution, metaphor and similarity. And, to a degree, this is so. Social space, the space of social practice, the space of the social relations of production and of work and non-work (relations which are to a greater or lesser extent codified) – this space is indeed condensed in monumental space. The notion of 'social condenser', as proposed by Russian architects in the 1920s, has a more general application. The 'properties' of a spatial texture are focused upon a single point: sanctuary, throne, seat, presidential chair, or the like. Thus each monumental space becomes the metaphorical and quasi-metaphysical underpinning of a society, this by virtue of a play of substitutions in which the religious and political realms symbolically (and ceremonially) exchange attributes – the attributes of power; in this way the authority of the sacred and the sacred aspect of authority are transferred back and forth, mutually reinforcing one another in the process. The horizontal chain of sites in space is thus replaced by vertical superimposition, by a hierarchy which follows its own route to the locus of power, whence it will determine the disposition of the sites in question. Any object – a vase, a chair, a garment – may be extracted from everyday practice and suffer a displacement which will transform it by transferring it into monumental space: the vase will become holy, the garment ceremonial, the chair the seat of authority. The famous bar which, according to the followers of Saussure, separates signifier from signified and desire from its object, is in fact transportable hither and thither at the whim of society, as a means of separating the sacred from the profane and of repressing those gestures which are not prescribed by monumental space – in short, as a means of banishing the obscene.

All of which has still not explained very much, for what we have said applies for all 'monumentality' and does not address the question of what particular power is in place. The obscene is a general category of social practice, and not of signifying processes as such: exclusion from the scene is pronounced silently by space itself.

XVI

Analysis of social space – in this case, of monumental space – brings out many differences: what appeared simple at first now emerges as full of complexities. These are situated neither in the geometrically objectified space of squares, rectangles, circles, curves and spirals, nor in the mental space of logical inherence and coherence, of predicates bound to substantives, and so on. For they also – indeed most importantly – involve levels, layers and sedimentations of perception, representation, and spatial practice which presuppose one another, which proffer themselves to one another, and which are superimposed upon one another. Perception of the entrance to a monument, or even to a building or a simple cabin, constitutes a chain of actions that is no less complex than a linguistic act, utterance, proposition or series of sentences. Yet, whatever analogies or correlations may legitimately be made between course and discourse, so to speak, these complexities cannot be said to be mutually defining or isomorphic: they are truly *different*.

 1 The level of *singularities* stretches outwards around bodies: that is, around each body and around the connections between bodies, and extends them into places affected by opposing quali- ties – by the favourable and the unfavourable, say, or by the feminine and the masculine. These qualities, though dependent on the places in question, are also what confer symbolic power on them. This level is governed, though at times in an inverted manner, by the laws of symmetry and asymmetry. Places so affected – and hence affect-laden, valorized – are not scattered through a mental space, indeed they are not separated from one another. What bind them together are rhythms – semiological differentials.

2 Singularities reappear transformed at another level, at the level of *generality*, in the space of political speech, of order and prescription, with its symbolic attributes, which are often religious, but sometimes simple symbols of power and violence. This is the space of activity, and hence the space of labour divided according to sex, age or group, and the space of communities (village or town). Here rhythms, bodies and words are subordinated to principles of coexistence dictated from above, and indeed often written down.

3 Lastly, the level of singularities also reappears, again modified, in the *particularities* attributed to groups, especially families, in spaces defined as permitted or forbidden.

XVII

This analysis leads back to buildings, the prose of the world as opposed, or apposed, to the poetry of monuments. In their pre-eminence, buildings, the homogeneous matrix of capitalistic space, successfully combine the object of control by power with the object of commercial exchange. The building effects a brutal condensation of social relationships, as I shall show later in more (economic and political) detail. It embraces, and in so doing reduces, the whole paradigm of space: space as domination/appropriation (where it emphasizes technological domination); space as work and product (where it emphasizes the product); and space as immediacy and mediation (where it emphasizes mediations and mediators, from technical *matériel* to the financial 'promoters' of construction projects). It reduces significant oppositions and values, among them pleasure and suffering, use, and labour. Such condensation of society's attributes is easily discernible in the style of administrative buildings from the nineteenth century on, in schools, railway stations, town halls, police stations or ministries. But displacement is every bit as important here as condensation; witness the predominance of 'amenities', which are a mechanism for the localization and 'punctualization' of activities, including leisure pursuits, sports and games. These are thus concentrated in specially equipped 'spaces' which are as clearly demarcated as factories in the world of work. They supply 'syntagmatic' links between activities within social spaces as such – that is, within a space which is determined economically by capital, dominated socially by the bourgeoisie, and ruled politically by the state.

It may be asked whether global space is determined by architectonics (our discussion of which is about to come to an end and debouch onto other analytical perspectives). The answer must be no – and this for several reasons. First of all, the global level is dependent upon dialectical processes which cannot be reduced to binary oppositions, to contrasts and complementarities, or to mirage effects and reduplications, even though such effects or oppositions may well be integral – and integrative – components thereof. They are, in other words, necessary but not sufficient conditions. The global level mobilizes triads, tripartite conflicts or connections. It will do no harm to recall the most essential of these connections now: capitalism cannot be analysed or explained by appealing to such binary oppositions as those between proletariat and bourgeoisie, wages and profit, or productive labour and parasitism; rather, it is comprised of three elements, terms or moments – namely land, labour and capital, or in other words rent, wages and profit – which are brought together in the global unity of surplus value.

The global level, moreover, has its own mode of existence, and its effects are qualitatively different from partial effects. Like language, global space (as, for example, that between monuments and buildings, the space of street or square) produces effects, along with that of communication, which are contradictory: effects of violence and persuasion, of (political) legitimation and delegitimation. Inasmuch as global space bears the inscriptions and prescriptions of power, its effectiveness redounds upon the levels we have been discussing – the levels of the architectural (monument/building) and the urban. Where global space contrives to signify, thanks to those who inhabit it, and for them, it does so, even in the 'private' realm, only to the extent that those inhabitants accept, or have imposed upon them, what is 'public'.

And this leads us into another area, another discussion.

4

From Absolute Space to Abstract Space

I

To recapitulate: social space, which is at first biomorphic and anthropological, tends to transcend this immediacy. Nothing disappears completely, however; nor can what subsists be defined solely in terms of traces, memories or relics. In space, what came earlier continues to underpin what follows. The preconditions of social space have their own particular way of enduring and remaining actual within that space. Thus primary nature may persist, albeit in a completely acquired and false way, within 'second nature' – witness urban reality. The task of *architectonics* is to describe, analyse and explain this persistence, which is often evoked in the metaphorical shorthand of strata, periods, sedimentary layers, and so on. It is an approach, therefore, which embraces and seeks to reassemble elements dispersed by the specialized and partial disciplines of ethnology, ethnography, human geography, anthropology, prehistory and history, sociology, and so on.

Space so conceived might be called 'organic'. In the immediacy of the links between groups, between members of groups, and between 'society' and nature, occupied space gives direct expression – 'on the ground', so to speak – to the relationships upon which social organization is founded. Abstraction has very little place in these relationships, which remain on the level of sex, age, blood and, mentally, on that of images without concepts (i.e. the level of *speech*).

Anthropology has shown us how the space occupied by any particular 'primitive' group corresponds to the hierarchical classification of the group's members, and how it serves to render that order always actual, always present.[1] The members of archaic societies obey social norms without knowing it – that is to say, without recognizing those norms as such. Rather, they live them spatially: they are not ignorant of them, they do not misapprehend them, but they experience them immediately. This is no less true of a French, Italian or Turkish village, provided always that note is taken of the role played by external factors – by markets, by social abstractions (money, etc.), or by outside political authorities. The near order, that of the locality, and the far, that of the state, have of course long ceased to coincide: they either clash or are telescoped into one another.[2] It is in this sense that 'architectonic' determinants, along with the space that they comprehend, persist in society, ever more radically modified but never disappearing completely. This underlying continuity does not exist solely in spatial reality, but also at the representational level. Pre-existing space underpins not only durable spatial arrangements but also *representational spaces* and their attendant imagery and mythic narratives – i.e. what are often called 'cultural models', although the term 'culture' gives rise to a good deal of confusion.

Knowledge falls into a trap when it makes representations of space the basis for the study of 'life', for in doing so it reduces lived experience. The *object* of knowledge is, precisely, the fragmented and uncertain connection between elaborated representations of space on the one hand and representational spaces (along with their underpinnings) on the other; and this 'object' implies (and explains) a *subject* – that subject in whom lived, perceived and conceived (known) come together within a spatial practice.

'Our' space thus remains qualified (and qualifying) beneath the sediments left behind by history, by accumulation, by quantification. The qualities in question are qualities *of space*, not (as latter-day representation suggests) qualities embedded *in space*. To say that such qualities constitute a 'culture', or 'cultural models', adds very little to our understanding of the matter.

Such qualities, each of which has its own particular genesis, its own particular date, repose upon specific spatial bases (site, church, temple, fortress, etc.) without which they would have disappeared. Their ultimate foundation, even where it is set aside, broken up, or localized, is

---

[1] See for example M. Fortes and E. E. Evans-Pritchard, *African Political Systems* (London: Oxford University Press, 1940).
[2] See 'Perspectives de la sociologie rurale', in my *Du rural à l'urbain* (Paris: Anthropos, 1970).

nature; this is an irreducible fact, though nature is hard to define in this role as the absolute within – and at the root of – the relative.

From Rome and the ancient Romans the Christian tradition inherited, and carried down into the modern world, a space filled with magico-religious entities, with deities malevolent or benevolent, male or female, linked to the earth or to the subterranean (the dead), and all subject to the formalisms of rite and ritual. Antiquity's representations of space have collapsed: the Firmament, the celestial spheres, the Mediterranean as centre of the inhabited earth. Its representational spaces, however, have survived: the realm of the dead, chthonian and telluric forces, the depths and the heights. Art – painting, sculpture, architecture – has drawn and continues to draw on these sources. The high culture of the Middle Ages (equivalent to the low culture of the modern world) had its epic space, the space of the *romanceros* or of the Round Table, which straddled dream and reality; the space of cavalcades, crusades and tourneys, where the distinction between war and festival becomes unclear. This space, with its continual appeals to minor local deities, is hard to disentangle (though it is in fact distinct) from the organizational and juridical space inherited from the Roman world. As for the lyrical space of legend and myth, of forests, lakes and oceans, it vies with the bureaucratic and political space to which the nation states have been giving form since the seventeenth century. Yet it also completes that space, supplying it with a 'cultural' side. This romantic representational space was derived, via the Romantic movement, from the Germanic barbarians who overthrew the Roman world and carried out the West's first great agrarian reform.

The process whereby an existing form leads back to immediacy via 'historical' mediations is a reverse repetition of the original formative process. Conflict is not rare between representational spaces and the symbolic systems they encompass, and this is notably true as between the imaginary realm of the Graeco-Roman (or Judaeo-Christian) tradition and a Romantic imagery of nature. This is in addition to the conflicts which ordinarily exist between the rational and the symbolic. Even today urban space appears in two lights: on the one hand it is replete with places which are holy or damned, devoted to the male principle or the female, rich in fantasies or phantasmagorias; on the other hand it is rational, state-dominated and bureaucratic, its monumentality degraded and obscured by traffic of every kind, including the traffic of information. It must therefore be grasped in two different ways: as *absolute* (apparent) within the *relative* (real).

What is the fantasy of art? To lead out of what is present, out of what is close, out of representations of space, into what is further off, into nature, into symbols, into representational spaces. Gaudí did for architecture what Lautréamont did for poetry: he put it through the bath of madness. He pushed the Baroque as far as it would go, but he did not do so on the basis of accepted doctrines or categorizations. As locus of a risible consecration, one which makes a mockery of the sacred, the Sagrada Familia causes modern space and the archaic space of nature to corrupt one another. The flouting of established spatial codes and the eruption of a natural and cosmic fertility generate an extraordinary and dizzying 'infinitization' of meaning. Somewhere short of accepted symbolisms, but beyond everyday meanings, a sanctifying power comes into play which is neither that of the state, nor that of the Church, nor that of the artist, nor that of theological divinity, but rather that of a naturalness boldly identified with divine transcendence. The Sagrada Familia embodies a modernized heresy which disorders representations of space and transforms them into a representational space where palms and fronds are expressions of the divine. The outcome is a virtual eroticization, one based on the enshrinement of a cruel, sexual–mystical pleasure which is the opposite, but also the reverse, of joy. What is obscene is modern 'reality', and here it is so designated by the staging – and by Gaudí as stage-manager.

In the extensions and proliferations of cities, housing is the guarantee of reproductivity, be it biological, social or political. Society – that is, capitalist society – no longer totalizes its elements, nor seeks to achieve such a total integration through monuments. Instead it strives to distil its essence into buildings. As a substitute for the monumentality of the ancient world, housing, under the control of a state which oversees both production and reproduction, refers us from a cosmic 'naturalness' (air, water, sun, 'green space'), which is at once arid and fictitious, to *genitality* – to the family, the family unit and biological reproduction. Being commutable, permutable and interchangeable, spaces differ in their degree of 'participation' in nature (they may also reject or destroy nature). Familial space, linked to naturalness through genitality, is the guarantor of meaning as well as of social (spatial) practice. Shattered by a host of separations and segregations, social unity is able to reconstitute itself at the level of the family unit, for the purposes of, and by means of, generalized reproduction. The reproduction of production relations continues apace amid (and on the basis of) the destruction of social bonds to the extent that the symbolic space of 'familiarity' (family life, everyday life), the only such space to be 'appropriated', continues to hold sway. What makes this possible is the way in which 'familiar' everyday practice is constantly referring from *representations of space* (maps and plans, transport and communications systems, information conveyed by images and signs) to *representational space* (nature, fertility). Reference from the one to the other, and back again, constitutes an oscillation which plays an ideological role but replaces any clear-cut ideology. In this sense space is a trap – and all the more so in that it flees immediate consciousness. This may help account for the passivity of the 'users' of space. Only a small 'elite' see the trap and manage to sidestep it. The elitist character of some oppositional movements and social critiques should perhaps be viewed in this light. Meanwhile, however, the social control of space weighs heavy indeed upon all those consumers who fail to reject the familiarity of everyday life.

Still, that familiarity tends to break apart. Absolute and relative are themselves prone to dissolution. Familiarity is misdirected and/or fetishized, alternately hallowed and profaned, at once power's proxy and a form of powerlessness, and a fictitious locus of gratification. Nor does it have any great immunity to all these contradictions.

Residua in space thus make possible not just dual ideological illusions (opacity/transparency) but also much more complex references and substitutions. It is for this reason that social space may be described and explained, at least partially, in terms of an intentional signifying process, in terms of sequential or stratified codes and in terms of imbricate forms. Dialectical movements 'superclassify' and 'supercode' overlapping categorizations and logical connections. (The movements of this kind which concern us for the moment are immediacy/mediation and/or relative/absolute.)

Symbols and symbolisms are much-discussed topics, but they are rarely discussed intelligently. It is too often forgotten that some if not all symbols had a material and concrete existence before coming to symbolize anything. The labyrinth, for instance, was originally a military and political structure designed to trap enemies inextricably in a maze. It served too as palace, fortification, refuge and shelter before coming to stand for the womb. And it was even later that the labyrinth acquired a further symbolic role as modulator of the dichotomy between presence and absence. Another example is the zodiac, which represents the horizon of the herder set down in an immensity of pasture: a figure, then, of demarcation and orientation. Initially – and fundamentally – absolute space has a relative aspect. Relative spaces, for their part, secrete the absolute.

II

The cradle of absolute space – its origin, if we are to use that term –
is a fragment of agro-pastoral space, a set of places named and exploited
by peasants, or by nomadic or semi-nomadic pastoralists. A moment
comes when, through the actions of masters or conquerors, a part of
this space is assigned a new role, and henceforward appears as transcendent, as sacred (i.e. inhabited by divine forces), as magical and cosmic.
The paradox here, however, is that it continues to be perceived as part
of nature. Much more than that, its mystery and its sacred (or cursed)
character are attributed to the forces of nature, even though it is the
exercise of political power therein which has in fact wrenched the area
from its natural context, and even though its new meaning is entirely
predicated on that action.

Around this nucleus of an organic coherence, which is the centre of
time because it is the centre of space, is distributed, more or less
'harmoniously', an already dense population. Actually, however, harmony between the nucleus and its surroundings only occurs if the
circumstances are right, only by the grace of 'historical' chance. In nearly
all cases, however, the political and religious centre is marked by the
conflict between town and country, between urban space and agrarian
space. The very rites of prohibition and protection that confer religious
and magical power upon central spaces are responses to real threats
from without.

The town and its site live off the surrounding country, exacting tribute
therefrom both in the form of agricultural produce and in the form of
work in the fields. The town has a two-sided relationship to the country,
however: first as an entity which draws off the surplus product of rural
society, and secondly as an entity endowed with the administrative and
military capacity to supply protection. Sometimes one of these roles
predominates, sometimes the other: by appropriating rural space the
town takes on a reality which is sometimes 'maternal' (it stores, stocks
or profitably exchanges a portion of the surplus product, later returning
a lesser or greater fraction of it to the original producers) and sometimes
'masculine' (it protects while exploiting – or exploits while protecting;
it holds the power; it oversees, regulates and on occasion – as in the
East – organizes agriculture, taking responsibility for major projects of
dyke construction, irrigation, drainage, etc.).

Thus the town – urban space – has a symbiotic relationship with that rural space over which (if often with much difficulty) it holds sway. Peasants are prone to restlessness, and as for herders, nomadic or semi-nomadic, the towns have always found it hard to contain them – they are, in fact, ever potential conquerors of the town.

The city state thus establishes a fixed centre by coming to constitute a hub, a privileged focal point, surrounded by peripheral areas which bear its stamp. From this moment on, the vastness of pre-existing space appears to come under the thrall of a divine order. At the same time the town seems to gather in everything which surrounds it, including the natural and the divine, and the earth's evil and good forces. As image of the universe (*imago mundi*), urban space is reflected in the rural space that it possesses and indeed in a sense *contains*. Over and above its economic, religious and political content, therefore, this relationship already embodies an element of symbolism, of image-and-reflection: the town perceives itself in its double, in its repercussions or echo; in self-affirmation, from the height of its towers, its gates and its campaniles, it contemplates itself in the countryside that it has shaped – that is to say, in its *work*. The town and its surroundings thus constitute a *texture*.

As guardian of civic unity and hence of the bond between all members of the city, including the country people, absolute space condenses, harbours (or at any rate seems to harbour) all the diffuse forces in play. Do the forces of death precede the forces of life, or vice versa? The question is a purely abstract one, for the two go hand in hand. Civic unity binds the living to the dead just as it binds the living to one another, especially in those instances, which are frequent, where the city as concentrated wealth is concretely embodied by a monarch. Absolute space is thus also and above all the space of death, the space of death's absolute power over the living (a power of which their sole sovereign partakes). Tombs and funerary monuments belong, then, to absolute space, and this in their dual aspect of formal beauty and terrifying content. A pre-eminence of formal beauty in such spaces leads to the mausoleum, the prestigious but empty monument; that of a terrorizing political content, on the other hand, gives rise to haunted places, places peopled by the living dead. The Christian cemetery is just such a place – though it must be said for cemeteries that they do democratize immortality.

Here and there, in every society, absolute space assumes meanings addressed not to the intellect but to the body, meanings conveyed by threats, by sanctions, by a continual putting-to-the-test of the emotions.

This space is 'lived' rather than conceived, and it is a representational space rather than a representation of space; no sooner is it conceptualized than its significance wanes and vanishes.

Absolute space does have dimensions, though they do not correspond to dimensions of abstract (or Euclidean) space. Directions here have symbolic force: left and right, of course – but above all high and low. I spoke earlier of three levels: surface, heights, depths – or, in other words, the earth, as worked and ruled by humanity; the peaks, the heavens; and abysses or gaping holes. These levels enter the service of absolute space, but each does so in its own way. Altitude and verticality are often invested with a special significance, and sometimes even with an absolute one (knowledge, authority, duty), but such meanings vary from one society or 'culture' to the next. By and large, however, horizontal space symbolizes submission, vertical space power, and subterranean space death. These associations offer unequivocal responses to demands for meaning, but they need tempering by some notion of ambiguity: nowhere is death perceived as 'pure death', or as 'pure' nothingness; nor are power, submission, knowledge, wisdom, and so forth, ever apprehended as 'pure'. Thus the very concept of abstract space is self-correcting. Even in this mitigated form, though, abstract space retains its essential traits. For those in its vicinity, this is the *true space*, the space of truth, and of truth's sudden eruptions (which destroy appearances – that is to say, other times and other spaces). Whether empty or full, absolute space is therefore a highly activated space, a receptacle for, and stimulant to, both social energies and natural forces. At once mythical and proximate, it generates times, cycles. Considered in itself – 'absolutely' – absolute space is located nowhere. It has no place because it embodies *all* places, and has a strictly symbolic existence. This is what makes it similar to the fictitious/real space of language, and of that mental space, magically (imaginarily) cut off from the spatial realm, where the consciousness of the 'subject' – or 'self-consciousness' – takes form. Absolute space is always at the disposal of priestly castes. It consecrates, and consecration metaphysically identifies any space with fundamentally holy space: the space of a sanctuary *is* absolute space, even in the smallest temple or the most unpretentious village church. The space of tombs, for its part, unless it contains a god or a monarch, is analogous merely to the spaces of birth, death or oblivion. Absolute space, being by definition religious as well as political, implies the existence of religious institutions which subject it to the two major mechanisms of *identification* and *imitation*. These mental categories, destined to become those of imagination and reflective thought, first appear as spatial forms. The material extension of absolute space occurs by virtue of these processes, to the benefit of priestly castes and the political power they exercise or serve.

Being ritually affixable to any place and hence also detachable therefrom, the characteristic 'absolute' requires an identifying mark. It therefore generates forms, and forms accommodate it. Such forms are microcosms of the universe: a square (the mandala), a circle or sphere, a triangle, a rational volume occupied by a divine principle, a cross, and so on.

In its ancient Greek version, absolute space may contain nothing. The temple (the Parthenon, say) is divided up into the portico or *naos*, the sanctuary or *pronaos*, and the *opisthodomos*, which is the secret dwelling-place of divinity – and of thought. It has aspects but no façade. The frieze girdles the entire edifice. Visitors may walk all the way around, but the place is not an 'object' that can be grasped otherwise than by means of a thought-process capable of perceiving it as a totality, and hence as endowed with meaning. Curves appear – intentionally – to be straight: the lines of the columns, as of the entablature, have a curvature which is 'imperceptible' because the eye compensates for it. Thus for the Greeks curves are as it were reabsorbed by straight lines, which in the process lose their rigidity, and are softened, while continuing to obey the dictates of the Logos. For it must be remembered that these adjustments called for meticulous calculation.[3]

Volume perceived and conceived, clarified by the light of the sun as by the light of understanding, is the Cosmos in epitome. This, whether that volume is vacant or occupied by thought. Consider the agora. It is part of absolute space, both religious and political – and it concentrates that space. The agora is empty – and must remain empty so that the *ecclesia*, or assembly of free citizens, may be held there. The Roman Forum, by contrast, contains state monuments, the tribune, temples, rostra, and later a prison: it is a place occupied and filled by objects and things, and as such it stands in contradiction to the space of the Greeks.

Though we have arrived at it by another path, here we may once more discern and identify an idea which is the key to the Greek 'miracle' – the simple idea of *unity*. 'Among the Greeks,' wrote Viollet-le-Duc, 'Construction and Art are one and the same thing; the form and the

---

[3] See Vitruvius, *The Ten Books on Architecture*, tr. Morris Hickey Morgan (1914; New York: Dover, 1960), book III, ch. 3, section 6 *et seq.* (pp. 80ff), along with the accompanying 'Vitruvian tables'.

structure are intimately connected . . . .' In the space of the Romans, by contrast, there was a separation, a rift: 'we have the construction, and we have the form which clothes that construction, and is often independent of it'.[4] The Romans organized volumes in such a way as to fulfil some particular function, whether in the basilica or in the baths; the use of constructed masses was clearly distinct from the presentation of surfaces or decoration – the elements of which were ornamental additions to heavy masses of bricks or rubble (i.e. cement and a sort of concrete). The 'orders' invented by the Greeks (Doric, Ionic and Corinthian) *were* the structure itself; the notion of 'order' embraced that of structure, so that the external appearance and the composition (or structure) of Greek buildings are indistinguishable from each other: each contains and reveals the other. It was impossible, according to Viollet-le-Duc, who brought a technician's viewpoint to the development of Hegel's ideas on Greek art and architecture, to strip a Greek temple of its 'order' without destroying the monument itself. The order was not decorative, nor were the columns and the capitals. 'The Greek orders are none other than the structure itself, to which that form was given which was most appropriate to its function. In the orders adopted from the Greeks the Romans saw only a decoration which might be removed, omitted, displaced, or replaced by something else.'[5]

In the West, therefore, absolute space has assumed a strict form: that of volume carefully measured, empty, hermetic, and constitutive of the rational unity of Logos and Cosmos. It embodies the simple, regulated and methodical principle or coherent stability, a principle operating under the banner of political religion and applying equally to mental and to social life. This assumes material form in monuments which govern time by means of well co-ordinated materials whose objective ordering – in terms of vertical pressures and physical mass – successfully achieves both a natural and a rational equilibrium.

To the extent that the Greek mind perceives space in order to shape it, perhaps the ancient Greeks were essentially sculptors. As Hegel pointed out, they were able to take natural materials, first wood and then stone, and endow them with meanings which rendered concrete and practical such social abstractions as assembly, shelter and protection. The shaping of nature, and hence of space (which Hegel still saw as

---

[4] Eugène Emmanuel Viollet-le-Duc, *Entretiens sur l'architecture*, 4 vols (Paris: A. Morel, 1863–72), vol. I, p. 102. Eng. tr. by Benjamin Bucknall: *Lectures on Architecture*, 2 vols (Boston, Mass.: Ticknor, 1889), vol. I, p. 101.

[5] Ibid., vol. I, p. 212; Eng. tr., vol. I, p. 210.

external to mental or social acts) so as to represent and symbolize gods, heroes, kings and leaders – such is the basic sense of Greek art. And this is especially true of sculpture, whether under its inorganic (architectural) aspect or under its organic one (the work of the sculptor).

Do we then have here the founding principle of Western culture? In part, yes – but in part only. The Greek unification of form with function and structure precluded any separation. But the Romans split up what had thus been unified, reintroducing difference, relativity, and varying (and hence civil) aims into a Greek space in which the fusion of politics and religion on the one hand and mathematical rationality on the other had been able to effect a metaphysical (eternal) closure. The city state, at once beautiful, true and good, identified mental with social, higher symbolisms with immediate reality, and thought with action, in a way that was destined to degenerate. The apotheosis of ancient Greek civilization pointed the way for its decline, as Nietzsche clearly saw. By contrast, did Roman diversity, governed as it was by an external constraining principle rather than by an internal unity, contain the seeds of further growth? It seems reasonable to suppose so.

Was the Greek spatial *habitus*, inseparably social and mental, a sufficient basis for the formulation of the essential concepts of form, function and structure? Undoubtedly, since Greek philosophy essayed such a formulation explicitly, since the philosophers took it in hand. This is even truer of Aristotle than of Plato: whereas in Plato the unity in question shines with the brightness of ontological transcendence, in Aristotle it becomes a theory of discourse, of classification, of coherence. No sooner have they crossed the threshold of their formulation than these concepts detach themselves from one another: the conceived separates from the lived, the *habitus* from the *intuitus*, and their presupposed unity is broken. In the Roman *intuitus*, on the other hand, unity in a sense enjoys a certain leeway, in that in each instance – baths being the perfect example – form, structure and function are subordinated to a principle both material (answering a need) and juridical (or civic), which dictates social use. Roman space, though encumbered by objects (as in the Forum), was a productive space. It was also a freer space, as witness the greater use of curves. The unity of the law, of property, or of the city state, being lived and perceived rather than conceived, was never immediately shattered. In the case of ancient Rome, need appears to have been an almost total determinant: both the baths and the villa incorporate responses to every demand of the bodies and minds of free – and rich – citizens.

It is indisputable that slaves made it possible for the city state to exist.

The claim made on the basis of this fact alone by a self-proclaimedly Marxist philosophy of history which posits a specific 'mode of production' founded on slavery succeeds only, however, in rendering unintelligible the realities of that city state, of Athens or Rome, of the Logos/Cosmos, and of Roman Law.

Is there a link between the space the Greeks invented and their inventions in respect of the alphabet, alphabetic script, graphics, arithmetic, geometry, and so on? Perhaps so, but this can in any event represent only a subsidiary aspect of their *habitus*. Furthermore, it would surely be unjust and specious to restrict Greek invention to the invention of a cosmological space. Absolute space always gives rise to diverse forms, and it is not at all clear that some of these may be attributed to reason and the rest to myth (or unreason). One response to the Greek Logos/Cosmos, for instance, was the labyrinth, whose symbolism restores (at a local level) the priority of the original mystery, of the maternal principle, of a sense of envelopment, and of temporal cycles.[6]

In short, absolute (religious and political) space is made up of sacred or cursed locations: temples, palaces, commemorative or funerary monuments, places privileged or distinguished in one way or another. Locations, therefore, governed by a good many prohibitions. In extreme cases, such places may be merely indicated, suggested or signified, as for example by a stone, or by a post whose verticality confers supreme dignity upon a point in space, or by a hole, or simply by a hollow. More commonly, however, the site is circumscribed, demarcated by a perimeter, and characterized by an assigned and meaningful form (square, curve, sphere, triangle, etc.). Everything in the societies under consideration was situated, perceived and interpreted in terms of such places. Hence absolute space cannot be understood in terms of a collection of sites and signs; to view it thus is to misapprehend it in the most fundamental way. Rather, it is indeed a space, at once and indistinguishably mental and social, which *comprehends* the entire existence of the group concerned (i.e. for our present purposes, the city state), and it must be so understood. In a space of this kind there is no 'environment', nor even, properly speaking, any 'site' distinct from the overall texture. Is there a distinction here between signifier and signified? Certainly not if what is meant thereby is a differentiation performed by an *intellectus*. Secret space, the space of sanctuary or palace, is entirely 'revealed' by the spatial order that it dominates. The thing signified, political in

---

[6] Cf., on Aegean palaces, Charles Le Roy, *Le monde égéen* (Paris: Larousse, 1969); also Gustav René Hocke, *Labyrinthe de l'art fantastique* (Paris: Gonthier, 1967).

nature, resides in the religious signifier. Are there grounds for discriminating between the two? No – because at the time with which we are concerned symbolisms and signs had not yet separated. The 'decoding' of space by means of its associated time was still brought about by acts, by ceremonial – specifically, by the Greeks' processions and 'theories'. Being ritual, gestural and 'unconscious' – but also real – decoding was part of the use of a space of this kind and of its image. We must avoid attributing to an ancient Greek climbing up to the Parthenon the attitude of a tourist 'reading' or 'decoding' the prospect before him in terms of his feelings, knowledge, religion or nationality. Here, at the dawn of Western civilization, time contained the spatial code, and vice versa. There was as yet no possibility of displacement into aestheticism, of co-optation of emotions or of 'lived experience' by morality, or of any such 'decodings' imposed upon works which were still experienced and perceived in an unmediated fashion. The concepts of *intuitus* and *habitus* are used here in order to avoid an anachronistic application of categories of a later time, generated subsequently by the *intellectus*, and hence to obviate misunderstandings and misapprehensions.[7] So long as time and space remain inseparable, the meaning of each was to be found in the other, and this *immediately* (i.e. without intellectual mediation).

Absolute space did not govern the private space of family and individual. But this did not mean that private space was left a great deal of freedom. Absolute space entertained no distinction between public and private, and only included the so-called private realm to the degree that this had its own religious or political status (home, household). Its freedom was a weak one – the freedom of houses or dwellings to cluster, with varying degrees of humility, around places invested with high (or low) significance.

In this respect too the Roman organization of space left more room for diversity. But at what cost?

III

The poets in their noble expatiations have neglected neither chasms and abysses nor their corollaries, summits and peaks. At the dawn of Western culture, Dante dealt in an incomparably powerful manner with the

[7] For these concepts of philosophical origin, see F. Gaboriau, *Nouvelle initiation philosophique*, vol. II (Paris: Casterman, 1963), pp. 65ff; also, of course, Aquinas's *Summa theologica*.

themes of the depths (Inferno) and the heights (Paradise), although in doing so he displayed a measure of disdain for surfaces, for the superficial – a bias which had to be corrected later (by Nietzsche). Evocation of the dichotomy between the shadows and the light, between diabolical and divine, continued right down to Hugo's sublime rhetoric. Relationships of this sort between space and language have indeed undergone vicissitudes which are still little known.

First among philosophers to do so, Heidegger, in *Sein und Zeit*, subjected the *mundus* to examination as image, as symbol, as myth. And – as place. He approached the 'world' more as a philosopher than as a historian, an anthropologist, or an analyst of societies.

The *mundus*: a sacred or accursed place in the middle of the Italiot township. A pit, originally – a dust hole, a public rubbish dump. Into it were cast trash and filth of every kind, along with those condemned to death, and any newborn baby whose father declined to 'raise' it (that is, an infant which he did not lift from the ground and hold up above his head so that it might be born a second time, born in a social as well as a biological sense). A pit, then, 'deep' above all in meaning. It connected the city, the space above ground, land-as-soil and land-as-territory, to the hidden, clandestine, subterranean spaces which were those of fertility and death, of the beginning and the end, of birth and burial. (Later, in Christian times, the cemetery would have a comparable function.) The pit was also a passageway through which dead souls could return to the bosom of the earth and then re-emerge and be reborn. As locus of time, of births and tombs, vagina of the nurturing earth-as-mother, dark corridor emerging from the depths, cavern opening to the light, estuary of hidden forces and mouth of the realm of shadows, the *mundus* terrified as it glorified. In its ambiguity it encompassed the greatest foulness and the greatest purity, life and death, fertility and destruction, horror and fascination. 'Mundus est immundus.'

Might a psychoanalysis of space account for this strange and powerful presence–absence? Undoubtedly, but does it not make more sense, instead of engaging in *a posteriori* rationalizations of that kind, to envision a slow process of 'historical' secretion, a laying-down and superimposition of strata of interpretation, along with their attendant rites and myths, occurring as the Italiots localized and focused their fears in the abyssal realm? That a void should be placed at the centre, and indeed at the centre of the conception of the 'world', is surely too strange a fact to be explained solely in terms of psychic realities –

particularly when one thinks of the future whose seeds this representational space contained.

Rome was itself the exorcist of the forces of the underworld, challenging those forces by representing them in a graspable manner. The Eternal City thus incorporated nature into its (military, juridical and political) order by means of a figurative process. The notion of the citizen–soldier, chief and father, did not exclude a role for femininity in the space of the city, either in representation or in reality. If the *mundus* played a part in the formation of the Roman mind it was an inverse and corollary one: the figure of the Father. The Father predominated; he became what he was: chief, political soldier, and hence Law or Right (as imposed on the vanquished in the ordering of victory: the sharing-out of booty and the reassignment of places – primarily land). The Pater–Rex did not have a passive relationship to the world; rather, he reorganized it according to his power and rights, Property and Patrimony, *jus utendi et abutendi* – the limits of which were set not by the 'being' of others but rather by the rights of those among the others who partook of the same power. The Pater–Rex, later Imperator, at once magistrate and priest, thus reconstituted the space around him as the *space of power*.

In this way arose the spatial (social) and mental arrangements which would give rise to Western society (and its ideologies) – to wit, (Roman) law, the notion of the Law, and the notions of Patrimony and of juridical and moral Paternity.

Paternity's imposition of its juridical law (the Law) on maternity promoted abstraction to the rank of a law of thought. Abstraction was introduced – and presupposed – by the Father's dominion over the soil, over possessions, over children, over servants and slaves, and over women. Assigned to the feminine sphere were immediate experience, the reproduction of life (which was, to begin with, inextricably bound up with agricultural production), pleasure and pain, the earth, and the abyss below. Patriarchal power was inevitably accompanied by the imposition of a law of signs upon nature through writing, through inscriptions – through stone. The shift from a maternal principle (which would retain its importance in the sphere of kinship relations) to the rule of paternity implied the establishment of a specific mental and social space; with the rise of private ownership of the land came the need to divide it up in accordance with abstract principles that would govern both property lines and the status of property-holders.

Rome: *orbis* and *urbs*. The ancient city was understood and perceived as an *imago mundi*, assembling and integrating elements in the vicinity which would otherwise have had discrete existences. Inserted into nature, occupying its own site, in a well-defined situation clearly distinguishable from the surroundings, it gave rise to a particular representation of space. The way citizens 'thought' their city was not as one space among others but instead as something vaster: the city constituted their representation of space as a whole, of the earth, of the world. *Within* the city, on the other hand, representational spaces would develop: women, servants, slaves, children – all had their own times, their own spaces. The free citizen – or political soldier – envisioned the order of the world as spatially embodied and portrayed in his city. The military camp, being an instrumental space, answered to a different order (a rectangular, strictly symmetrical space, organized according to *cardo* and *decumanus*).

The founding of Rome – in the traditional account at least – was effected in a distinctly ritual manner. The founder, Remus, described a circle with his plough, thus subtracting a space from nature and investing it with a political meaning. Everything in this foundation story – the details of which are immaterial for our purposes – is at once symbolic and practical; reality and meaning, the immediate and the abstract, are one.

Everything suggests that the space of the Romans was apprehended and constructed in accordance with a directing *intuitus*. *Orbis* and *urbs*: always the circular, non-geometric form. The resulting rationality, whether spatial or juridical, is detectable everywhere in the essential and most concrete creations of the Roman mind: vault, arch, circle (circus, *circulus*) – even the Roman toga, which, in some periods at any rate, was cut by simply opening a hole for the head in a round piece of material. *Intuitus* here – as opposed to *habitus* – does not designate a theoretical intuition of a basically intellectual nature, but rather a practice, a spatial practice, mobilized by (equally spatial) representations.

A visitor to Rome curious about the genesis of this space would do well to consider not only the Rome of marble but also the Rome of brick; to inspect not only the Coliseum and the Forum, for all that they are rich in significance, but to pay careful attention too to the Pantheon, and this without lingering before the marble façade. The interior reproduces the world itself, as it emerges in and through the city, opening to the celestial powers, welcoming all gods and embracing all places. The visitor should ignore his guidebook long enough to analyse the construction of this space, with its prodigiously interlaced curves and entangled archwork (load-bearing or not). What Rome offers is an image that engenders (or produces) space. What space? Specifically, the space of power. Political space is not established solely by actions (with material violence generating a place, a legal order, a legislation): the genesis of a space of this kind also presupposes a practice, images, symbols, and the construction of buildings, of towns, and of localized social relationships.

The paradoxical fact is that this *intuitus*, in a sophisticated and impoverished form, was destined to become a *habitus*. A representation of space embodied in stone, in the city, in paternalistic law, in the Empire, would be transformed into a representational space, submerged into a rediscovered, degenerate *mundus* – a subterranean and hellish abyss. And this representational space would in turn become Christianity's 'foundation' – and its basic resource. This occurred during the long decline of the Empire and the city. As Augustine, that barbarian of genius, would put it, 'Mundus est immundus.'

Here, in summary, are those aspects of Rome and the Roman spirit that an analytic approach enables us to discern.

1 *Spatial practice*, dual in character: the Roman road, whether civil or military, links the *urbs* to the countryside over which it exercises dominion. The road allows the city, as people and as Senate, to assert its political centrality at the core of the *orbis terrarum*. The gate, through which the imperial way proceeds from *urbs* to *orbis*, marks the sacrosanct enceinte off from its subject territories, and allows for entrance and exit. At the opposite pole – the pole of 'private' life, juridically established in the heart of 'political' society, and according to the same principles, those of property – we find the Roman house, a response to clearly defined needs.[8]

2 The *representation of space*, dual in character: on the one hand the *orbis* and the *urbs*, circular, with their extensions and implications (arch, vault); on the other hand the military camp with its strict grid and its two perpendicular axes, *cardo* and *decumanus* – a closed space, set apart and fortified.

3 *Representational space*, dual in character: the masculine principle, military, authoritarian, juridical – and dominant; and the feminine, which, though not denied, is integrated, thrust down into the 'abyss' of the earth, as the place where seeds are sown and the dead are laid, as 'world'.

---

[8] See Vitruvius's precise description in *Ten Books*, book VI, chs 7–8 (pp. 185ff.).

These three levels of determination correspond, within an overall unity, to the perceived, the conceived, and the directly experienced (or 'lived'). In and through a spatial practice, refined in the course of a history, an *intuitus* was transformed into a *habitus* by means of a process first of consolidation, then of degeneration. During this process too, and after it, the *intellectus* made its appearance in the conceptualizing discourse of Vitruvius, as also of a variety of other authors (for instance, Cicero, Seneca). The triad perceived–conceived–lived, along with what is denoted and connoted by these three terms, contributes to the production of space through interactions which metamorphose the original *intuitus* into a quasi-system: the vault and its magic, the arch, or the aqueduct. In the case of Rome, organization, thought and the production of space went together, indeed almost hand in hand. And they did so not under the sign of the Logos but under the sign of the Law.

IV

Christianity was to thrive on a play on words: 'Mundus est immundus' (which was closely bound up with another, just as celebrated and just as sophisticated, the play on the Logos and the Word). As for the philosophy of later times, the philosophy of Christian society, it thrived on the Augustinian dichotomy between time and space (or between subject and object), with its devaluation of the latter.[9]

Closer to modernity, and thanks to Marx's influence, a tendency emerged to overestimate the economic sphere, either by identifying it with history (so-called historical materialism) or else by opposing it to history (ordinary economism). In either case, history as the precondition and underpinning of the economic realm was misapprehended. What then of the Logos, and the logic, of the Greeks? What of the Romans' Law, and laws? Their status remained blurred, fetishized by some and discredited by others. And yet they continued to engender practice, for they were not mere ideologies. Logic is an integral part of knowledge, as law is of praxis. To confine these categories to anthropology, or purely and simply to historicity, is hardly a satisfactory solution. Their ambiguity would be diminished, however, if reflective thought were to take space into account. By 'space', however, I mean to say 'real' space – not an abstract, purified, or emptied-out space, but space in its concrete modalities. Were logic and law not originally forms of spatial

---

[9] See St Augustine, *Confessions*, book X.

organization, forms which presupposed and embodied representations of space and representational spaces?

It is indeed curious in more than one respect that 'we' Westerners, inheritors of an exhausted tradition, and members of a society, culture and civilization that 'we' scarcely know how to characterize (is it capitalism? Judaeo-Christianity? both of these? a 'culture of non-body'? a society at once – and contradictorily – permissive and repressive? a system of bureaucratically managed consumption?) should consider ourselves closer to the Logos and Cosmos of the Greeks than to the Roman world – a world by which, nevertheless, we are deeply haunted.

The Greek polis, with its acropolis and agora, came into being through a synoecism, a unification of villages, upon a hilltop. Its birth was attended by the clear light of day. The sea was never far away, with all its resources. The unknown, the far-off, dangerous but not inaccessible, were stimulants at once, and inextricably, to curiosity, imagination and thought.

Something which resulted, here as elsewhere, from an encounter and a practice, had enigmatic and marvellous qualities bestowed upon it by a later rhetoric. The Greek city did not exorcize the forces of the underworld; rather, it rose above them and so surmounted them. Occasionally it captured them: Eleusis. For the citizen and city-dweller, representational space and the representation of space, though they did not coincide, were harmonious and congruent.[10] A unity was achieved here between the order of the world, the order of the city and the order of the house – between the three levels of segments constituted by physical space, political space (the city along with its domains), and urban space (i.e. within the city proper). This unity was not a simple or a homogeneous one, but rather a unity of composition and of proportion, a unity embracing and presupposing differences and hierarchy. By the same token knowledge and power, social theory and social practice, were commensurate with each other. And time, the rhythm of days and feasts, accorded with the organization of space – with household altars, with centres of collective activity, with the *boule* in the agora (a free and open citizens' assembly), with temples and with stadia.

All historical societies have diminished the importance of women and

[10] As demonstrated from his own particular perspective – that of a psychological history – by Jean-Pierre Vernant. See his *Mythe et pensée chez les Grecs, études de psychologie historique* (Paris: François Maspéro, 1965); Eng. tr.: *Myth and Thought among the Greeks* (London and Boston, Mass.: Routledge and Kegan Paul, 1983). Vernant's interpretation of the Greek mind, though more precise than Nietzsche's, and more firmly grounded in philology, lacks the poetic breadth of the Nietzschean view.

restricted the influence of the female principle. The Greeks reduced the woman's station to that of the fertility of a field owned and worked by her husband. The female realm was in the household: around the shrine or hearth; around the *omphalos*, a circular, closed and fixed space; or around the oven – last relic of the shadowy abyss. Women's social status was restricted just as their symbolic and practical status was – indeed, these two aspects were inseparable so far as spatiality (spatial practice) was concerned.

The underworld had thus not disappeared. In daytime, Zeus and reason had vanquished the shadowy or chthonian forces. But in the depths of the infernal world, their defeat notwithstanding, the Titans were still active. In the country of the dead the shades had drunk the waters of Lethe. Greek genius was able to localize the underworld, to specify and name it, and in so doing to subordinate it to the surface world – to the mountains with their grazing flocks, to the cultivated fields, to the sea ploughed by ships laden with riches. Instead of dominating and appropriating the netherworld after the fashion of the Romans, the Greeks set that world apart and *situated* it (as at Delphi, or in the revels of the Bacchantes). The meaning of such images is not to be found in literary works. On the contrary, rites and mythic narratives (from Hesiod to Plato) tell in images and symbols what is occurring in social space. Conceptual rationalizations were indeed offered by the Greeks themselves – but only much later (along with philosophy), towards the end of their civilization.

V

If most societies have followed this same route, how are we to account for their differences? How is it that different societies assign different roles to the male principle and its dominant form, and that this dominant form itself is differently formulated from one society to another? Greece, for example, which took Athens as its model, and Italy, which took Rome, differ so radically that the one produced and transmitted the Logos (logic and knowledge) while the other produced and transmitted the Law.

Psychoanalysis might on the face of it be expected to find problems such as these easy to tackle, but in practice the triangular Oedipal model can support only a very mechanistic and homogenizing causal explanation. The 'Oedipal triangle' is supposedly to be found everywhere, and is said to be a structure having explanatory force, but if it is an unchanging structure how does it give rise to such diverse outcomes?

In any event, our present approach to the question is a quite different one, for our aim is to treat social practice as an extension of the body, an extension which comes about as part of space's development in time, and thus too as part of a historicity itself conceived of as *produced*.

There is surely an argument for drawing a distinction within this history between manliness and masculinity. In Rome the masculine virtues and values, those of the military man and the administrator, were in command. Manliness, by contrast, was a Greek attribute – the kind of manliness that dictates constant defiance towards one's enemies and constant rivalry with one's friends, that cultivates performance, whether in brutal or subtle form, as its basic *raison d'être* and goal, and that aspires above all to *excel*; this is an aspiration, however, which despises everyday tasks yet, capriciously, confuses matters when long-term decisions are called for. Manliness so understood, and elevated to the cosmic level, to the level of the gods, conserves the traits of small groups in competition.

In their cult of manliness and rivalry, the Greeks distinguished between good and bad approaches to the eristic or agonistic. The bad sought the destruction of the adversary, while the good meant respecting the adversary even while seeking to outdo him.[11] *Dike*, or justice, discriminated between these two aspects of challenge and defiance, a distinction which is not implied in the idea of hubris. Whereas, in the Roman case, there are grounds for contrasting an initial *intuitus* with a final *habitus*, no such division is called for apropos of the Greeks.

The founding image of Greek space was a space already fully formed and carefully populated; a space in which each focal point, whether that of each house or that of the polis as a whole, was ideally placed upon a well-chosen, well-situated eminence, sunlit and close to an abundant source of water. The Greek city, as a spatial and social hierarchy, utilized its meticulously defined space to bring demes, aristocratic clans, villages, and groups of craftsmen and traders together into the unity of the polis. At once means and end, at once knowledge and action, at once natural and political, this space was occupied by people and monuments. Its centre – the agora – served as focus, as gathering-place. At the highest

---

[11] Cf. Nietzsche's reprise of the concept of *eris* in *Thus Spoke Zarathustra*, tr. R. J. Hollingdale (Harmondsworth, Middx: Penguin, 1961): 'Of the Friend' in part I, and 'Of the Compassionate' in part II; also, 'You should always be the first . . . – this precept made the soul of a Greek tremble' ('Of the Thousand and One Goals', ibid., part II, p. 85). For the dual aspect of *eris*, see Vernant, *Mythe et pensée*.

point of the acropolis, the temple presided over and rounded out the city's spatio-temporal space. Built in no image, the temple was simply *there*, 'standing in the rocky valley'. It arranged and drew about itself (and about the god to which it was devoted) the grid of relations within which births and deaths, adversity and good fortune, victories and defeats came about (Heidegger). There was nothing decorative here, and nothing functional. The space, the cut of the stones, the geometry of the masses, the overall scheme – none of these could be separated from the others. The beams and lintels with their supports and props determined the disposition of space and the distribution of volumes. Hence the stress on the 'orders' and their significance. These 'orders', as defined by Doric, Ionic and Corinthian columns, refer both to construction and to decoration. The Cosmos, like a fine head of hair above a noble brow, deployed its glory without separating the good from the beautiful.

What, then, of difference? Difference was *produced*. Not produced as such, however – not conceptually, nor by virtue of an idea. Difference was never part – except perhaps much later, and then indirectly – of a body of knowledge, of a sequence of propositions, or of an epistemological field, whether or not associated with a core of knowledge. A difference conceptualized is surely already *reduced*, solely by virtue of the fact that the two elements in question are now governed by the same comparison, are part of the same thought, part of the same intellectual act. Even if this act is followed by an action, a practical action which realizes it, the difference is still merely *induced*.

Between the Cosmos and the 'world', difference arose as part of a 'historical' process, each side of the dichotomy being ignorant of, or misapprehending, the other. It might be claimed, with the benefit of hindsight, that a particular image or concept of space must have been informed by the low or by the high – by the abyss below or the summits above – and that it emphasized such and such a direction, such and such an orientation. Fair enough. But neither of the opposing images was constituted specifically against the other, in contradistinction to it. Rather, the difference occurred spontaneously, which is what distinguishes *produced* difference from difference which is *induced*, and generally *reduced*.

VI

What is the mode of existence of absolute space? Is it imagined or is it real?

To phrase the question in this manner makes any coherent answer impossible. Faced with such an alternative, one can only oscillate indefinitely between the choices offered. Imaginary? Of course! How could an 'absolute' space have a concrete existence? Yet it must also be deemed real, for how could the religious space of Greece or Rome *not* possess political 'reality'?

There is thus a sense in which the existence of absolute space is purely mental, and hence 'imaginary'. In another sense, however, it also has a social existence, and hence a specific and powerful 'reality'. The 'mental' is 'realized' in a chain of 'social' activities because, in the temple, in the city, in monuments and palaces, the imaginary is transformed into the real. What the above formulation of the question ignores or fails to grasp is the existence of these works, an existence which certainly transgresses and in all likelihood transcends such trivialized and latter-day categories as 'the imaginary' and 'the real'. When asked whether a temple and its surroundings are imaginary or real, the realist will naturally see only stones, whereas the metaphysician will see a place consecrated in the name of a divinity. But of course there must be more to it than this.

Absolute space has not disappeared. Nor does it survive only in churches and cemeteries. The Ego takes refuge in a pit – in its 'world' – whenever it falls from its perch on some crag of the Logos. Its voice may emerge from an often mephitic and sometimes inspired cavern. Is this perhaps the space of speech? Both imaginary and real, it is forever insinuating itself 'in between' – and specifically into the unassignable interstice between bodily space and bodies-in-space (the forbidden). Who speaks? And where from? As it becomes more and more familiar, this question serves increasingly to conceal the paradox of absolute space – a mental space into which the lethal abstraction of signs inserts itself, there to pursue self-transcendence (by means of gesture, voice, dance, music, etc.). Words are in space, yet not in space. They speak of space, and enclose it. A discourse on space implies a truth of space, and this must derive not from a location within space, but rather from a place imaginary and real – and hence 'surreal', yet concrete. And, yes – conceptual also.

Might not this space, extracted from nature yet endowed with properties just as natural as those of sculptures hewn from wood and stone, be also the space of art?

VII

As part of the protracted decline of the Roman state–city–Empire (as defined by its political power, and by that power's basis in the earth, in landownership), the city gradually ceased to be. The villa of a latifundiary landowner retained not a trace of the sacred. It was the concretization, within agro-pastoral space, of a codified, law-bound spatial practice, namely private ownership of the land. The villa thus combined in a single unit of material production the general traits of Roman society (an order grounded in juridical principles), a refined – albeit not very creative – aesthetic taste, and a search for the comforts of life. Testimony to this is to be found as early as the classical period, in the writings of Cicero, Pliny and others. The resulting diversification of space, along with the legal predominance of the private realm, meant the loss of Greek order and a rupture of the unity of form, structure and function; it also meant a split, within buildings themselves, between decorative and functional elements, between the treatment of masses and the treatment of surfaces, and hence between construction and composition, architecture and urban reality. Consequently the Roman villa (of the Lower Empire and the decadent era) emerges as the generator of a new space, a space with a great future in Western Europe. Herein lies the secret of the Roman world's survival despite its decline. It is not just that the villa gave rise to many of our towns and villages; it also introduced a conception of space the characteristics of which would continue to manifest themselves in later times: the dissociation of component elements, and a consequent practical diversification; subordination to the unifying but abstract principle of property; and the incorporation into space of this same principle, which is in itself impossible to live, even for the landowner, because it is juridical in nature, and hence external, and supposedly superior, to 'lived experience'.

This, then, was the road taken by the Roman spirit on the way to its demise. (A long road, in point of fact, because in the twentieth century the end has still not been reached.) Once unshackled, the principle of private property did not remain sterile: rather, it gave birth to a space. The centuries-long silence of the state is portrayed in official history, and indeed in the work of most historians, as a void, a complete hiatus in historical existence. Nothing could be further from the truth. The

Gallo-Roman West preserved the most valuable Roman achievements: the art of building, the art of irrigation systems and dykes, the great roads, agricultural advances (to which the Gauls had made their own contribution), and, last and most important, the right of (private) property. This 'right' should not, any more than money or the commodity, be looked upon as the root of all evil. It is not intrinsically bad. The property principle, by dominating space – and this in the literal sense of subjecting it to its *dominion* – put an end to the mere contemplation of nature, of the Cosmos or of the world, and pointed the way towards the mastery which transforms instead of simply interpreting. It may be asked, however, whether in the society which it dominated this principle did not reach an impasse. Inasmuch as it was taken in isolation and erected into an absolute, it certainly did. Which is why the barbarians' arrival on the scene had a salutary effect, for in violating the sanctity of property these intruders fertilized it. For this to happen, of course, it was also necessary that they be accepted and given a chance to establish themselves, to turn the villae to good account, and to get the Gallo-Roman settlers to work by subordinating them to leaders of village communities who had now become lords. So far as space was concerned, the barbarians might be said to have rejuvenated it by rediscovering the old markers of agro-pastoral (and in fact primarily pastoral) times.

It may thus be seen that during the supposed emptiness of the late imperial or early medieval period a new space was established which supplanted the absolute space, and secularized the religious and political space, of Rome. These changes were necessary though not sufficient conditions for the subsequent development of a historical space, a space of accumulation. The 'villa', now either a lordly domain or a village, had durably defined a *place* as an establishment bound to the soil.

VIII

Rendered more sophisticated by (Augustinian) theology, the *imago mundi* we have been discussing survived the decline of the Roman Empire and state, the rise of the latifundia, and their dramatic clash with the barbarian innovators. Viewed in this light, the year 1000 appears as a truly pregnant moment. Within an apparent void, a new departure was being prepared. Contemporaries were overwhelmed by anxiety because they could perceive only the past. But space had already been transformed, and was already the birthplace and cradle of what was to come.

Christianity, whatever institutional ups and downs it was experiencing, was a great worshipper of tombs. Its holiest places, those stamped by divinity – Rome, Jerusalem, Santiago de Compostela – were all tombs: St Peter's, Christ's, St James's. The great pilgrimages drew the crowds to shrines, to relics, to objects sanctified by death. The 'world' held sway. This was a religion which 'coded' death, ritualizing, ceremonializing and solemnizing it. The monks in their cloisters contemplated death, and could contemplate only death: they had to die in the 'world' so that the 'world' might be fulfilled. Essentially cryptic in nature, religion revolved around those underground places, church crypts. Lying beneath each church or monastery, the crypt always held the bones or a portion of the remains of a consecrated figure – sometimes mythical, sometimes historical. Historical figures were generally martyrs, who had borne witness with their lives, and continued to bear witness from their catacombs – from 'depths' which no longer had anything in common with the ancient world's realm of shadows. The saint's presence in the crypt was supposed to concentrate there the life and death forces diffused throughout the 'world'; absolute space was identified with subterranean space. Such was the dismal religion which waxed as Rome, its city and state, waned. It paralleled an agricultural society of mediocre productivity where agriculture itself (except around the monasteries) was degenerating, where famine threatened, and where whatever fertility did exist was attributed to occult forces. It was to this context that the syncretic unity of an Earth-Mother, a cruel God-the-Father and a benevolent mediator was applied. Crypts and tombs always held traces and representations of saints, but, it would seem, hardly ever sculptures. Paintings, yes – paintings remarkable in that they were never seen, except perhaps, on the saint's feast day, by clergy entering the crypt with lighted candles. At such moments of intensity the images came to life, the dead made their appearance. Cryptal art of this kind has nothing visual about it, and for those who think in the categories of a later time, projecting them into the past, it poses an insoluble problem. How can a painting remain out of sight, condemned to a purely nocturnal existence? What is the *raison d'être* of Lascaux's frescoes, or of those in the crypt of St-Savin? The answer is that these paintings were made not to be seen, but merely to 'be' – and so that they might be known to 'be' there. They are magical images, condensing subterranean qualities, signs of death and traces of the struggle against death, whose aim is to turn death's forces against death itself.

Consider the Church. What a narrow, indeed mistaken, view it is which pictures the Church as an entity having its main 'seat' in Rome and maintaining its presence by means of clerics in individual 'churches' or villages and towns, in convents, monasteries, basilicas, and so forth. The fact is that the 'world' – that imaginary–real space of shadows – was inhabited, haunted by the Church. This underworld broke through here and there – wherever the Church had a 'seat', from that of the lowliest country priest to that of the Pope himself; and wherever it thus pierced the earth's surface, the 'world' emerged. The 'world' – that of religious agitation, of the Church suffering and militant – lay and moved below the surface. This space, the space of Christendom, was a space that could in the twelfth century be occupied by the powerful personality of a Bernard of Clairvaux. Indeed, without its magico-mystical, imaginary–real unity, it would be impossible to account for the influence of this genius, who controlled two kings and told the sovereign pontiff, 'I am more Pope than you.' Just as something new was appearing on the horizon, Bernard of Clairvaux revalorized the space of the signs of death, of desperate contemplation, of asceticism. The masses rallied about him – and not only the masses. His poor-man's bed epitomized his space.

What exactly happened in the twelfth century? According to the received wisdom of the historians, history suddenly resumed after a long interruption. Only now were certain 'factors' created which would mould the modern epoch – and the job of tracking these down makes for a good deal of suspense. The restraint displayed for so long by History is only rivalled by that of the historians, flailing about in this crepuscular dawn, unravelling some facts but few causes. They demonstrate admirable prudence too, in hesitating to speak of revolution apropos of the great movements of the twelfth century[12] – the more so since to do so might oblige them to consider the peasant revolution (or 'revolt of the serfs'), which challenged the state of servility, in its connections with the urban revolution, which overthrew the existing social arrangements as a whole. Who would profit from these transformations? The monarch, certainly, and his authority, and the state, which to begin with had a feudal and military character. But of course those changes which first hove into view in the twelfth century did not occur immediately. What was the precise mix of happenstance and determinism that made possible the careers of such exceptional men as Bernard of Clairvaux, Suger and Abelard? There is no way for us to grasp retrospectively what came about at that time if we cannot form a clear

[12] See for example Charles-Edmond Petit-Dutaillis, Les Communes françaises (Paris: Albin Michel, 1947), and even Georges Duby in some of his more recent contributions.

perception of the locus and cradle of these events. That the towns came once more to the fore is beyond dispute. To the question 'What did the towns contribute that was new – what did they produce?' we are inclined to answer, 'A new space.' Does this obviate the methodological and theoretical difficulties that arise when we consider time – historical, or supposedly historical, time – in isolation? Perhaps. The rise of the medieval town has to be viewed along with its implications and consequences. It presupposed a surplus production in the countryside sufficient to feed the urban population, both because the town was organized as a market and because urban craftsmen worked with materials (wool, leather) produced by agricultural labour. This was what led to the setting-up of corporate bodies of communitarian inspiration within the urban collectivity. Although the members of such corporations had nothing 'proletarian' about them, it is true to say that the advent of these associations heralded the arrival of a collective worker, a worker able to produce 'socially' – that is to say, for society, and in this instance for the town.

The Papacy sought to defend itself against these developments, counter-attacking and scoring some measure of success. Its grand design, however – namely, the replacement of the imperial state, whose mantle the Roman Church wished to assume, by a vast ecclesiastical state – was doomed to failure. The nations, the nation states, were now about to appear. Monastic culture was on the ebb. What was about to disappear was absolute space; it was already crumbling as its supports gave way. What then was about to emerge? The space of a secular life, freed from politico-religious space, from the space of signs of death and of non-body.

The urban landscape of the Middle Ages turned the space which preceded it, the space of the 'world', upon its head. It was a landscape filled with broken lines and verticals, a landscape that leapt forth from the earth bristling with sculptures. In contrast to the maleficent utopia of the subterranean 'world', it proclaimed a benevolent and luminous utopia where knowledge would be independent, and instead of serving an oppressive power would contribute to the strengthening of an authority grounded in reason. What do the great cathedrals say? They assert an inversion of space as compared with previous religious structures. They concentrate the diffuse meaning of space onto the medieval town. They 'decrypt' in a vigorous (perhaps more than a rigorous) sense of the word: they are an emancipation from the crypt and from cryptic space. The new space did not merely 'decipher' the old, for, in deciphering it, it surmounted it; by freeing itself it achieved illumination and elevation. The field now remained decidedly, and decisively, in possession of what has been called 'white communication'.[13] The other sort, the black, was not, however, annihilated. Merely, it took refuge in the subterranean parts of society, in places hidden away from face-to-face communication.

An extraordinary trio mobilized and resisted this great movement of emergence: Bernard of Clairvaux, Suger and Abelard. These three cannot be understood separately. Bernard, the perfect 'reactor', had the ear of the powerful yet knew how to hold the attention of the masses. Suger, who served the state – a state which was royal, military, and already 'national' because territorial – conceived and carried out political projects. And Abelard – the heretic – was at the cutting edge of the possible, part of the sort of thinking which by questioning basic assumptions shook the edifice to its foundations. He was also the most effective of the three, his apparent failure notwithstanding. Despite a persecution which spared him no humiliation, which seized upon a romantic intrigue as a stick with which to beat him, Abelard would later be recognized as the 'most modern' figure of his time.

The crypt at St-Savin holds the now symbolic 'earthly dust' and images of St Gervase and St Protase, and of their edifying lives and martyrdom. The church vault, however, features scenes from the Scriptures, from the Old and New Testaments – painted imagery diametrically opposed to cryptic/cryptal space. The vault 'decrypts' by exposing the contents of the underground chambers to the light of day. St-Savin's counterposed images thus perfectly crystallize the moment of emergence that I have been describing.

In his book *Gothic Architecture and Scholasticism*, Erwin Panofsky is not content, when attempting to discover the links between the various aspects of the twelfth century, with an appeal to a Hegelian *Zeitgeist* – to the notion of a pervasive and thus banal spirit of the times. The idea of an analogy between architecture and philosophy has nothing paradoxical or new about it *per se*,[14] but Panofsky goes beyond the identification of a fruitful encounter between technique and symbol[15] – beyond an approach that itself transcended Viollet-le-Duc's rationalistic interpretation (which, despite his sophisticated analysis of social and historical processes, remained mechanistic, technicistic and functionalist

[13] See Georges Bataille, *Le Coupable* (Paris: Gallimard, 1961), p. 81.
[14] See Karl Hampe, *Le Haut Moyen Age* (Paris: Gallimard, 1943) [tr. of *Das Hoch-mittelalter* (Berlin: Propyläen-Verlag, 1932)], where this idea is clearly set forth (pp. 212–28, especially p. 228 on Gothic script).
[15] Cf. Emile Mâle, *L'art religieux du XIIᵉ au XIIIᵉ siècles* (Paris, 1896).

in character[16]). Cathedrals can be accounted for neither by intersecting rib vaults, nor by buttresses and flying buttresses – even if such features are necessary conditions thereof. The same goes for the soul's yearning heavenwards, the youthful ardour of a new generation, and other such considerations. Panofsky posits a homology (not just an analogy) between philosophy and architecture, arguing that each, though complete in its own way, partakes with the other of a unity of which it is a 'manifestation' – an elucidation, in the sense in which faith may be said to be elucidated by reason. To the question which of the two has priority, Panofsky's answer is philosophy. For priority there has to be. Scholasticism produced a mental habit or *habitus*, and hence a *modus operandi* derived from a *modus essendi*, from a *raison d'être*. The *habitus* of architecture was directly descended from the providential reason which at the time presided over the unity of truth – over that unity of reason and faith whose culminating expression was the *Summa theologica*.[17] The spatial arrangement of the Gothic church corresponds for Panofsky to that great work – or rather it 'reproduces' it, embodying as it does a reconciling of opposites, a tripartite totality, and the organizational equilibrium of a system whose component parts are themselves homologues.[18] Thus Panofsky sees nothing problematic about deriving a mental space, that of a speculative construction, the *Summa theologica*, from an abstract representation, that of a unity of homologous parts itself analogous to the unity of the Divinity (one-in-three and three-in-one); nor in further deriving from that mental space a social one – the space of the cathedral. What is really being engendered and produced (or reproduced) here, however, is the divine act of creation itself. One would indeed have to be a person of great religious faith to see nothing objectionable in such an argument, which is in reality a fine example of the abuse of a concept – the concept of *production* in the event – by blindly divorcing it from all content and all context. The identification of thought with the productive activity of God is justified by the adduction of would-be scientific concepts such as structural affinity – or the supposed 'search for the geometrical location of the symbolic expression specific to a society and an epoch'.[19] It is as though simply replacing

---

[16] See Pierre Francastel, *Art et technique aux XIXe et XXe siècles* (1956; Paris: Denoël, 1964) pp. 83–4 and 92ff.

[17] Erwin Panofsky, *Gothic Architecture and Scholasticism* (1951; New York: New American Library, 1976), pp. 44ff.

[18] Cf. ibid., p. 45, citing the *Summa theologica*.

[19] Cf. Pierre Bourdieu's 'Postface' to Panofsky's book in French translation: *Architecture gothique et pensée scholastique* (Paris: Editions de Minuit, 1967), p. 135.

the word 'create' by the word 'produce' sufficed to validate this extra-ordinary leap – and with it an idealism and spiritualism of the most irresponsible and facile variety. The thesis is hardly persuasive.

Panofsky was in search of a principle of unity. Why did he opt for a *habitus* rather than an *intuitus*? And was he even speaking of a *habitus* as defined by Aquinas, for humanity, as a 'mode of being' implying a 'power of use and enjoyment',[20] and hence as a quality which is a basic attribute of a person (consider the connection with *habere* and *habitare*)? This is what distinguishes *'habitus'* from 'habit'. How could a doctrine contain a *habitus* (or mental habit) and a *modus operandi* capable – barring miracles – of giving rise to several such different frameworks as those of writing, art, music, and so on? Still, this spiritualistic nonsense does conceal the concrete intuition of a certain unity, a certain *pro-duction*. What Panofsky discovered – or, at least, what emerged from his work – is the idea of a 'visual logic'.[21] What does he mean by this? That the religious edifice, by rising higher, receives more light; that its naves no longer have the compact and sombre atmosphere of so-called Romanesque churches; that its walls become less massive now that they no longer bear all the weight, and that the pillars, small columns and ribbing rise with slender elegance towards the vault; that stained-glass windows make their appearance and the making of them becomes an art. More than this, too: that the Scholastic mind accepts and even demands a double clarification – the 'clarification of function through form' and the 'clarification of thought through language'.[22]

Panofsky does not take his thinking as far as it will go, however. The full implication of his 'visual logic' is that all should be revealed. All? Yes – everything which was formerly hidden, the secrets of the world. Even demonic and evil forces. Even natural beings – plants and animals. Even living bodies. As they burst up into the light, bodies took their revenge; the signs of non-body[23] became subordinate to those of the body – including the resurrected body of the living God, of Christ. This was the new alliance of the 'world', opening up now to the light, with

[20] See Gaboriau, *Nouvelle initiation*, vol. II, pp. 62, 97. There is nothing intrinsically wrong with the introduction of these philosophical (scholastic) concepts, but their specula-tive use, without any point of reference besides the Thomist system itself, opens the way to some very questionable manoeuvres.

[21] See Panofsky, *Gothic Architecture*, p. 58.

[22] Ibid., pp. 59–60.

[23] In his *Conjunciones y disyunciones* (Mexico City: Joaquín Mortiz, 1969), tr. Helen R. Lane as *Conjunctions and Disjunctions* (New York: Viking, 1974), Octavio Paz attempts to paint a symmetrical picture of relationships – similarities and contrasts – between medieval Christian and Buddhist art (see pp. 51 ff.; and Eng. tr., pp. 45 ff.).

the Logos and the Cosmos. And it was a trend that encouraged the rediscovery of Greek thought, of Plato and Aristotle. The resurrection of the flesh, hitherto a peripheral matter, now became central; this is the meaning of the Last Judgements (though such works continued to induce terror by invoking death and the underworld). Once the under-world had come to the surface, while the surface world rose upwards and offered itself to view as occupier of space, sculpture had little difficulty routing cryptic/cryptal painting. Whence the profusion of capitals, and of statues on façades. Freed from their former weightiness, surfaces now carried decoration glorifying the body (even if the idea of sin still managed here and there to bring minds back to putrefaction, to the *immundus*, to the 'world'). Sculpture was once more, as in ancient Greece, the primordial, the leading art. Painting retained a measure of dignity only as an art of lighting (i.e. as stained glass).

To limit this new creative force to an 'architectural composition' that made it possible to 're-experience the very processes of cogitation' (in the *Summa theologica*) is to frame a hypothesis so reductionist as to be startling in the extreme.[24] It has a double advantage, however, for it gets us up to the *aggiornamento* of Scholasticism while denigrating all the reforming, subversive and exemplary aspects of the medieval revolution in the West. Does it make sense, then, to speak of a 'visual logic'? Certainly: an emergence from darkness and a coming out into the light. The point is, though, that this goes far beyond Gothic architecture and involves the towns, political action, poetry and music, and thought in general. The role of Abelard, his thought and life, can only be understood in terms of a revolt of the body which certainly went beyond any 'visual logic' — which went as far, in fact, as to anticipate a reconciliation between flesh and spirit effected thanks to the intervention of the Third Person, the Holy Spirit.

What is involved, therefore, is a *production* — the production of a space. Not merely a space of ideas, an ideal space, but a social and a mental space. An *emergence*. A decrypting of the space that went before. Thought and philosophy came to the surface, rose from the depths, but life was decrypted as a result, and society as a whole, along with space. If one were of a mind to distinguish, after the fashion of textual analysis,[25] between a genotype and a phenotype of space, it would be from this 'emergence' that the 'genospatial' would have to be derived.

Of an originality and revolutionary force such that it spread through-

---

[24] See Panofsky, *Gothic Architecture*, p. 59.
[25] See Julia Kristeva, *Semeiotike: recherches pour une sémanalyse* (Paris: Seuil, 1969).

out the West from its starting-point in the Ile-de-France with extraordinary (relative) speed, the 'production' with which we are concerned may correctly be described as tending towards the 'visual'. The importance taken on by façades confirms this – indeed it is in itself sufficient proof of the fact. Organized with the greatest care, these high and highly worked surfaces were strictly governed by the Church's commands: Law, Faith, Scripture. The living, naked body had a very limited role: Eve, Adam, and occasional others. One finds few female bodies aside from those of ascetics and those of the damned. The façade rose in affirmation of prestige; its purpose was to trumpet the associated authorities of Church, King and city to the crowds flocking towards the porch. Despite the efforts of medieval architects to have the exterior present the interior and render it visible, the mere existence of the façade sufficed to destroy any such concordance.

The production of a luminous space and the emergence of that space did not as yet, in the thirteenth century, entail either its subordination to the written word or its mounting as 'spectacle'.[26] Still, to the extent that he is accurate, Panofsky is describing a threatening gambit. The trend towards visualization, underpinned by a strategy, now came into its own – and this in collusion on the one hand with abstraction, with geometry and logic, and on the other with authority. Social space, even as early as this, was already being affected by this alchemical formula with its disturbing ingredients and surprising effects. Admittedly, the (loose) threshold beyond which realization becomes reification, and vitality becomes alienated vitality, was not as yet crossed. But portents of that step were certainly present. The negative and lethal magic of signs – that magic which by means of a painting can immobilize a bird in full flight, in perfect mimicry of the hunter's mortal strike – carried the day. The other kind of magic, by contrast, the magic of the spoken word, whose symbolisms (the breath of the Spirit, the bird of prophecy, the act of creation) infused even the realm of death with life, could only retreat before the intense onslaught of visualization. As for sculpture, it is more eloquent than painting in the three dimensions of space; but what it says it says all at once, and once and for all. There is no appeal.

The verticality and political arrogance of towers, their feudalism, already intimated the coming alliance between Ego and Phallus. Uncon-

---

[26] The first development, as it occurred from the fifteenth century on, has been described by Marshall McLuhan in *The Gutenberg Galaxy* (Toronto: University of Toronto Press, 1962). The second is the subject of Guy Debord's *La société du spectacle* (1967; Paris: Champ Libre, 1973); Eng. tr.: *The Society of the Spectacle*, rev. edn (Detroit: Black and Red, 1977).

sciously, of course – and all the more effectively for that.

The Phallus is seen. The female genital organ, representing the world, remains hidden. The prestigious Phallus, symbol of power and fecundity, forces its way into view by becoming erect. In the space to come, where the eye would usurp so many privileges, it would fall to the Phallus to receive or produce them. The eye in question would be that of God, that of the Father or that of the Leader. A space in which this eye laid hold of whatever served its purposes would also be a space of force, of violence, of power restrained by nothing but the limitations of its means. This was to be the space of the triune God, the space of kings, no longer the space of cryptic signs but rather the space of the written word and the rule of history. The space, too, of military violence – and hence a *masculine* space.[27]

IX

Consider the demise of a society able to spend its surplus in sumptuous fashion on festivals, monuments and wars waged for mere show or mere prestige: how and when did the non-accumulative and the non-historical make their joint disappearance?

The development of a theory of accumulation, initiated by Marx, remains unfinished. What made primitive accumulation possible? And what were its implications, aside from the feasibility of investing wealth rather than saving or squandering it, and aside from the rationality attending that change (cf. Max Weber)?

The accumulation of money for investment, and productive investment itself, are hard to conceive of without a parallel accumulation of techniques and knowledge. Indeed these are all really aspects of an indivisible accumulation process. So, if the Middle Ages saw a growth in the productive forces and in production (first of all in agriculture, the precondition of the rise of towns), this is attributable to the diffusion of techniques and to the fact that they were adopted in one place or another. The documentary evidence confirms this.

[27] It is hard to imagine a less convincing or murkier thesis than the claim of some psychoanalysts that speech is linked to the penis; see for example C. Stein, *L'enfant imaginaire* (Paris: Denoël, 1971), p. 181. As for a phallus that allegedly castrates the clitoris and diminishes the vagina, it hardly seems unjust that it should subsequently be emasculated by the 'eye of God', see S. Viderman, *La construction de l'espace analytique* (Paris: Denoël, 1970), pp. 126ff. I cannot help feeling that something essential is being overlooked amidst all this exchanging of low blows.

The problem that has not yet been satisfactorily resolved is this:

In many societies, notably in Western antiquity, a number of the preconditions for the accumulation process were present, including a commodity and money-based economy, scientific thought and knowledge, and the existence of towns. How is it then that this process was not set in motion at that time and place, and that, in so far as we can assign it a historical origin, it dates only from medieval Europe? What conditions were not met earlier? What obstacles stood in the way?

Many answers to these questions have been proposed – slavery, constant wars, extravagance, the parasitism of the ruling classes (or even of the Roman *plebes*) – but none is theoretically satisfying; any or all of these historical 'factors' may have had a part in the interdiction or elimination of a trend towards accumulation, yet none can fully account for it. One is almost ready to hear that spiritual or political authorities, in their profound wisdom, took measures to prevent such a development – a hypothesis that would amount to endowing castes, priests, warlords or political leaders not so much with a profound as with a superhuman wisdom.

I propose the following answer: the space that emerged in Western Europe in the twelfth century, gradually extending its sway over France, England, Holland and Italy, was the space of accumulation – its birthplace and cradle. Why and how? Because this secularized space was the outcome of the revival of the Logos and the Cosmos, principles which were able to subordinate the 'world' with its underground forces. Along with the Logos and logic, the Law too was re-established, and contractual (stipulated) relationships replaced customs, and customary exactions.

With the dimming of the 'world' of shadows, the terror it exercised lessened accordingly. It did not, however, disappear. Rather, it was transformed into 'heterotopical' places, places of sorcery and madness, places inhabited by demonic forces – places which were fascinating but tabooed. Later, much later, artists would rediscover this ferment of sacred and accursed. At the time when it held sway, however, no one could represent this 'world'; it was simply there. Space was ridden with hidden powers, more often malign than well-disposed. Each such place had a name, and each denomination also referred to the relevant occult power: *numen–nomen*. Place-names (*lieux-dits*) dating from the agro-pastoral period had not been effaced during the Roman era. On the contrary, the Romans' innumerable minor superstitions relating to the earth, carried down via the villae and tied into the great maledictions of Christianity, could only sustain the profusion of sacred/cursed sites scattered across the face of the land. In the twelfth century a metamorphosis occurred, a displacement, a subversion of signifiers. More precisely, what had formerly signified, in an immediate manner, that which was forbidden, now came to refer solely to itself *qua* signifier – stripped of any emotional or magical referential charge. Few places, it would seem, were deprived of their names, but, typically, new names were superimposed upon the old, thus creating a web of place-names innocent of any religious overtones. Common examples of such names are Château-neuf, Ville-Franche, Les Essarts, and Bois-le-Roi. May such reference to groups of words and signs which as signifiers have been stripped of meaning be legitimately considered part of a great subversive current? Certainly. To deny this, in fact, one would have to be the sort of fetishizer of signs who takes them for the immutable foundation of knowledge and the unvarying basis of society. Besides being decrypted, medieval space was also cleared. Social practice – which did not know where it was going – made space available for something else, made it vacant (though not empty). As part of the same process, the 'libido' was freed – that tripartite libido which was denounced by Augustinian theology and which founded the secular world: *libido sciendi, dominandi, sentiendi*: curiosity, ambition, sensuality. Thus liberated, libido mounted an assault upon the space open before it. This space, deconsecrated, at once spiritual and material, intellectual and sensory, and populated by signs of the body, would become the recipient, first of an accumulation of knowledge, then of an accumulation of riches. Its source, to locate it precisely, was less the medieval town envisaged as a community of burghers than that town's marketplace and market hall (along with their inevitable companions the campanile and the town hall).

In this connection – apropos of the marketplace and the market hall – it bears repeating that the degradation of money and the baleful character of the commodity manifested themselves only later. At the time which concerns us, the exchangeable 'thing', the object produced to be sold, was still a rarity – and had a liberating function. It was an iconoclastic force and a scandal to the spirit of religious devotion promoted by the likes of Bernard of Clairvaux – who was the founder of a kind of Cistercian state, and an apologist, on the one hand for poverty, asceticism and contempt for the world, and on the other hand for the absolute hegemony of the Church.

Money and commodities, still *in statu nascendi*, were destined to bring with them not only a 'culture' but also a space. The uniqueness of the marketplace, doubtless on account of the splendour of religious and political structures, has tended to be overlooked. We should therefore remind ourselves that antiquity looked upon trade and tradespeople as external to the city, as outside its political system, and so relegated them to the outskirts. The basis of wealth was still real property, ownership of the land. The medieval revolution brought commerce inside the town and lodged it at the centre of a transformed urban space. The marketplace differed from the forum as from the agora: access to it was free, and it opened up on every side onto the surrounding territory – the territory the town dominated and exploited – and into the countryside's network of roads and lanes. The market hall, an inspired invention, was for its part as far removed from the portico as it was from the basilica; its function was to shelter the transaction of business while permitting the authorities to control it. The cathedral church was certainly not far away, but its tower no longer bore the symbols of knowledge and power; instead the freestanding campanile now dominated space – and would soon, as clock-tower, come to dominate time too.

Historians, though loth to acknowledge the subversive character of this period, have nevertheless shed light on the unevenness of the process involved. The seaboard towns of the Mediterranean easily won municipal freedoms, as did the old cities of the south of France and the cloth-towns of Flanders. In northern France, on the other hand, it was only by violence that towns were able to wrest concessions, franchises, charters and municipal constitutions from the bishops and barons. This unequal development – unequal in the degree of violence, unequal in terms of success or failure – only serves to underscore the rapidity of the spread and the extension achieved by the new space. By the fourteenth century this space, known and recognized now, and hence representable, was able to generate purely symbolic towns, founded for the purposes of commerce in regions which were still exclusively agro-pastoral, and where consequently no commercial activity was as yet taking place. Take for example the *bastides* of south-western France: spaces commercial in the strictest sense, egalitarian and abstract in nature; townships isolated and sleepy from the start, though glorified by names such as Grenade, Barcelone, Florence, Cologne or Bruges.

These places can only be understood as offshoots of the great subversive movement of the twelfth century. Yet towns such as Montauban are perfect epitomes of the commercial town – representations of an ideal type, complete with a variety of corresponding implications and extensions, among them their secular character, their civic and civil organization, their later adoption, first of Protestantism, then of Jacobinism, and so on.

This space which established itself during the Middle Ages, by whatever means it did so, whether violent or no, was by definition a space of exchange and communications, and therefore of networks. What networks? In the first place, networks of overland routes: those of traders, and those of pilgrims and crusaders. Traces of the imperial (Roman) roads were still discernible, and in many cases these roads survived intact. The new networks may be described, specifically, as hydraulic in character. The role of the ports and seaboard cities did not diminish – far from it. The 'thalassocracy' did not retain its hegemony everywhere, however, and a gradual shift towards the North Sea and Atlantic ports tended to put the Mediterranean at a disadvantage. Rivers and later canals, together with roads, constituted the new hydraulic web. The importance of the part played by inland water transport is well known. It linked up the local, regional and national markets that were already operating and those still in the process of development (Italy, France, Flanders, Germany). This communications network was simply the physical reflection – the natural mirror as it were – of the abstract and contractual network which bound together the 'exchangers' of products and money.

It would be a mistake, though, to define the new space solely in terms of these networks: we must not fall back into the one-way determinism of specialized scientific disciplines – of geography or geopolitics. Social space is multifaceted: abstract and practical, immediate and mediated. Religious space did not disappear with the advent of commercial space; it was still – and indeed would long remain – the space of speech and knowledge. Alongside religious space, and even within it, there were places, there was room, for other spaces – for the space of exchange, for the space of power. Representations of space and representational spaces diverged, yet the unity of the whole was not shattered.

Medieval space has something miraculous about it. There is no need to cross-section it theoretically – longitudinally, transversely or vertically – to identify orders and estates, ranks and hierarchies. The social edifice itself resembled a cathedral, and indeed is arguably a better candidate for homology with the *Summa theologica*. The top of the social pyramid, it might be objected, did not reach to the heavens, so there is no analogy here. But the point is rather that *one and the same illusion* sustained the belief that the tops of the city's towers grazed the vault of Heaven and embodied the celestial virtues; the belief that those at the tip of the social pyramid rubbed shoulders with divinity; the belief that reason at the zenith of its speculative constructions held out a hand to the faith directly dispensed by divine grace; and the belief, finally, that poetry may go down into the Inferno and then reascend to Paradise.

This was a society which, if not utterly transparent, certainly had a great limpidity. The economic sphere was subordinate to relationships of dependence; violence itself had a sovereign clarity; everyone knew how and why death overtook them, how and why they suffered, and why chance bestowed their few joys upon them. Society as a whole was emerging into the light. Unfortunately, money, though it had helped dispel the shadows, would soon usher in the most opaque and impenetrable relations imaginable. Medieval space raised itself above the earth; it was not yet by any means an abstract space. A large – though diminishing – portion of 'culture', of impressions and representations, was still *cryptic*, still attached to places that were holy or damned, or haunted – to caverns, grottoes, dark vales, tombs, sanctuaries and underground chambers. Whatever started to emerge was raised further into the light by the movement of the times. Such 'decrypting' was not read, or said, but lived; the process aroused terror or joy, but was generally persuasive rather than violent. When painting reasserted its priority in the Quattrocento it fell to the artists to proclaim this general transition from the cryptic to the decrypted. This was not the art of the visible *per se*, however. Knowledge was still knowledge. Decrypting in this sense had little to do with the deciphering of a text. Emergence from obscurity was an irreversible proceeding, and what emerged did so not as a sign but 'in person'.

Thus time was not separated from space; rather it oriented space – although a reversal of roles had begun to occur with the rise of medieval towns, as space tended to govern those rhythms that now escaped the control of nature (or of nature's space). Where was the connection or bond between space and time? Beyond the acquired knowledge of the period, no doubt, yet below the level grasped by its theory of knowledge: in a praxis, an 'unconscious' praxis, which regulated the concordance of time and space by limiting clashes between representations and countering distortions of reality. Time was punctuated by festivals – which were celebrated in space. These occasions had both imaginary (or mythical) and real (or practical) 'objects', all of which would appear, rise, fall, disappear and reappear: the Sun, the Christ, saints male and female, the Great Virgin Mother. As places diversified, so did social time: business time (the time of the market hall) ceased to coincide with the time of the Church, for its secularization proceeded hand in hand with that of the space to which it related. And the time of communal councils likewise ceased to coincide with the time of private life.

X

In the Western Europe of the sixteenth century 'something' of decisive importance took place. This 'something', however, was not a datable event, nor an institutional change, nor even a process clearly measurable by some economic yardstick, such as the growth of a particular form of production or the appearance of a particular market. The West was nevertheless turned upside down. The town overtook the country in terms of its economic and practical weight, in terms of its social import-ance; landownership lost its former absolute primacy. Society underwent a global change, but one uneven in its effects, as becomes apparent as soon as we consider particular sectors, elements, moments or insti-tutions.

Nowhere was there an absolute rupture with what had gone before. According to one's perspective as one views these few decades, it can seem as if everything changed or as if everything went on as before.

Perhaps examining space may help us solve the methodological and theoretical problem embodied in the question 'What changed in this crucial period?' Transition implies mediation. The historical mediation between medieval (or feudal) space and the capitalist space which was to result from accumulation was located in urban space – the space of those 'urban systems' which established themselves during the transition. In this period the town separated from the countryside that it had long dominated and administered, exploited and protected. No absolute rift between the two occurred, however, and their unity, though riven with conflict, survived. The town, in the shape of its oligarchy, continued to exercise control over its domains. From the height of their towers, 'urbanites' continued to contemplate their fields, forests and villages. As for what peasants 'are', the town-dwellers conceived these recently converted pagans either as fantasy or as objectors, and accordingly treated them with embarrassment or contempt, as something out of a fairytale or out of a tale of terror. The urbanites located themselves by reference to the peasants, but in terms of a distantiation from them:

there was therefore duality in unity, a perceived distance and a conceived unity. The town had its own rationality, the rationality of calculation and exchange – the Logos of the merchant. In taking over the reigns of power from the feudal lords, it seized control of what had been their monopoly: the protection of the peasants and the extraction of their surplus labour. Urban space was fated to become the theatre of a compromise between the declining feudal system, the commercial bourgeoisie, oligarchies, and communities of craftsmen. It further became *abstraction in action* – active abstraction – vis-à-vis the space of nature, generality as opposed to singularities, and the universal principle *in statu nascendi*, integrating specificities even as it uncovered them. Urban space was thus a tool of terrifying power, yet it did not go so far as to destroy nature; it merely enveloped and commandeered it. Only later, in a second spiral of spatial abstraction, would the state take over: the towns and their burghers would then lose not only control of space but also dominion over the forces of production, as these forces broke through all previous limits in the shift from commercial and investment capital to industrial capital. Surplus value would no longer have to be consumed where it was produced; rather, it would be susceptible of realization and distribution far away from its source, far beyond the local boundaries which had thus far hemmed it in. The economic sphere was destined to burst out of its urban context; that context would itself be overturned in the process, although the town would survive as a centre, as the locus of a variety of compromises.

The emergence of the new in Europe that we have been discussing occurred at a privileged moment, the moment of relative equipoise between a declining countryside (i.e. landownership, agricultural production) and a town (i.e. commerce, movable property, urban crafts) on the ascendant. This was the point at which the town was conceptualized, when representations of space derived from the experience of river and sea voyages were applied to urban reality. The town was given written form – described graphically. Bird's-eye views and plans proliferated. And a language arose for speaking at once of the town and of the country (or of the town in its agrarian setting), at once of the house and of the city. This language was a *code of space*.[28]

---

[28] In his investigation of the 'open work' and the 'absent structure', Umberto Eco embraces error and even delusion when, without a shred of supporting evidence, he accepts the notion that, thanks to a favourable historical development and the increasing rationality of society, art, culture and material reality, this whole complex has in the second half of the twentieth century become susceptible of coding and decoding. According to Eco, this superior rationality takes the form of communication. The communicable is presumably

Truth to tell, the first formulation of such a unitary code dates back to antiquity, and specifically to Vitruvius. The work of the Roman architect contains an elaborate attempt to establish term-by-term correspondences between the various elements of social life in the context of a particular spatial practice, that of a builder working in a city that he knows from the inside. The books of Vitruvius open with an explicit statement that is a sort of premonitory exposure of the naïvety of all who evoke Saussure's signifier–signified distinction and make it the cornerstone of their 'science':

In all matters, but particularly in architecture, there are these two points: the thing signified, and that which gives it its significance. That which is signified is the subject of which we may be speaking; and that which gives significance is a demonstration on scientific principles.[29]

The Vitruvian books implicitly embody all the elements of a code.

1 A complete alphabet and lexicon of spatial elements: water, air, light, sand, bricks, stones, conglomerates and rubbles, colouring materials, apertures and closures (doors, windows), etc.; also an inventory of the materials and *matériel* (tools) used.

2 A grammar and a syntax: description of the way the above-mentioned components are combined into wholes – into houses, basilicas, theatres, temples or baths; and directions for their assembly.

3 A style manual: recommendations of an artistic or aesthetic kind concerning the proportions, 'orders' and effects to be sought.

What is missing from the Vitruvian spatial code? On casual inspection, nothing. Everything is apparently covered in this dictionary of use value

---

decipherable, and consequently everything in the culture – each element or aspect of it – is said to constitute a semiological system. This evolutionistic rationalism and this sanguine view of the nature of communication (reading/writing) are typical embodiments of Eco's almost charming ideological naïvety. [See Eco, 'La funzione e il segno: semiologia dell'architettura', in *La struttura assente* (Milan: Bompiani, 1968). Eng. tr.: 'Function and Sign: Semiotics of Architecture', in M. Gottdiener and Alexandros Ph. Lagopoulos (eds), *The City and the Sign: An Introduction to Urban Semiotics* (New York: Columbia University Press, 1986.) – *Translator.*]

[29] 'Cum in omnibus enim rebus, tum maxime etiam in architectura haec duo insunt: quod significatur et quod significat. Significatur proposita res de qua dicitur: hanc autem significat demonstratio rationibus doctrinarum explicata' (book I, ch. 1, section 3; Eng. tr.: *Ten Books*, p. 5).

in which exchange value has simply no part. Vitruvius supplies us with a fine analytic tool with which to understand the spatial practice of the ancient Greek and Roman city, with its elaborate representations of space (astronomy, geonomy) and its magico-religious representational spaces (astrology).[30] Vitruvius goes into considerable detail: in connection with modules and mouldings – that is to say, with orders and ordonnance – he offers a methodical study, a true systematization of both terminology and objects (or 'things signified').

Nevertheless, for all its thoroughness, the approach initiated by Vitruvius's would-be comprehensive treatise on spatial semiology continued for many centuries to overlook something essential – namely, the analysis and explanation of the 'urban effect'. The city in Vitruvius is conspicuous by its absence/presence; though he is speaking of nothing else, he never addresses it directly. It is as though it were merely an aggregation of 'public' monuments and 'private' houses (i.e. those owned by the place's notabilities). In other words, the paradigm of civic space is barely present even though its 'syntagmatic' aspects – the connections between its component parts – are dealt with at length. As early as Vitruvius, then, emphasis on the technical and empirical already implied that operational considerations were paramount.

Only in the sixteenth century, after the rise of the medieval town (founded on commerce, and no longer agrarian in character), and after the establishment of 'urban systems' in Italy, Flanders, England, France, Spanish America, and elsewhere, did the town emerge as a unified entity – and as a *subject*. By the time it thus asserted itself, however, its eclipse by the state was already imminent. Still, the town became the basic premise of a discourse which offered a glimpse of a harmonious transcendence of the ancient conflict between nature, the world, and the 'rural animal' (Marx) on the one hand, and the artificial, the acquired, and the 'urban animal' on the other. At this unique moment, the town appeared to found a history having its own inherent meaning and goal – its own 'finality', at once immanent and transcendent, at once earthly (in that the town fed its citizens) and celestial (in that the image of the City of God was supplied by Rome, city of cities). Together with its territory, the Renaissance town perceived itself as a harmonious whole, as an organic mediation between earth and heaven.

---

[30] Vitruvius's plan and discussion of the Roman theatre show how the 'musical harmony of the stars' governed the sounds of musical instruments just as it governed zodiacal fortunes (book V, ch. 6, section 1; Eng. tr., p. 146). Similarly, he asserts that the pitch of the human voice is regulated by its position relative to a harp or 'sambuca' clearly discernible in the heavens (book VI, ch. 1, sections 5–8; Eng. tr., pp. 171–3).

The urban effect is linked to the architectural effect in a unity of composition and style. While it may be true to say that in the sixteenth and seventeenth centuries, beginning with Galileo, the 'human being' lost his place in the 'world' and Cosmos on account of the collapse of the Greek unity of 'action, time and space',[31] the fact remains that this 'Renaissance' being continued to situate himself in his town. Spatial practice and architecture-as-practice were bound up with each other, and each expressed the other. So the architect was effective and architecture was 'instrumental'. The Renaissance town ceased to evolve 'after the fashion of a continuous narrative', adding one building after another, an extension to a street, or another square to those already in existence. From now on each building, each addition, was politically conceived; each innovation modified the whole, and each 'object' − as though it had hitherto been somehow external − came to affect the entire fabric.[32] The centre–periphery split that would occur later, as cities fell apart under the impact of industrialization and statification, was not yet in the offing. The dominant contrast for the moment was between 'inside' and 'outside' within the unity of the architectural effect and the urban effect,[33] the unity of the country villa and the town house. This was the time of Palladio. Owing to a substantialistic or naturalistic fallacy, the space of the Renaissance town has occasionally been described as 'organic', as though it had a coherence akin to that of an organism, defined by a natural goal-directedness, with the whole governing the parts.

Such a unity, to the extent that it may ever be said to have existed in an urban space, as a 'purposiveness without purpose', may most appropriately be ascribed to the cities of antiquity. The concept of the organic denotes and connotes a blind development leading from birth to the decline of life. Can it be said of the medieval town with its burghers that it developed 'organically' − and hence blindly? Possibly − but certainly only until that moment when political power, the power of an oligarchy, of a prince or of a king, asserted itself. At that point space was necessarily transformed. If political power controlled the 'whole', this was because it knew that change to any detail could change

[31] Cf. Alexandre Koyré, *From the Closed World to the Infinite Universe* (Baltimore: Johns Hopkins University Press, 1957), pp. 2–3.
[32] See Manfredo Tafuri, *Teorie e storia dell'architettura* (Rome and Bari: Laterza Figli, 1968), pp. 25–6. Eng. tr. by Giorgio Verrecchia: *Theories and History of Architecture* (New York: Harper and Row, 1980), pp. 15–16.
[33] Cf. *La Città di Padova: saggio di analisi urbana* (Rome: Officina, 1970), pp. 218ff. (This is a remarkable collection of essays on Padua.)

that whole; the organic surely now relinquished the field to the political principle. Note, however, that there is as yet no need to evoke any abstract and detached category of the 'functional'.

For a good many 'positive' minds, nothing could be clearer or more susceptible of empirical verification than the supposed 'needs' and 'functions' of a social reality conceived of as organic in nature. In point of fact, however, nothing could be more obscure. Whose needs? By whom are such needs formulated? And by what are they satisfied or saturated? We are told that the baths of Diocletian answer to the 'needs' and 'functions' of the bathroom. To the contrary, the baths were in the highest degree multifunctional, and they met 'social needs' far more than they did 'private' ones; they were, in short, part of a *different* urban reality.

Façade and perspective went hand in hand. Perspective established the line of façades and organized the decorations, designs and mouldings that covered their surfaces. It also drew on the alignment of façades to create its horizons and vanishing-points.

The façade tells us much – and much that is surprising. It is curious, in view of its artificial and studied character, that the façade is arguably the basis for the 'organic' analogy. The notion of façade implies right and left (symmetry), and high and low. It also implies a front and a back – what is shown and what is not shown – and thus constitutes a seeming extension into social space of an asymmetry which arose rather late in the evolution of living organisms as a response to the needs of attack and defence. Inasmuch as the prestigious surface of the façade is decorative and decorated, and thus in some sense fraudulent, can we take a non-disparaging view of it? Certainly it has often been viewed otherwise – for example, as a face or countenance perceived as expressive, and turned not towards an ideal spectator but towards the particular viewer. By virtue of this analogy with a face or countenance, the façade became both eloquent and powerful. It was called upon to create ensembles, to become master of the internal (structured) disposition of space as well as of its own function (which it both fulfilled and concealed). From this 'perspective', *everything was façade*. And, if perspective governed the arrangement of component elements, of houses or other structures, the inverse was equally true, for these could also be said, by virtue of their alignment and grouping, to give rise to a perspective. It seems natural enough in this connection to draw an analogy between various (pictural and architectural) artistic forms. A picture, as a painted surface, privileges one dimension, orienting itself towards the viewer and grouping its subjects, whether inanimate or living, according to the same logic. It is a sort of face and a sort of façade. A painting turns in the direction of anyone approaching it – that is, in the direction of the public. A portrait looks *out* before, while and after it is looked *at*. A canvas, or a painted wall, has a countenance, one which actively invites scrutiny. Both face and façade have something of the gift about them, something of favour and fervour. Can the façade effect become predominant? Undoubtedly. Expressiveness emanates from the face; dissimulation, therefore, likewise. Virtues are presumed to derive from it, and it is the subject of much proverbial wisdom – consider the expression 'saving face'. It is not just buildings, but also manners and customs, and everyday life with its rites and festivals, that can fall under the sway of the prestige thus generated.

Papal Rome furnishes a rather good example of a space where the façade was master, where everything was face and façade. By virtue of an easily understood reciprocal relationship, the façade in this context was cause as well as effect: each building, house or church imposed the supremacy of the façade; every monument was at the same time the result of that supremacy. The basic configuration of space applied equally to the whole and to each detail. Symbolism infused meaning not into a single object but rather into an ensemble of objects presented as an organic whole. St Peter's in Rome is the Church itself: the Church 'entire and whole' – body and countenance – 'fastening upon her prey'. The prestigious dome represents the head of the Church, while the colonnades are this giant body's arms, clasping the piazza and the assembled faithful to its breast. The head thinks; the arms hold and contain. It seems that one might justifiably speak here, without over-generalizing, of a culture of the façade and of the face. As a principle more concrete than the 'subject' of the philosophers, the countenance, along with its complements (masks) and supplements (dress), may certainly be said to determine ways of life.

Seductive as this hypothesis may be, it is liable to oust a fundamental concept, that of production, in favour of an ideological account of generation. When an institution loses its birthplace, its original space, and feels threatened, it tends to describe itself as 'organic'. It 'naturalizes' itself, looking upon itself and presenting itself as a body. When the city, the state, nature or society itself is no longer clear about what image to present, its representatives resort to the easy solution of evoking the body, head, limbs, blood or nerves. This physical analogy, the idea of an organic space, is thus called upon only by systems of knowledge or power that are in decline. The ideological appeal to the organism is by extension an appeal to a unity, and beyond that unity (or short of it) to an *origin* deemed to be known with absolute certainty, identified beyond any possible doubt – an origin that legitimates and justifies. The notion of an organic space implies a myth of origins, and its adduction eliminates any account of genesis, any study of transformations, in favour of an image of continuity and a cautious evolutionism.

The façade and façade effects have had an eventful history, one that traverses the periods of the Baroque, of exoticism and of a variety of mannerisms. Only with the rise of the bourgeoisie and of capitalism was this principle thoroughly developed. And even then this was done in a contradictory way. Fascism sought to enthrone an organic fantasy of social life based on the notions of blood, race, nation, and an absolute national state. Hence its use of the façade, a democratic parody of which is to be found in the detached, suburban house, with its front and back – its face, as it were, and its obscene parts.

XI

Between the twelfth and the nineteenth centuries wars would revolve around accumulation. Wars used up riches; they also contributed to their increase, for war has always expanded the productive forces and helped perfect technology, even as it has pressed these into the service of destruction. Fought over areas of potential investment, these wars were themselves the greatest of investments, and the most profitable. Cases in point are the Hundred Years War, the Italian wars, the Wars of Religion, the Thirty Years War, Louis XIV's wars against the Dutch and against the Holy Roman Empire, and the wars of the French Revolution and Empire. The space of capitalist accumulation thus gradually came to life, and began to be fitted out. This process of animation is admiringly referred to as history, and its motor sought in all kinds of factors: dynastic interests, ideologies, the ambitions of the mighty, the formation of nation states, demographic pressures, and so on. This is the road to a ceaseless analysing of, and searching for, dates and chains of events. Inasmuch as space is the locus of all such chronologies, might it not constitute a principle of explanation at least as acceptable as any other?

Industry would pitch its tent in a space in which the communitarian traditions of the countryside had been swept away and urban institutions brought to ruin by wars (though the links between towns, the 'urban system', had not disappeared). This was the space, piled high with the rich spoils of years of rapine and pillage, which was to become the industrial space of the modern state.

To summarize: before the advent of capitalism, the part played by violence was extra-economic; under the dominion of capitalism and of the world market, it assumed an economic role in the accumulation process; and in consequence the economic sphere became dominant. This is not to say that economic relations were now *identical* to relations of power, but merely that the two could no longer be separated. We are confronted by the paradoxical fact that the centuries-old space of wars, instead of sinking into social oblivion, became the rich and thickly populated space that incubated capitalism. This is a fact worth pondering. What followed was the establishment of the world market, and the conquest and plunder of the oceans and continents by Europeans – by Spain, England, Holland and France. Far-ranging expeditions of this kind called for material resources as much as for goals and fantasies (not that the one excluded the other). Where was this historical process concentrated? Where was its point of combustion? From what crucible did all these creative and catastrophic forces flow? The answer is, those regions which to this day are the most industrialized of Europe and the most subject to the imperatives of growth: England, northern France, the Netherlands, the lands lying between the Loire and the Rhine. The philosophical abstractions of negation and negativity take on a distinctly concrete form when we 'think' them in the context of social and political space.

In the wake of Marx, many historians have tried to account in economic terms for the violence we have been discussing, but in so doing they have merely projected a schema applicable enough to the imperialist era back onto an earlier time. They have made no attempt to understand how the economic sphere achieved its predominance – a development which (along with other factors: surplus value, the bourgeoisie and its state) defines capitalism itself. Indeed, they have failed to understand Marx's thinking on this score – his idea that the historical with its categories was predominant during a specific period, but that it was subordinated to the economic sphere in the nineteenth century.

Does this mean that the 'economistic' explanation of history should be replaced by a 'polemological' one? Not exactly. War has been unfairly classed, however, as a destructive and evil force as opposed to a good and creative one: whereas economics could lay claim (at least as the economists saw it) to being positively and peacefully 'productive', the historians adjudged wars nothing but evil-hearted actions, the outcome of harmful passions – of pride, ambition and excess. The trouble with this apologetic kind of thinking, which is still fairly widespread, is that it ignores both the role of violence in capitalist accumulation and the part played by war and armies as productive forces in their own right. This was something that Marx had pointed out, and underlined briefly but firmly. What did war produce? The answer is: Western Europe – the space of history, of accumulation, of investment, and the basis of the imperialism by means of which the economic sphere would eventually come into its own.

Violence is in fact the very lifeblood of this space, of this strange body. A violence sometimes latent, or preparing to explode; sometimes unleashed, and directed now against itself, now against the world; and a violence everywhere glorified in triumphal arches (Roman in origin), gates, squares and prospects.

It was in this space of earth and water, a space which it had produced and sustained, that war, in Western Europe, deployed its contradictory – destructive and creative – forces. The Rhine, the North Sea, or the canals of Flanders had as great a strategic importance as the Alps, the Pyrenees, or the plains and the mountains. A single rationality may be discerned in seventeenth-century France in the actions of a Turenne, a Vauban and a Riquet – warrior, strategist and engineer respectively. It is a rationality usually associated with Cartesian philosophy, but it differs from that philosophy in the way in which a social practice does differ from an ideology, the correspondence between the two being somewhat loose and uncertain.

Did those who made history – simple soldier or field marshal, peasant or emperor – work consciously in the service of accumulation? Of course not. Now that historical time is collapsing, it is surely incumbent upon us to distinguish here, more subtly than was done at the moment when that time was first analysed, between motives, reasons, causes, aims and outcomes. Pride and ambition were certainly often motives, for example: dynastic conflicts clearly helped cause wars. As for results, they became evident only after the fact. We are thus led back to a dialectical formula which is far more acceptable than the historical verities with which the dogmatists assail us – I refer to Marx's well-known assertion that men make their history and do not know that they are making it.

To hold the conception of a whole – in the event, of a specific space – does not release us from the obligation to examine the details. The period we have been considering witnessed the glory and the decline of the town. As we have seen, society in the sixteenth century stood at a watershed. Space and time were urbanized – in other words, the time and space of commodities and merchants gained the ascendancy, with their measures, accounts, contracts and contractors. Time – the time appropriate to the production of exchangeable goods, to their transport, delivery and sale, to payment and to the placing of capital – now served to measure space. But it was space which regulated time, because the movement of merchandise, of money and of nascent capital, presupposed places of production, boats and carts for transport, ports, storehouses, banks and money-brokers. It was now that the town recognized itself and found its image. It no longer ascribed a metaphysical character to itself as *imago mundi*, centre and epitome of the Cosmos. Instead, it assumed its own identity, and began to represent itself graphically; as already noted, plans proliferated, plans which as yet had no reductive function, which visualized urban reality without suppressing the third – the divine – dimension. These were true tableaux, bird's-eye views; the town was putting itself in perspective, like a battlefield, and indeed a siege in progress was often depicted, for war often raged around the towns, and they were forever being taken, violated and despoiled. The towns were the location of wealth, at once threatening (and threatened) 'objects' and 'subjects' of accumulation – and hence too 'subjects' of history.

Throughout these conflicts, despite and because of them, the towns achieved a dazzling splendour. As the reign of the *product* began, the *work* reached the pinnacle of its achievement. These towns were in effect works of art themselves, subsuming a multitude of particular works: not only paintings, sculptures and tapestries, but also streets, squares, palaces, monuments – in short, architecture.

XII

Some theories of the state consider it to be the work of political geniuses; others deem it the result of history. The second thesis, provided it is not based on the conclusions of specialists extrapolating from their particular areas of competence (from law, from political economy, or even from political organizations themselves), and provided it achieves a certain level of generality, rejoins Hegelianism.

It is doubtful whether Marx had a fully worked-out theory of the state, although he promised both Lassalle (letter of 22 February 1848) and Engels (letter of 5 April 1848) that he would provide one. Certainly he left no complete account of the state, any more than he left a complete theory of dialectical thought. He did, however, leave a number of fragments on the subject, and a number of not unimportant suggestions.

Marx contested Hegel's theory all his life long, dismantling it, appropriating scraps of it, and replacing parts of it. Thus he proposed that a social and industrial rationality be substituted for the state and political rationality that Hegel had elevated to the status of an absolute; he viewed the state as a superstructure and not as the essence and crowning achievement of society; and he introduced the idea of the working class as the basis of a transformation that would lead to the withering-away of the state.

I suggest that the weakness not only of Hegelianism but also of the critique of Hegelianism may lie in a misapprehension of the role of space and of the corollary role of violence. For Hegel space brought historical time to an end, and the master of space was the state. Space perfected the rational and the real – simultaneously. As for violence, Hegel made it part of his speculative categories: struggle, active negativity, war, the expression of contradictions. Marx and Engels for their part showed that there could be no such thing as 'pure' and absolute violence existing apart from a class struggle, in the absence of any 'expression' of an economically dominant class, for the state could not establish itself without calling upon material resources, without a goal and that goal's repercussions upon the productive forces and upon the relations of production. Violence was indeed the midwife, but only the midwife, of a progeny conceived without its help. Neither Marx and Engels nor Hegel clearly perceived the violence at the core of the accumulation process (though Marx did consider pirates and corsairs, the sixteenth-century traffic in gold, etc.), and thus its role in the production of a politico-economic space. This space was of course the birthplace and cradle of the modern state. It was here, in the space of accumulation, that the state's 'totalitarian vocation' took shape, its tendency to deem political life and existence superior to other so-called 'social' and 'cultural' forms of practice, while at the same time concentrating all such political existence in itself and on this basis proclaiming the principle of *sovereignty* – the principle, that is to say, of its own sovereignty. It was here that the state was constituted as an imaginary and real, abstract–concrete 'being' which recognized no restraints upon itself other than those deriving from relations based on force (its relations with its own internal components, and those with its congeners – invariably rivals and virtual adversaries). The concept of sovereignty, as we have seen, enabled the monarchic state to assert itself against the Church and the Papacy, and against the feudal lords. It treated the state and its henchmen as 'political society', dominating and transcending civil society, groups and classes. Even if, like Marx, one proves to one's own satisfaction that the state and its constitution are not independent of the relations of production, of classes and their contradictions, the fact remains that the state with its sovereignty rises above these factors and reserves the right to resolve contradictions by force. The state legitimates the recourse to force and lays claim to a monopoly on violence.

Sovereignty implies 'space', and what is more it implies a space against which violence, whether latent or overt, is directed – a space established and constituted by violence. Beginning in the sixteenth century, the accumulation process exploded the framework of small medieval communities, towns and cities, fiefdoms and principalities. Only by violence could technical, demographic, economic and social possibilities be realized. The spread of sovereign power was predicated on military domination, generally preceded by plunder. In time, states became empires – the empire of Charles V and the Hapsburgs, the empire of the tsars, then Napoleon's empire and the empire which had Bismarck as its strategist. These empires, which antedated imperialism, were themselves destined sooner or later to collapse, falling victim to a space which now escaped their control. The nation state, based on a circumscribed territory, triumphed both over the city state – though this did survive into the nineteenth century, witness Venice and Florence – and over the imperial state, whose military capabilities were eventually overwhelmed. The centre–periphery relationship, though it existed on a scale as yet by no means worldwide, already suggested the limitations of centralized state power, the vulnerability of a 'sovereign' centre.

None of which changes the fact that every state is born of violence, and that state power endures only by virtue of violence directed towards a space. This violence originated in nature, as much with respect to the sources mobilized as with respect to the stakes – namely, wealth and land. At the same time it aggressed all of nature, imposing laws upon it and carving it up administratively according to criteria quite alien to the initial characteristics of either the land or its inhabitants. At the same time too, violence enthroned a specific rationality, that of accumulation, that of the bureaucracy and the army – a unitary, logistical, operational and quantifying rationality which would make economic growth possible and draw strength from that growth for its own expansion to the point where it would take possession of the whole planet. A founding violence, and continuous creation by violent means (by fire and blood, in Bismarck's phrase) – such are the hallmarks of the state. But the violence of the state must not be viewed in isolation: it cannot be separated either from the accumulation of capital or from the rational and political principle of *unification*, which subordinates and totalizes the various aspects of social practice – legislation, culture, knowledge, education – within a determinate space; namely, the space of the ruling class's hegemony over its people and over the nationhood that it has arrogated. Each state claims to produce a space wherein something is accomplished – a space, even, where something is brought to perfection: namely, a unified and hence homogeneous society. In fact, and in practice, what state and political action institutes, and consolidates by every available means, is a *balance of power* between classes and fractions of classes, as between the spaces they occupy. What, then, is the state? According to the 'politicologists', it is a framework – that of a power which makes decisions in such a way as to ensure that the interests of certain minorities, of certain classes or fractions of classes, are imposed on society – so effectively imposed, in fact, that they become indistinguishable from the general interest. Fair enough, but we must not forget that the framework in question is a *spatial* one. If no account is taken of this spatial framework, and of its strength, we are left with a state that is simply a rational unity – in other words, we revert to Hegelianism. Without the concepts of space and of its production, the framework of power (whether as reality or concept) simply cannot achieve concreteness. We are speaking of a space where centralized power sets itself above other power and eliminates it; where a self-proclaimed 'sovereign' nation pushes aside any other nationality, often crushing it in the process; where a state religion bars all other religions; and where a class in power claims to have suppressed all class differences. The relationship between institutions other than the state itself (for instance, university, tax authority, judiciary) and the effectiveness of those institutions has no need of the mediation of the concept of space to achieve self-representation, for the space in which they function is governed by statutes (and regulations for their enforcement) which fall *within* the political space of the state. By contrast the state framework, and the state *as* framework, cannot be conceived of without reference to the *instrumental* space that they make use of. Indeed each new form of state, each new form of political power, introduces its own particular way of partitioning space, its own particular administrative classification of discourses about space and about things and people in space. Each such form commands space, as it were, to serve its purposes; and the fact that space should thus become *classificatory* makes it possible for a certain type of non-critical thought simply to register the resultant 'reality' and accept it at face value.

An effective examination of space – of political space and of the politics of space – ought to enable us to dissolve the antithesis between 'liberal' theories of the state, which define it as the embodiment of the 'common good' of its citizens and the impartial arbiter of their conflicts, and 'authoritarian' theories, which invoke the 'general will' and a unifying rationality as justification for the centralization of power, a bureaucratico-political system, and the existence and importance of an *apparat*.

To the aforementioned facets of the production of abstract space may be added a general *metaphorization* which, applied to the historical and cumulative spheres, transfers them into that space where violence is cloaked in rationality and a rationality of unification is used to justify violence. As a result, the trend towards homogeneousness, instead of appearing as such, is perceived only through such metaphors as 'consensus', parliamentary democracy, hegemony, or *raison d'état*. Or even as the 'spirit of enterprise'. In a very particular kind of 'feedback', exchanges between knowledge and power, and between space and the discourse of power, multiply and are regularized.

In this way the capitalist 'trinity' is established in space – that trinity of land–capital–labour which cannot remain abstract and which is assembled only within an equally tri-faceted institutional space: a space that is first of all *global*, and maintained as such – the space of sovereignty, where constraints are implemented, and hence a fetishized space, reductive of differences; a space, secondly, that is *fragmented*, separating, disjunctive, a space that locates specificities, places or localities, both in order to control them and in order to make them negotiable; and a space, finally, that is *hierarchical*, ranging from the lowliest places to the noblest, from the tabooed to the sovereign.

But this is to run ahead, and we must return to the point we had reached in our exposition.

XIII

The work of Rabelais reveals a surprising relationship between readable and non-readable, between what appears and what remains hidden. What is said is apprehended in the mode of something appearing or emerging. The 'seen' (as opposed to appearances) refers neither to the seer nor to the visible, but rather to a nocturnal invisibility about to be exposed to daylight. Hardly are words written down than they announce this birth of each thing and preside over it. 'But had you opened that box, you would have found inside a heavenly and priceless drug . . . .'[34] And what indeed is the content of the box – what is it that comes thus into the light of day? The whole of the past, certainly, which has been buried by memory and forgetfulness; but the reality of the flesh is also being actualized here. The living body is present, as a place of transition between the depths and the surface, the threshold between hiding-place and discovery; meanwhile the writer – 'with much help from [his] spectacles, following that art by which letters can be read that are not apparent'[35] – uses his magic words to draw secrets from the drear realm of Dionysus into the realm of Apollo, from the crypts and caverns of the body into the clarity of dream and reason. The most immediate experience, and the test of the 'physical', serve as lessons to the highest form of knowledge. The emergence of the world thus continues apace as the Logos achieves its concrete realization. Texts refer neither to other texts nor to their contexts; rather, they refer to non-texts. So true is this that Rabelais, inspired master of the word that he was, ends up attacking those 'transporteurs de noms' (or 'jugglers-with-names') who replace thought with plays on words or with attributions based on colours. Such is his frustration that he hails the Egyptians' wisdom in using hieroglyphs 'which none understood who did not understand' – a veritable call to arms of listening, of aural understanding, against the visual.[36]

For Descartes and the Cartesians, God never rested. Creation was continuous. What is the meaning of this thesis of Descartes's, which was adopted by Spinoza and Leibniz before being taken to the point of absurdity by Malebranche?

1 The material world, i.e. space, continues to exist only inasmuch as it is sustained by divine thought, and contained in that thought: *produced* by it, continually and literally secreted by it – an organic mirror of the infinite.

2 The laws of space, which are mathematical laws, are laid down by God and upheld by him; nothing escapes them, and mathematical calculation reigns in nature because such calculation is coextensive with the space produced by God.

3 Novelty is constantly occurring in nature, even though the elements of nature (*natures*) are perfectly simple – so simple, in fact, that there is really only one, namely geometrical space.

[34] François Rabelais, *Gargantua and Pantagruel*, tr. J. M. Cohen (Harmondsworth, Middx: Penguin, 1955), I, 'Author's Prologue', p. 38.
[35] Ibid., I.1, p. 42.
[36] Ibid., I.9, p. 58.

Divine action, like human action, proceeds after the fashion of a lacemaker creating extraordinarily complex figures from a single thread. This metaphor is proposed in perfect seriousness by Descartes himself (in the *Meditations*). Indeed, when Descartes says that everything in nature is merely figures and movement, these terms should be taken not metaphorically but literally. The Cartesian God produces, works, and strives to create just as finite beings do, even if he does not become exhausted as a result.

It is in the context of space that productive labour is thus integrated into the essence of the divine. For the Cartesians, God embodies a sort of transcendent unity of labour and nature. Human activity imitates divine creative activity: on the one hand there is the work of craftsmen, as they make themselves masters of nature; on the other hand there is knowledge (*connaissance*) – the knowledge called for by the creative (productive) process, no longer the contemplation of antiquity or of the Middle Ages, but the Cartesian form of theoretical thought, destined to be developed, and transformed, by Hegel and Marx. The time of knowledge dominates a spatial order constituted according to the logical laws of homogeneity, under the gaze of the Lord and before the eyes of the thinking 'subject'.

The predominance of the visual (or more precisely of the geometric–visual–spatial) was not arrived at without a struggle.

In the eighteenth century music was in command. It was the pilot of the arts. On the basis of physical and mathematical discoveries, it advanced from the fugue to the sonata and thence to grand opera and the symphony. It also gave birth to an idea with infinite repercussions – the idea of harmony. Musical controversies engaged popular opinion; they had philosophical and hence universal implications. The *philosophes* concerned themselves with music, listened to music, and wrote about music.

The space of the eighteenth century, already politicized, already visual–geometric in character, and buttressed by painting and by monumental architecture (Versailles), thus suffered the onslaught of music. This onslaught stood also for the revenge of the body and the signs of the body upon the non-body and its signs – a campaign commonly known as 'eighteenth-century materialism'. The superiority of the visual over the other senses and sense organs was seriously challenged by Diderot, who pointed out that a blind person knew as much, had as many ideas, and lived as 'normally' as someone with sight. This allowed the philosopher to ask what purpose might be served by sight, which was apparently just a sort of luxury, agreeable but not really necessary. The significance of this philosophical criticism cannot be properly grasped unless it is placed in the context of the great eighteenth-century debates over music and the attendant rise of a powerful concept which united the Cosmos and the World: the concept of Harmony.

XIV

We already know several things about abstract space. As a product of violence and war, it is political; instituted by a state, it is institutional. On first inspection it appears homogeneous; and indeed it serves those forces which make a *tabula rasa* of whatever stands in their way, of whatever threatens them – in short, of differences. These forces seem to grind down and crush everything before them, with space performing the function of a plane, a bulldozer or a tank. The notion of the instrumental homogeneity of space, however, is illusory – though empirical descriptions of space reinforce the illusion – because it uncritically takes the instrumental as a given.

Critical analysis, by contrast, is immediately able to distinguish three aspects or elements here, aspects which might better be described – to borrow a term from the study of musical sounds – as 'formants'. These formants are unusual (though not unique) in the following respect: they imply one another and conceal one another. (This is not true of bipartite contrasts, the opposing terms of which, by reflecting each other in a simple mirror effect, illuminate each other, so to speak, so that each becomes a signifier instead of remaining obscure or hidden.) What, then, are these three elements?

1 The geometric formant   This is that Euclidean space which philosophical thought has treated as 'absolute', and hence a space (or representation of space) long used as a space of *reference*. Euclidean space is defined by its 'isotopy' (or homogeneity), a property which guarantees its social and political utility. The reduction to this homogeneous Euclidean space, first of nature's space, then of all social space, has conferred a redoubtable power upon it. All the more so since that initial reduction leads easily to another – namely, the reduction of three-dimensional realities to two dimensions (for example, a 'plan', a blank sheet of paper, something drawn on that paper, a map, or any kind of graphic representation or projection).

2 *The optical (or visual) formant* The 'logic of visualization' identified by Erwin Panofsky as a strategy embodied in the great Gothic cathedrals now informs the entirety of social practice. Dependence on the written word (Marshall McLuhan) and the process of spectacularization (Guy Debord) are both functions of this logic, corresponding respectively to each of its two moments or aspects: the first is metaphoric (the act of writing and what is written, hitherto subsidiary, become essential – models and focal points of practice), and the second is metonymic (the eye, the gaze, the thing seen, no longer mere details or parts, are now transformed into the totality). In the course of the process whereby the visual gains the upper hand over the other senses, all impressions derived from taste, smell, touch and even hearing first lose clarity, then fade away altogether, leaving the field to line, colour and light. In this way a part of the object and what it offers comes to be taken for the whole. This aberration, which is normal – or at least normalized – finds its justification in the social importance of the written word. Finally, by assimilation, or perhaps by simulation, all of social life becomes the mere decipherment of messages by the eyes, the mere reading of texts. Any non-optical impression – a tactile one, for example, or a muscular (rhythmic) one – is no longer anything more than a symbolic form of, or a transitional step towards, the visual. An object felt, tested by the hands, serves merely as an 'analogon' for the object perceived by sight. And Harmony, born through and for listening, is transposed into the visual realm; witness the almost total priority accorded the arts of the image (cinema, painting).

The eye, however, tends to relegate objects to the distance, to render them passive. That which is merely *seen* is reduced to an image – and to an icy coldness. The mirror effect thus tends to become general. Inasmuch as the act of seeing and what is seen are confused, both become impotent. By the time this process is complete, space has no social existence independently of an intense, aggressive and repressive visualization. It is thus – not symbolically but in fact – a purely visual space. The rise of the visual realm entails a series of substitutions and displacements by means of which it overwhelms the whole body and usurps its role. That which is merely seen (and merely visible) is hard to see – but it is spoken of more and more eloquently and written of more and more copiously.

3 *The phallic formant* This space cannot be completely evacuated, nor

entirely filled with mere images or transitional objects. It demands a truly full object – an objectal 'absolute'. So much, at least, it contributes. Metaphorically, it symbolizes force, male fertility, masculine violence. Here again the part is taken for the whole; phallic brutality does not remain abstract, for it is the brutality of political power, of the means of constraint: police, army, bureaucracy. Phallic erectility bestows a special status on the perpendicular, proclaiming phallocracy as the orientation of space, as the goal of the process – at once metaphoric and metonymic – which instigates this facet of spatial practice.

Abstract space *is not* homogeneous; it simply *has* homogeneity as its goal, its orientation, its 'lens'. And, indeed, it renders homogeneous. But in itself it is multiform. Its geometric and visual formants are complementary in their antithesis. They are different ways of achieving the same outcome: the reduction of the 'real', on the one hand, to a 'plan' existing in a void and endowed with no other qualities, and, on the other hand, to the flatness of a mirror, of an image, of pure spectacle under an absolutely cold gaze. As for the phallic, it fulfils the extra function of ensuring that 'something' occupies this space, namely, a signifier which, rather than signifying a void, signifies a plenitude of destructive force – an illusion, therefore, of plenitude, and a space taken up by an 'object' bearing a heavy cargo of myth. The use value of a space of this kind is political – exclusively so. If we speak of it as a 'subject' with such and such an aim and with such and such means of action, this is because there really is a subject here, a political subject – power as such, and the state as such.

Thus to look upon abstract space as homogeneous is to embrace a representation that takes the effect for the cause, and the goal for the reason why that goal is pursued. A representation which passes itself off as a *concept*, when it is merely an image, a mirror, and a mirage; and which, instead of challenging, instead of refusing, merely *reflects*. And what does such a specular representation reflect? It reflects the result sought. 'Behind the curtain there is nothing to see', says Hegel ironically somewhere. Unless, of course, 'we' go behind the curtain ourselves, because someone has to be there to see, and for there to be something to see. In space, or behind it, there is no unknown substance, no mystery. And yet this transparency is deceptive, and everything is concealed: space is illusory and the secret of the illusion lies in the transparency itself. The apparatus of power and knowledge that is revealed once we have 'drawn the curtain' has therefore nothing of smoke and mirrors about it.

Homogeneous in appearance (and appearance is its strength), abstract space is by no means simple. In the first place, there are its constitutive dualities. For it is both a result and a container, both produced and productive – on the one hand a representation of space (geometric homogeneity) and on the other a representational space (the phallic). The supposed congruence of the formants of this duality serves, however, to mask its *duplicity*. For, while abstract space remains an arena of practical action, it is also an ensemble of images, signs and symbols. It is unlimited, because it is empty, yet at the same time it is full of juxtapositions, of proximities ('proxemics'), of emotional distances and limits. It is thus at once lived and represented, at once the expression and the foundation of a practice, at once stimulating and constraining, and so on – with each of these 'aspects' depending on (without coinciding with) its counterpart. What emerges clearly, all the same, are the three elements of the perceived, the conceived and the lived (practice, and representations in their dual manifestation).

The individual's orientation to abstract space is accomplished socially. For individuals, for example, the location of the instruments of labour, and of the places where labour is performed (as well, naturally, as the ways of getting there), is not separate from the representation by means of signs and symbols of the hierarchy of functions. On the contrary, the one includes the other. The underpinnings of a way of life embody and fashion that way of life. And position (or location) with respect to production (or to work) *comprehends* the positions and functions of the world of production (the division of labour) as well as the hierarchy of functions and jobs. The *same* abstract space may serve profit, assign special status to particular places by arranging them in the hierarchy, and stipulate exclusion (for some) and integration (for others). Strategies may have multiple 'targets', envisaging a specific object, putting specific stakes into play and mobilizing specific resources. The *space of work* has two complementary aspects: productive activity and position in the mode of production. Any relationship to things in space implies a relationship to space itself (things in space dissimulate the 'properties' of space as such; any space infused with value by a symbol is also a *reduced* – and homogenized – space).

Spatial practice thus simultaneously defines: places – the relationship of local to global; the representation of that relationship; actions and signs; the trivialized spaces of everyday life; and, in opposition to these last, spaces made special by symbolic means as desirable or undesirable, benevolent or malevolent, sanctioned or forbidden to particular groups. We are not concerned here with mental or literary 'places', nor with philosophical *topoi*, but with places of a purely political and social kind.

The upshot is certain global phenomena affecting space as a whole (exchange, communications, urbanization, the 'development' of space), as well as a number of compartmentalizations, disintegrations, reductions and interdictions. *The space of a (social) order is hidden in the order of space.* Operating-procedures attributable to the action of a power which in fact has its own location in space appear to result from a simple logic of space. There are beneficiaries of space, just as there are those excluded from it, those 'deprived of space'; this fact is ascribed to the 'properties' of a space, to its 'norms', although in reality something very different is at work.

How is this possible? How could such capabilities, such efficacy, such 'reality' lie hidden within abstraction? To this pressing question here is an answer whose truth has yet to be demonstrated: *there is a violence intrinsic to abstraction*, and to abstraction's practical (social) use.

Abstraction passes for an 'absence' – as distinct from the concrete 'presence' of objects, of things. Nothing could be more false. For abstraction's *modus operandi* is devastation, destruction (even if such destruction may sometimes herald creation). Signs have something lethal about them – not by virtue of 'latent' or so-called unconscious forces, but, on the contrary, by virtue of the forced introduction of abstraction into nature. The violence involved does not stem from some force intervening aside from rationality, outside or beyond it. Rather, it manifests itself from the moment any action introduces the rational into the real, from the outside, by means of tools which strike, slice and cut – and keep doing so until the purpose of their aggression is achieved. For space is also instrumental – indeed it is the most general of tools. The space of the countryside, as contemplated by the walker in search of the natural, was the outcome of a first violation of nature. The violence of abstraction unfolds in parallel with what we call 'history' – the 'history' that I have reviewed in the preceding discussion, while trying to lay the emphasis on this often overlooked side of things.

Was a precise threshold crossed in the course of the transition which I have been outlining in brief? Was there an exact moment when phallic–visual–geometric space vanquished earlier perceptions and forms of perception?

Even if one takes a pro-revolutionary stance, it is no longer easy to look upon all results of the great revolutions as 'beneficial'. The French Revolution, for example, gave birth (contradictorily) to the nation, the state, law (modern law, i.e. Roman law revised and 'appropriated'), rationality, compulsory military service, the unpaid soldier, and permanent war. To this list may be added the disappearance of forms of community control over political authorities that had been enjoyed since antiquity. To say nothing of the bourgeoisie, capitalism – in short, the advent of generalized violence.

Also among revolution's effects, direct and indirect, was the definitive constitution of abstract space, with its phallic, visual and geometric formants. It goes without saying that this effect did not make its appearance as such: it was not exactly laid down in the articles of the Napoleonic Code. But, as Hegel says, the most creative periods of history were (and are) the most agonizing. After production, however, comes a time for taking stock and (to use a typographical analogy) for imposition. A time too, sometimes, for happiness, which is recorded only on history's 'blank pages'. The appearance and 'imposition' of abstract space cannot be dated: we are not concerned here with events or institutions in any clearly defined sense – even though, by the late twentieth century, the results are there plain to see. The formative process involved cannot be grasped without transcending the familiar categories of the 'unconscious' and the 'conscious'. Nothing could be more 'conscious' than the use of metaphors, for metaphors are an intrinsic part of discourse, and hence of consciousness; but nothing could be more 'unconscious' either, if one considers the *content* that emerges subsequently, in the course of usage (whether of words or of concepts). Textual criticism, in the sense of the careful and slow amassing of a body of critical knowledge, could play an important part here. Might not Romanticism be said to have lived through – even if it misunderstood – the transitional moment that separated abstract spatiality from a more unmediated perception? Was the Romantic movement not in fact shot through – and hence actuated – by this particular antagonism, even if it has been ignored in favour of more dramatic ones? Here, in brief, are a few suggestive questions in this connection.

1 Is there not a certain Romantic poetry that exists precisely on this threshold?
2 Is this poetry not the way across the threshold – or at any rate the ornament on the great portal?
3 Does not the poetry of a Victor Hugo portend the triumph of the visual, of the phallic, and of the now-consecrated geometric realm?

Hugo the 'visionary' evokes the abyss, the depths, the 'mouth of darkness'. He gives utterance (to words). He wants the light to rout the shadows. He envisions the victory of the Logos. Every possible visual metaphor is trundled out with maximum fanfare. The eye (of God, of the eternal Father) takes up residence in the tomb. The sounds of a fife make lace in the air. The bleeding hog rises from the dirt where he is lying in agony and is suddenly found balancing the scales of eternity, face to face with God: 'Le porc sanglant et Dieu se regardèrent'. The eye is master of the field. Is this stupidity or genius? A false problem. The tone is epic indeed: Vision and Sight, Truth and the Heavens sweep to triumph. As for the enemy, it fades away before this onslaught. All those twilight peoples, those denizens of the night, genies, ancestors or demons, are dispersed with the coming of the day. But what will that day be like? Into what shadows have they disappeared? What science has chased them hence? Before God, reaper of eternal summer.

Was this not the threshold? And has it not been crossed?

5

Contradictory Space

I

No *science* of space (geometry, topology, etc.) can brook contradictions in the nature of space. If social space itself were constituted by dualities (or dual properties), these could not embody contradictions in the nature of space, for duality does not imply antagonism – on the contrary. If it were true that space was the location – or set of locations – of coherence, and if it could be said to have a *mental* reality, then space could not contain contradictions. From Heraclitus to Hegel and Marx, dialectical thinking has been bound up with time: contradictions voice or express the forces and the relationships between forces that clash within a history (and within history in general).

The illusion of a transparent, 'pure' and neutral space – which, though philosophical in origin, has permeated Western culture – is being dispelled only very slowly. We have already seen how complex it is by looking at it from many viewpoints – historical, physical, physiological, linguistic, and so on. Social space embodies distinct and distinctive 'traits' which attach to the 'pure' mental form of space, without, however, achieving a separate existence as its external superadded content. Their analysis tells us what it is that confers a concrete (practical) existence upon space instead of leaving it confined within (mental) abstraction.

II

Should we be content simply to introduce the idea of a 'plural', 'polyscopic', or 'polyvalent' space? No – our analysis needs to be taken further than that. First of all, we ought to ask (in fully worked-out terms) the following questions.

1 Is there a logic of space? If so, how is it to be defined and what is its scope?
2 Does it have limits, and if so what are they? If not, where exactly does whatever is irreducible to logical form begin?
3 Where does thought, starting from the 'pure' form, encounter its first obstacle – and what is that obstacle? Opacity and compactness? Complexity? Sensory content and an irreducible practice? A residue resistant to every analytic effort?

A critique of the Cartesian concept of space dealing with its extensions into modern philosophy does not *ipso facto* entail a critique of spatial logic. The fact is that Cartesian space is open to an *intuitus*. Perfectly defined, born as an already adult and mature consciousness of self, and hence somewhat separated from the 'real', from the 'world', the Cartesian subject nevertheless miraculously, thanks to divine intervention, grasps an 'object' – space – which is the result neither of intellectual construction nor of sensory elaboration but which is, rather, given *en bloc* as suprasensory purity, as infinitude. In contrast to such a Cartesian intuition, a logic merely determines a network of relationships constitutive of the 'object'.

Much effort has been expended in contemporary thinking on attempts to bring entire sectors of reality under the rule of logic, or, to put it another way, to treat specific domains as determined and defined in accordance with a logical thesis about coherence and cohesiveness, equilibrium and regulation. There has thus been a good deal of discussion of the logic of life, the logic of the social, the logic of the market, the logic of power, and so on – without any preliminary definition of the *logical*, or of its bounds. A desire to avoid dialectical thought is what lies at the root of this search for one 'logic' after another; the result is a threat to logic itself.

III

Logical relationships are relationships of inclusion and exclusion, conjunction and disjunction, implication and explication, iteration and reiteration, recurrence and repetition, and so forth. Thus propositions, judgements, concepts or chains of concepts may include one another, and result from inclusions, or else they may be mutually exclusive. Such logical relationships imply neither a pre-existing 'reality' nor a pre-existing 'truth'. They may be represented by geometic figures; thus circles, larger ones including smaller, may serve to symbolize concepts. Such representation merely illustrates relations which have no basic need of representation, since they are themselves of a strictly formal nature. Logical relations embody the (necessary and sufficient) rationality of mathematical relations – that is, relations between figures, sets or groups (associativity, commutability, etc.).

It is beyond dispute that relations of inclusion and exclusion, and of implication and explication, obtain in practical space as in spatial practice. 'Human beings' do not stand before, or amidst, social space; they do not relate to the space of society as they might to a picture, a show, or a mirror. They know that they *have* a space and that they *are* in this space. They do not merely enjoy a vision, a contemplation, a spectacle – for they act and situate themselves in space as active participants. They are accordingly situated in a series of enveloping levels each of which implies the others, and the sequence of which accounts for social practice. For anthropology, as it examines a so-called archaic or peasant society, there is the body ('proxemics'); the dwelling with its 'rooms'; and the vicinity or community (hamlet or village) along with its dependent lands (fields under cultivation or fallow, pasture, wood and forest, game preserves, etc.). Beyond these spheres lies the strange, the foreign, the hostile. Short of them, the organs of the body and of the senses. Like (supposedly) primitive peoples, the child, who, doubtless on account of its unproductive and subservient role, is mistakenly viewed as a simple being, must make the transition from the space of its body to its body in space. And, once that operation is complete, it must proceed to the perception and conceptualization of space. According to our present analysis, these successive achievements start and end with objective 'properties' – with material symmetries and duplications upon which inclusions/exclusions are superimposed. For such inclusions embody exclusions: there are places that are prohibited (holy or damned heterotopias) for various reasons, and others that are open of access, or to which access is encouraged; in this way parts or subdivisions of space are dramatically defined in terms of the opposition between beneficent and maleficent, both of which are also clearly distinguished from neutral space.

Relationships of this kind may be figuratively represented by means of rectangles or squares: some are included by others, but at the same time they include – or are excluded by – yet others. Circles can perform an analogous representational function. Such figures help us understand the importance of grids and of the so-called radial–concentric form – and hence too, at a higher level of complexity, the importance of the cylinder and the cube. To understand their importance, though, is to limit that importance – as indeed we did earlier in showing how the form is transfigured by whatever aspects of it are apprehended in the course of the so-called 'historical' process.

The theme of iteration or repetition and its consequences (combinations of elements, differences induced within wholes) is encountered in a good many areas of study. The question is whether we are confronted here by a logical structure of such a kind that it may be described and grasped from two converging angles of approach, one starting out from what is implied, the other from what does the implying – the first from the smallest wholes discerned, the second from the vastest and most comprehensive. If so, we might reasonably be expected to arrive at an all-inclusive intelligibility. The first approach would enumerate parts of space, and thus objects in space (not just the tools of everyday life, of home and work, but also the containers of those tools – huts, cabins, houses, buildings, streets, squares, and so on, all duly marked for and by the needs of practical life). These elements could thus be inventoried in a concrete way. The second approach, by contrast, would describe space as a whole – the relations constituting society at the global level. Once an exact correspondence was attained between these two ways of apprehending space – i.e. between implication and explication – we would be in a position to grasp both the transformations brought about by the active elements within space and the genesis of space as an ensemble that is at once social and mental, abstract and concrete.

Anthropology would seem to have confirmed that such a hypothesis applies beyond the realm of 'pure' abstraction. Our knowledge of particular village communities, whether Dogon, Bororo or Basque, or of particular towns, be they ancient Greek or modern, indeed embraces surfaces and volumes bound by links of mutual implication and characterized by their overlappings, by more or less complex geometries that can be represented by figures. Here we indeed find objects and furnishings, along with 'rooms', shelters and family houses; we also find ampler places, named or designated (by means of common or proper names) as *topoi*. And all exhibit a duality that refers us back to the general properties of logico-mathematical entities, while at the same time – in practical terms – making possible multiple trajectories: outside–inside, inside–outside, and so on.

Whence the noticeable tendency of present-day anthropology to treat space as a means of classification, as a nomenclature for things, a taxonomy, on the basis of operations conceived of as independent of their content – that is, as independent of things themselves. This tendency converges with efforts to apply similar procedures, implying an identification of mental and social, to the family,[1] to exchange and communication, and to tools and objects themselves. A 'pure' self-sufficient knowledge is thus assigned a specific set of determinations: it is said to consist of categorizations implicit in its objects. We are dealing, therefore, with a hypothesis which presents itself not merely as a code capable of deciphering a given obscure message (in this instance, social space), but also as a thoroughgoing evacuation of the 'object'.

IV

An immediate objection may be made to any such reduction of content to its (formal) container. The fact is that this procedure abolishes differences from the outset, whereas a descriptive approach preserves differences in their discreteness and then plunges into the poorly charted realm of the specific.

In its most extreme form, reductionism entails the reduction of time to space, the reduction of use value to exchange value, the reduction of objects to signs, and the reduction of 'reality' to the semiosphere; it also means that the movement of the dialectic is reduced to a logic, and social space to a purely formal mental space.

What possible justification could there be for conflating an empty, Euclidean geometric space that is unaffected by whatever may fill it and a visual space with well-defined optical properties – both these spaces being treated in addition as indistinguishable from the space of a practice embracing morphologically privileged and hierarchically ordered places where actions are performed and objects are located? The thesis of an inert spatial medium where people and things, actions and situations,

---

[1] The prototype of this approach is Claude Lévi-Strauss's *Les structures élémentaires de la parenté* (Paris: Presses Universitaires de France, 1949); Eng. tr. by J. H. Bell, J. R. von Sturmer and R. Needham (ed.): *The Elementary Structures of Kinship*, rev. edn (Boston, Mass.: Beacon Press. 1969). This work contrives to deal with the family and with social relationships without once mentioning sex or eroticism. In this connection see Georges Bataille, *L'érotisme* (1957; Paris: 10/18, 1965), pp. 229–30; Eng. tr. by Mary Dalwood: *Eroticism* (1962; London and New York: Marion Boyars, 1987), pp. 210ff.

merely take up their abode, as it were, corresponds to the Cartesian model (conceiving of things in their extension as the 'object' of thought) which over time became the stuff of 'common sense' and 'culture'. A picture of mental space developed by the philosophers and epistemologists thus became a transparent zone, a logical medium. Thenceforward reflective thought felt that social space was accessible to it. In fact, however, that space is the seat of a practice consisting in more than the application of concepts, a practice that also involves misapprehension, blindness, and the test of lived experience.

Is there such a thing as a logic of space? Yes and no. In a way mathematics as a whole constitutes a logic of space. Space conceived of in its 'purity', however, as Leibniz clearly showed, has neither component parts nor form. Its parts are indiscernible, in which respect it closely resembles 'pure' identity – itself empty because of its 'purely' formal character. Before any determination can exist here, some content must come into play. And that content is the act which recognizes parts, and, within those recognized parts, an order – and hence a time. Otherwise, differences could not be *thought* – only thought *about*. To the question whether symbolic logic can be given expression without appealing to a before and an after, to a left and a right, or to symmetries and asymmetries, Lewis Carroll, for one, has shown that the answer is 'no'.[2] A logician of genius, Carroll clearly points up all the steps between pure form and the diversity of ranked contents; the latter he presents one by one along the way, fully aware of each's import and *raison d'être*. He links the mental to the social in terms of the mediating role of words, signs, doubles or shadows, and games (Alice, the looking-glass, etc.). The extension of these mediations is very great, irreducible yet conceivable (representable). Logic, so far from sitting in judgement over the confusion of orders, dimensions and levels, in fact only achieves concreteness in the process of discriminating between them. By pointing out and labelling the work of metaphor, logic effectively hinders its operation. The most pernicious of metaphors is the analogy between mental space and a blank sheet of paper upon which psychological and sociological determinants supposedly 'write' or inscribe their variations or variables. This is a metaphor used by a large number of authors, many of them

---

[2] See Lewis Carroll, *Symbolic Logic* and *The Game of Logic* (New York: Dover, 1955), 'The Biliteral Diagram' (p. 22), 'The Triliteral Diagram' (pp. 39ff), and the accompanying table of the classes and of the interpretation of spatial classes (pp. 54–5).

highly esteemed,[3] who often seek to lend it the authority either of philosophy in general or of particular philosophers.[4] What can be clearly seen by reading such authors is the way in which technicizing, psychologizing or phenomenologically oriented approaches displace the analysis of social space by immediately replacing it with a geometric – neutral, empty, blank – mental space. Consider for instance how Norberg-Schulz, a theoretician of space, defines a centre, namely as the point made by the pencil on a blank sheet of paper. From this perspective the marking-out of space has no aim or meaning beyond that of an *aide-mémoire* for the (subjective) recognition of places; Norberg-Schulz postulates an *Eigenraum* that is close (no pun intended) to the *proxemics* of the anthropologist Hall.[5] Thus objective space and the subjective image of space – the mental and the social – are simply identified.

The ultimate effect of descriptions of this kind is either that everything becomes indistinguishable or else that rifts occur between the conceived, the perceived and the directly lived – between representations of space and representational spaces. The true theoretical problem, however, is to relate these spheres to one another, and to uncover the mediations between them.

The emphasis thus comes to be laid on an *illusory space* deriving neither from geometrical space as such; nor from visual space (the space of images and photographs, as of drawings and plans) as such; nor even from practical and directly experienced social space as such; but rather from a telescoping of all these levels, from an oscillation between them or from substitutions effected among them. In this way, for example, the visual realm is confused with the geometrical one, and the optical transparency (or legibility) of the visual is mistaken for logico-mathematical intelligibility. And vice versa.

So what has to be condemned here, in the last analysis, is both a false consciousness of abstract space and an objective falseness of space itself. There is a 'common sense' for which the visual order that reduces objects to specular and spectacular abstraction is in no way distinct from scientific abstraction and its analytic (and hence reductive) procedures. A logic of reduction/extrapolation is applied to the blackboard as to the drawing-board, to the blank sheet of paper as to schemata of all kinds,

---

[3] See for example Christopher Alexander, *Notes on the Synthesis of Forms* (Cambridge, Mass.: Harvard University Press, 1964); also Christian Norberg-Schulz, *Existence, Space and Architecture* (New York: Praeger, 1971).

[4] Among them Heidegger, Merleau-Ponty, Bachelard and Piaget.

[5] See Edward T. Hall, *The Hidden Dimension* (Garden City, NY: Doubleday, 1966); and, again, Norberg-Schulz, *Existence, Space and Architecture*, pp. 18, 114.

to writing as to contentless abstraction. This *modus operandi* has even graver consequences inasmuch as the space of the mathematicians, like any abstraction, is a powerful means of action, of domination over matter – and hence of destruction. By itself, the visual realm does no more than sublimate and dissolve the body and natural energy as such; in combination, however, it acquires the disquieting ability to compensate for the impotence of pure looking by means of the power of technical agencies and of scientific abstraction.

Our present analysis will not attain its full meaning until political economy has been reinstated as the way to understand productive activity. But a new political economy must no longer concern itself with things in space, as did the now obsolete science that preceded it; rather, it will have to be a political economy of space (and of its production).

For the purposes of the present discussion, we may leave aside such considerations as accelerating technology, unfettered demographic expansion, and ecological dangers – all of which supply additional justification for such a foregrounding of space. Our approach here is a response to the impossibility of envisaging the pullulating humanity of the future (and, in some parts of our world, of the present) without at once raising the issue of space and its attendant problems. It should be emphasized *en passant* that this approach is to be sharply distinguished from a philosophy, or from a philosophical attitude, because it is founded on a practice, and a practice which is restricted neither to architecture nor to so-called town-planning, but which is broad enough to embrace overall social practice as soon as reflective thought comes to grips with the economic and political spheres.

At this stage in our investigation, what have we established? A few propositions, certainly. For mental and social to be reconnected, they first have to be clearly distinguished from one another, and the mediations between them re-established. *The concept of space is not in space.* Likewise the concept of time is not a time within time. Of this the philosophers have long been aware. The content of the concept of space is not absolute space or space-in-itself; nor does the concept contain a space within itself. The concept 'dog' does not bark. Rather, the concept of space denotes and connotes all possible spaces, whether abstract or 'real', mental or social. And in particular it has two aspects: representational spaces and representations of space.

Confusion has arisen, however, due to the fact that the philosophers, in their capacity as epistemologists, have envisaged spaces after the fashion of mathematicians: as Cartesian spaces for the classification of knowledge. They have thus proceeded as though the concept of space engendered or produced (mental) space. As a consequence, thought has been left in the unhappy position of having to plump either for a split between mental and social or else for a confused mixture of the two. The first choice meant accepting a chasm between the logical, mathematical, and epistemological realms on the one hand, and practice on the other. The second imposed an implacable systematizing and absolutely all-inclusive logic of society, of the social (and spatial) *res*, of the commodity, of capital, of the bourgeoisie, of the capitalist mode of production, and so on.

'True space' was thus substituted for the 'truth of space', and applied to such practical problems as those of bureaucracy and power, rent and profit, and so on, so creating the illusion of a less chaotic reality; social space tended to become indistinguishable from the space of planners, politicians and administrators, and architectural space, with its social constructed character, from the (mental) space of architects.[6]

V

Around 1910 academic painters were still painting 'beautiful' figures in an 'expressive' way: faces that were moving because they expressed emotions – the emotions of the painter – and desirable nudes giving voice to the desires of spectator and painter alike. The pictorial avant-garde, meanwhile, were busily detaching the meaningful from the expressive. They were not too clearly aware of this, however, for they were no great manipulators of concepts. Yet through their experimental activity these painters were acute witnesses to the beginnings of the 'crisis of the subject' in the modern world. In their pictorial practice they clearly apprehended a new fact, one bound up with the disappearance of all points of reference: the fact, namely, that only *signifying* elements could be communicated, because only they were independent of the 'subject' – that is, of the author, of the artist, and even of the spectator as an individual. This meant that the pictorial object, the painting, arose neither from the imitation of objective reality (all of whose points of reference – traditional space and time, common sense, perception of the 'real' defined by analogy with nature – were disappearing), nor from an 'expressiveness' bound up with emotions and feelings of a subjective kind. In their pictures these painters subjected the 'object' to the worst

---

[6] Cf. Philippe Boudon, *L'espace architectural, essai d'épistémologie* (Paris: Denoël, 1972).

– and before long the ultimate – atrocities. And they set about this work of breaking and dislocating with a will. Once the rift between 'subject' and 'object' had been opened, there were no limits. So wide did this rift become, indeed, that *something else* was able to emerge.

If we are to believe the most authoritative commentators, the turning-point was 1907.[7] It was at this time that Picasso discovered a new way of painting: the entire surface of the canvas was used, but there was no horizon, no background, and the surface was simply divided between the space of the painted figures and the space that surrounded them.[8] Whereas Matisse during the same period was perfecting the rhythmic treatment of the picture surface, Picasso bent his vigorous efforts to its structuring; indeed he went beyond structuring (to put it in the terms of a later date) and rendered it 'dialectical' through highly developed antagonism of line and plane rather than of colour, rhythm or background. He was not dismantling the picture surface alone, but objects too, so setting in train that paradoxical process whereby the third dimension (depth) was at once *reduced* to the painted surface and *restored* by virtue of the simultaneity of the multiple aspects of the thing depicted (analytical cubism). What we have therefore, all at once, are: the objectified end of points of reference (of Euclidean space, perspective, horizon line, etc.); a space at once *homogeneous* and *broken*; a space exerting *fascination* by means of its structure; a dialectical process initiated on the basis of antagonisms (paradigms) which does not go so far as to fracture the picture's unity; and an *absolute visualization* of things that supersedes that incipient dialectical framework.

The dissociation between the expressive and the meaningful and the liberation of the signifier had enormous consequences. The more so, because these developments were not confined to painting. Pride of place is given to painting here on account of its special relationship to space at the moment under consideration. In the first place, the liberation in question went so far as to affect the signification itself, in that the sign (the signifier) became detached from what is designated (the signified). The sign was now no longer the 'object' but rather the object on the canvas – and hence the treatment received by the objective realm as (at the same time and at one stroke) it was broken up, disarticulated, and made 'simultaneous'. As for the 'signified', it remained present – but

---

[7] Cf. Wilhelm Boeck and Jaime Sabartés, *Picasso* (New York and Amsterdam: Harry N. Abrams, 1955), p. 142: 'Unlike the many-figured paintings of 1906, *Les demoiselles d'Avignon* shows no space surrounding the figures.'

[8] '. . . the space they occupy and the space they leave unoccupied complement each other as the positive and the negative' (ibid.).

hidden. It was thus also (and above all) disquieting, evoking neither pleasure, nor joy, nor calm – only intellectual interest and most likely anxiety. Anxiety in face of what? In face of the shattered figures of a world in pieces, in face of a disjointed space, and in face of a pitiless 'reality' that cannot be distinguished from its own abstraction, from its own analysis, because it 'is' already an abstraction, already in effect an analytics. And to the question of what takes the place of subjectivity, of expressiveness, the answer is: the violence which is unleashed in the modern world and lays waste to what exists there.

To return to the case of Picasso, there is nothing simple about it, and we should indeed treat it as a 'case' rather than joining the pathetic chorus of the cultists. The notion that Picasso is a revolutionary artist ('revolutionary' because 'communist') who – his 'communism' notwith-standing – has conquered the bourgeois world and so achieved universal glory, is the product of a horrifying naïvety, if only on the grounds that the 'communist world' has in fact never accepted him. Picasso has in no sense conquered the world – nor has he been co-opted. Initially, he supplied the 'vision' that the existing world implied and awaited, and he did so just as the crisis broke, just as all the reference points were evaporating and violence was being unleashed. He did so in parallel with imperialism – and with the Great War, which was the first sign that a world market was at last becoming established, and the earliest figure of the 'world'. In parallel, too – and simultaneously – with the Bauhaus, or, in other words, with abstract space. Which, again, is not to say that Picasso was the cause of that space; he did, however, *signify* it.

Picasso's space *heralded* the space of modernity. It does not follow that the one *produced* the other. What we find in Picasso is an unreservedly visualized space, a dictatorship of the eye – and of the phallus; an aggressive virility, the bull, the Mediterranean male, a *machismo* (unquestionable genius in the service of genitality) carried to the point of self-parody – and even on occasion to the point of self-criticism. Picasso's cruelty toward the body, particularly the female body, which he tortures in a thousand ways and caricatures without mercy, is dictated by the dominant form of space, by the eye and by the phallus – in short, by violence. Yet this space cannot refer to itself – cannot acknowledge or admit its own character – without falling into self-denunciation. And Picasso, because he is a great and genuine artist, an artist who made of art an all-consuming fire, inevitably glimpsed the coming dialectical transformation of space and prepared the ground for it; by discovering and disclosing the contradictions of a fragmented space – contradictions which reside in him, and in all his works whether given form or not – the painter thus bore witness to the emergence of another space, a space not fragmented but differential in character.

VI

During this same period, Frank Lloyd Wright set out to abolish enclosing walls designed to separate the inside from the outside, interior from exterior. The wall was reduced to a surface, and this in turn to a transparent membrane. Light flooded into the house, from each of whose 'rooms' nature could be contemplated. From this moment on, the materiality of thick and heavy walls relinquished its leading architectural role. Matter was now to be no more than an envelope for space, ceding its hegemony to the light which inhabited that space. Following the tendency of philosophy, of art and literature, and of society as a whole, towards abstraction, visualization and formal spatial relations, 'architecture strove for immateriality'.[9]

Before long, however, a disjunction manifested itself that had not emerged at the outset. Walls having lost their importance (whether as walls or as curtains), interior space was liberated. The façade vanished (though it would reappear in the fascist era, with its pomp and brutality even more pronounced, its monumentality more oppressive than ever), and this led to a sundering of the street. The disarticulation of external space (façades, building-exteriors) may be clearly observed in Le Corbusier, as much in his written works as in his buildings. Le Corbusier claims to be concerned with 'freedom': freedom of the façade relative to the interior plan, freedom of the bearing structure relative to the exterior, freedom of the disposition of floors and sets of rooms relative to the structural frame. In actuality, what is involved here is a fracturing of space: the homogeneity of an architectural ensemble conceived of as a 'machine for living in', and as the appropriate habitat for a man–machine, corresponds to a disordering of elements wrenched from each other in such a way that the urban fabric itself – the street, the city – is also torn apart. Le Corbusier ideologizes as he rationalizes – unless perhaps it is the other way round. An ideological discourse upon nature, sunshine and greenery successfully concealed from everyone at this time – and in particular from Le Corbusier – the true meaning and

---

[9] Michel Ragon, *Histoire mondiale de l'architecture et de l'urbanisme modernes*, 3 vols (Tournai: Casterman, 1971–8), vol. II, p. 147.

content of such architectural projects. Nature was in fact already receding; its image, consequently, had become exalting.

VII

The belief that artists, plastic artists, are in some way the cause or *ratio* of space, whether architectural, urbanistic, or global, is the product of the naïvety of art historians, who put the social sphere and social practice in brackets and consider *works* as isolated entities. It is worth stressing this point, because what we are considering here was a *change of course*, not only in the history of art but also in the history of modern society and its space. That painters paved the way for the architectural space of the Bauhaus is indisputable. But how exactly did they do so? Just about the same time as Picasso, other great artists such as Klee and Kandinsky were inventing not merely a new way of painting but also a new 'spatiality'. It is possible that they went even further than Picasso in this direction – especially Klee. The object (painted on the canvas) was now apprehended in a perceptible – and hence readable and visible – relationship to what surrounded it, to the whole space of the picture. In Klee's work, as in Picasso's, space is detached from the 'subject', from the affective and the expressive; instead, it presents itself as meaningful. Picasso, however, projects the object's various aspects onto the canvas simultaneously, as analysed by eye and brush, whereas for Klee thought, guided by the eye and projecting itself onto the painted surface, actually revolves around the object in order to situate it. Thus the surroundings of the object become visible. And the object-in-space is bound up with a presentation of space itself.

It fell to the painters, then, to reveal the social and political transformation of space. As for the architecture of the period, it turned out to be in the service of the state, and hence a conformist and reformist force on a world scale. This despite the fact that its advent was hailed as a revolution – even as *the* anti-bourgeois revolution in architecture! The Bauhaus, just like Le Corbusier, expressed (formulated and met) the architectural requirements of state capitalism; these differed little, in point of fact, from the requirements of state socialism, as identified during the same period by the Russian constructivists. The constructivists displayed more imagination (in the utopian mode) than their Western counterparts; and, whereas they were characterized as reactionaries in their country, their Bauhaus contemporaries were dubbed subversives. This confusion has already persisted for half a century and is still far from having been dispelled: ideology and utopianism, inextricably bound up with knowledge and will, both remain vigorous. In the realm of nature rediscovered, with its sun and light, beneath the banner of life, metal and glass still rise above the street, above the reality of the city. Along with the cult of rectitude, in the sense of right angles and straight lines. The order of power, the order of the male – in short, the moral order – is thus naturalized.

There is nevertheless a strange contrast between the creative effervescence of the period we have been discussing, just before and just after the First World War, and the sterility of the second post-war era.

VIII

In the 'advanced' – i.e. the industrialized – countries, the inter-war years saw the beginnings of fragmentation in the kind of thinking about space that took place outside (or beyond) classical philosophy, as also outside the sphere of aesthetics proper – the kind of thinking, therefore, that sought some connection with 'reality'. In crude outline, theses were put forward on 'cultural space' which were then contested – on the face of it, at any rate – by theses on behavioural space. Culturalist anthropology was opposed not by the liberal humanism bequeathed by the nineteenth century, but rather by behaviourist psychology. And the two doctrines came together in the United States.

The ethnologists and anthropologists (among whom we should once again cite Mauss, Evans-Pritchard, and Rapoport) tended to project onto the present and future their often sophisticated analyses of societies as far removed and isolated as could be imagined from history, from cities, from industrial technologies. So far from relegating descriptions of peasant or tribal dwellings to the realm of folklore, this school of thought sought inspiration therein. The success enjoyed by this approach must be attributed to the fact that it evades modernity (in its capitalist form) and promotes mimesis, in the sense of a propensity to reason by analogy and to reproduce by means of imitation. Thus the theory of cultural space was transformed into a cultural model of space.

This static conception was countered by another – equally static – according to which space as directly experienced was indistinguishable from a set of conditioning factors and could be defined in terms of reflexes. At least this theory did not place a desiccated abstraction, namely culture, in the foreground. It even went so far as to assign the cultural sphere to the category of 'representational spaces', so indirectly

raising the question of the relationship between ideology and metaphysics. On the other hand, it suffered from all the shortcomings common to capitalist behaviourism and its 'socialist' competitor, Pavlovian
theory. Reductionistic in its core, this attitude excluded all inventiveness
and conjured away the need for a new space to be created as the
precondition of a new life (not that the mere invention of space is the
sufficient condition of a new life).

IX

What may be concluded from the foregoing considerations is the con-
verse of a Cartesian axiom: abstract space cannot be conceived of in
the abstract. It does have a 'content', but this content is such that
abstraction can 'grasp' it only by means of a practice that *deals with it*.
The fact is that abstract space contains contradictions, which the abstract
form seems to resolve, but which are clearly revealed by analysis. How
is this possible? How may a space be said to be at once homogeneous
and divided, at once unified and fragmented? The answer lies first of
all – and this has nothing whatsoever to do with any signifier–signified
relationship supposedly immanent to space – in the fact that the 'logic
of space', with its apparent significance and coherence, actually conceals
the violence inherent in abstraction. Just as violence is intrinsic to tools
in general (since tools cut, slice, assail and brutalize natural materials),
and to signs in general, it is also of necessity immanent to instrumental
space no matter how rational and straightforward this space may appear.
But at this point our analysis needs to be carried a step further.

Today it is easier for us to understand, since such notions have entered
the 'culture', that exchange value, the commodity, money and capital
are concrete abstractions, forms having a social existence (just like
language, which has caused so much ink to flow – and just like space)
but needing a content in order to exist socially. Capital inevitably
subdivides and disperses as individual 'capitals', but this does not mean
that it fails to retain its unity or ceases to constitute a whole – that
being a necessary condition of its operation (as capital market). Fractions
of capital enter into conflict with one another – commercial capital,
industrial capital, investment capital, finance capital – yet the formal
unity of capital subsists. The *form* persists, subsuming all such 'fractions'. And indeed the socially 'real' appearance it presents of itself is
that of unity, of capital *per se*. Its true heterogeneity, its conflicts and
contradictions, do not appear as such. Likewise in the case of property,

which is divided into immovable and movable property, landed property and money. As for the market, its fragmentation, with which we are quite familiar, is part of its very concept: there is the market in commodities (which a one-sided interpretation of Marxism places above all others), the capital market, the labour market, the market in land (construction, housing – and hence space), and the markets in works of art, in signs and symbols, in knowledge, and so on.

Abstract space can only be grasped *abstractly* by a thought that is prepared to *separate* logic from the dialectic, to *reduce* contradictions to a false coherence, and to *confuse* the residua of that reduction (for example, logic and social practice). Viewed as an instrument – and not merely as social appearance – abstract space is first of all the locus of nature, the tool that would dominate it and that therefore envisages its (ultimate) destruction. This same space corresponds to the broadening of that (social) practice which gives rise to ever vaster and denser networks on the surface of the earth, as also above and below it. It further corresponds, however, to *abstract labour* – Marx's designation for labour in general, for the average social labour that produces exchange value in general – and hence the general form of the commodity; abstract labour is in no way a mental abstraction, nor is it a scientific abstraction in the epistemological sense (i.e. a concept separated from practice so that it can be inventoried and incorporated into an absolute knowledge); rather, it has a *social* existence, just as exchange value and the value form themselves have. If one were to try and enumerate the 'properties' of abstract space, one would first have to consider it as a medium of *exchange* (with the necessary implication of interchangeability) tending to absorb *use*. This in no way excludes its *political* use, however – rather the opposite; the space of state domination and of (military) violence is also the space where strategies are put into effect. But its rationality (and it is a limited one) has something in common with the rationality of the factory – although one cannot go so far as to assume any precise parallelism between the technical and social divisions of labour. It is in this space that the world of commodities is deployed, along with all that it entails: accumulation and growth, calculation, planning, programming. Which is to say that abstract space is that space where the tendency to homogenization exercises its pressure and its repression with the means at its disposal: a semantic void abolishes former meanings (without, for all that, standing in the way of the growing complexity of the world and its multiplicity of messages, codes and operations). Both the vast metaphorization which occurs as history proceeds, and the metonymization which takes place by virtue of the process of accumulation, and which transports the body outside of itself in a paradoxical kind of alienation, lead equally to this same abstract space. This immense process starts out from physical truth (the presence of the body) and imposes the primacy of the written word, of 'plans', of the visual realm, and of a flattening tendency even within that realm itself. Abstract space thus simultaneously embraces the hypertrophied analytic intellect; the state and bureaucratic *raison d'état*; 'pure' knowledge; and the discourse of power. Implying a 'logic' which misrepresents it and masks its contradictions, this space, which is that of bureaucracy, embodies a successful integration of spectacle and violence (as distinct from 'pure' spectacle). Lastly, we find that abstract space so understood is hard to distinguish from the space postulated by the philosophers, from Descartes to Hegel, in their fusion of the intelligible *(res extensa)* with the political – their fusion, that is to say, of knowledge with power. The outcome has been an authoritarian and brutal spatial practice, whether Haussmann's or the later, codified versions of the Bauhaus or Le Corbusier; what is involved in all cases is the effective application of the analytic spirit in and through dispersion, division and segregation.

The space that homogenizes thus has nothing homogeneous about it. After its fashion, which is polyscopic and plural, it subsumes and unites scattered fragments or elements by force. Though it emerged historically as the plane on which a socio-political compromise was reached between the aristocracy and the bourgeoisie (i.e. between the ownership of land and the ownership of money), abstract space has maintained its dominance into the era of conflict between finance capital – that supreme abstraction – and action carried out in the name of the proletariat.

X

The space developed by avant-garde artists, by those artists who registered the collapse of the old points of reference, introduced itself into this fabric or tissue as a *legitimating ideology*, an ideology that justifies and motivates. These artists *presented* the object within the space of the dominant social practice. Meanwhile, the architects and city-planners offered – as an *ideology in action* – an empty space, a space that is primordial, a container ready to receive fragmentary contents, a *neutral* medium into which disjointed things, people and habitats might be introduced. In other words: incoherence under the banner of coherence, a cohesion grounded in scission and disjointedness, fluctuation and the ephemeral masquerading as stability, conflictual relationships embedded within an appearance of logic and operating effectively in combination.

Abstract space has many other characteristics also. It is here that desire and needs are uncoupled, then crudely cobbled back together. And this is the space where the middle classes have taken up residence and expanded – neutral, or seemingly so, on account of their social and political position midway between the bourgeoisie and the working class. Not that this space 'expresses' them in any sense; it is simply the space assigned them by the grand plan: these classes find what they seek – namely, a mirror of their 'reality', tranquillizing ideas, and the image of a social world in which they have their own specially labelled, guaranteed place. The truth is, however, that this space manipulates them, along with their unclear aspirations and their all-too-clear needs.

As a space where strategies are applied, abstract space is also the locus of all the agitations and disputations of mimesis: of fashion, sport, art, advertising, and sexuality transformed into ideology.

XI

In abstract space, where an anaphorization occurs that transforms the body by transporting it outside itself and into the ideal–visual realm, we also encounter a strange substitution concerning sex. In its initial, natural form, the sexual relationship implies a certain reciprocity; at a later stage this bond may be abstractly justified and legitimated in a way that changes it into a social reality (often wrongly described as 'cultural'). Physical reciprocity is legalized as contractual reciprocity, as a 'commitment' witnessed and underwritten by authority. During this process, however, the original bond undergoes a dangerous modification.

The space where this substitution occurs, where nature is replaced by cold abstraction and by the absence of pleasure, is the mental space of castration (at once imaginary and real, symbolic and concrete): the space of a metaphorization whereby the image of the woman supplants the woman herself, whereby her body is fragmented, desire shattered, and life explodes into a thousand pieces. Over abstract space reigns phallic solitude and the self-destruction of desire. The representation of sex thus takes the place of sex itself, while the apologetic term 'sexuality' serves to cover up this mechanism of devaluation.

Its natural status gone, its appeals for a 'culture' of the body unheeded, sex itself becomes no more than another localization, specificity or specialization, with its own particular location and organs – 'erotogenic zones' (as assigned by sexologists), 'organs' of reproduction, and the like. Now neither natural nor cultural, sexuality is apparently controlled as a coded and decodable system allotted the task of mediating between the 'real' and the imaginary, between desire and anxiety, between needs and frustration. Confined by the abstraction of a space broken down into specialized locations, the body itself is pulverized. The body as represented by the images of advertising (where the legs stand for stockings, the breasts for bras, the face for make-up, etc.) serves to fragment desire and doom it to anxious frustration, to the non-satisfaction of local needs. In abstract space, and wherever its influence is felt, the demise of the body has a dual character, for it is at once symbolic and concrete: concrete, as a result of the aggression to which the body is subject; symbolic, on account of the fragmentation of the body's living unity. This is especially true of the female body, as transformed into exchange value, into a sign of the commodity and indeed into a commodity *per se*.

Typically, the identification of sex and sexuality, of pleasure and physical gratification, with 'leisure' occurs in places specially designated for the purpose – in holiday resorts or villages, on ski slopes or sun-drenched beaches. Such leisure spaces become eroticized, as in the case of city neighbourhoods given over to nightlife, to the illusion of festivity. Like play, Eros is at once consumer and consumed. Is this done by means of signs? Yes. By means of spectacles? Certainly. Abstract space is doubly castrating: it isolates the phallus, projecting it into a realm outside the body, then fixes it in space (verticality) and brings it under the surveillance of the eye. The visual and the discursive are buttressed (or contextualized) in the world of signs. Is this because of what Schelsky calls 'the iron law of commercial terrorism'? Undoubtedly – but it is also, and most of all, because of the process of localization, because of the fragmentation and specialization of space within a form that is nevertheless homogeneous overall. The final stage of the body's abstraction is its (functional) fragmentation and localization.

The oddness of this space, then, is that it is at once homogeneous and compartmentalized. It is also simultaneously limpid and deceptive; in short, it is fraudulent. Falsely true – 'sincere', so to speak; not the object of a false consciousness, but rather the locus and medium of the generation (or production) of false consciousness. Appropriation, which in any case, even if it is concrete and effective, ought to be symbolizable – ought, that is, to give rise to symbols that *present* it, that render it present – finds itself *signified* in this space, and hence rendered illusory. Once this dilemma has been acknowledged, its implications and consequences are well-nigh inexhaustible. Abstract space *contains* much, but at the same time it masks (or denies) what it contains rather than indicating it. It contains specific imaginary elements: fantasy images, symbols which appear to arise from 'something else'. It contains representations derived from the established order: statuses and norms, localized hierarchies and hierarchically arranged places, and roles and values bound to particular places. Such 'representations' find their authority and prescriptive power in and through the space that underpins them and makes them effective. In this space, things, acts and situations are forever being replaced by representations (which, inasmuch as they are ideological in nature, have no principle of efficiency). The 'world of signs' is not merely the space occupied by space and images (by object-signs and sign-objects). It is also that space where the Ego no longer relates to its own nature, to the material world, or even to the 'thingness' of things (commodities), but only to things bound to their signs and indeed ousted and supplanted by them. The sign-bearing 'I' no longer deals with anything but other bearers of signs.

This homogenizing and fractured space is broken down in highly complex fashion into models of sectors. These models are presented as the product of objective analyses, described as 'systemic', which, on a supposedly empirical basis, identify systems of subsystems, partial 'logics', and so on. To name a few at random: the transportation system; the urban network; the tertiary sector; the school system; the work world with its attendant (labour) market, organizations and institutions; and the money market with its banking-system. Thus, step by step, society in its entirety is reduced to an endless parade of systems and subsystems, and any social object whatsoever can pass for a coherent entity. Such assumptions are taken for established fact, and it is on this foundation that those who make them (ideologues, whether technocrats or specialists, convinced of their own freedom from ideology) proceed to build, isolating one parameter or another, one group of variables or another. The logical consistency and practical coherence of a particular system will be asserted with no prior evaluation – even though the most cursory analysis would inevitably destroy the premise. (For example, is the 'urban network' exemplified by a particular city? or is it a representation of the city in general?) The claim is that specific mechanisms are being identified in this way which partake of a 'real' aspect of reality, and that these mechanisms will be clearly discernible once they, and some particular facet of the 'real', have been isolated. In actuality, all we have here is a tautology masquerading as science and an ideology masquerading as a specialized discipline. The success of all such 'model-building', 'simulation' and 'systemic' analysis reposes upon an unstated postulate – that of a space underlying both the isolation of variables and the construction of systems. This space validates the models in question precisely because the models make the space functional. And this works *up to a point* – the point at which chaos ensues.

XII

The visual–spatial realm – which, as I tried to show earlier, is not to be confused either with geometrical space, or with optical space, or with the space of natural immediacy – has a vast *reductive* power at its practical disposal. Though heir to history and to history's violence, this realm is responsible for the reduction of the space of earlier times, that of nature and that of history. Which means the destruction of the 'natural' as well as of the urban landscape. To say this is to evoke specific events, specific destructive decisions, and doubtless also certain displacements and substitutions that are more covert than events and decisions – and for that very reason more significant. When an urban square serving as a meeting-place isolated from traffic (e.g. the Place des Vosges) is transformed into an intersection (e.g. the Place de la Concorde) or abandoned as a place to meet (e.g. the Palais Royal), city life is subtly but profoundly changed, sacrificed to that abstract space where cars circulate like so many atomic particles. It has been noted time and again that Haussmann shattered the historical space of Paris in order to impose a space that was strategic – and hence planned and demarcated according to the viewpoint of strategy. The critics have perhaps paid insufficient attention, however, to the quality of the space Haussmann thus mortally wounded, a space characterized by the high and rare qualitative complexity afforded by its *double* network of streets and passageways. Is it conceivable that a complete correspondence could occur between a virtually total visualization (i.e. a 'visual logic' carried to the extreme) and a 'logic of society' in the sense of a strategy of the state bureaucracy? Such a concordance seems improbable – a coincidence too neat to be true. Yet Oscar Niemeyer's Brasilia clearly fits the bill. Nor has this fact gone unnoticed.[10] So faithfully is technocratic and state-bureaucratic society projected into the space of Brasilia that there is an almost self-consciously comic aspect to the process.

[10] See Charles Jencks, *Architecture 2000: Predictions and Methods* (New York: Praeger, 1971), pp. 10, 12.

The reduction with which we are concerned is directed towards the already reduced dimensions of Euclidean space; as we have already seen, this space is literally flattened out, confined to a surface, to a single plane. The steps in this flattening process, at once combined and disconnected, are worth recalling. The person who sees and knows only how to see, the person who draws and knows only how to put marks on a sheet of paper, the person who drives around and knows only how to drive a car – all contribute in their way to the mutilation of a space which is everywhere sliced up. And they all complement one another: the driver is concerned only with steering himself to his destination, and in looking about sees only what he needs to see for that purpose; he thus perceives only his route, which has been materialized, mechanized and technicized, and he sees it from one angle only – that of its functionality: speed, readability, facility. Someone who knows only how to see ends up, moreover, seeing badly. The reading of a space that has been manufactured with readability in mind amounts to a sort of pleonasm, that of a 'pure' and illusory transparency. It is hardly surprising that one soon seems to be contemplating the product of a coherent activity, and, even more important, the point of emergence of a discourse that is persuasive only because it is coherent. Surely this effect of transparency – so pleasing, no doubt, to lovers of the logical – is in fact the perfect booby trap. That, at any rate, is what I have been trying to show. Space is defined in this context in terms of the perception of an *abstract subject*, such as the driver of a motor vehicle, equipped with a collective common sense, namely the capacity to read the symbols of the highway code, and with a sole organ – the eye – placed in the service of his movement within the visual field. Thus space appears solely in its reduced forms. *Volume* leaves the field to *surface*, and any overall view surrenders to visual signals spaced out along fixed trajectories already laid down in the 'plan'. An extraordinary – indeed unthinkable, impossible – confusion gradually arises between space and surface, with the latter determining a spatial abstraction which it endows with a half-imaginary, half-real physical existence. This abstract space eventually becomes the simulacrum of a full space (of that space which was formerly full in nature and in history). Travelling – walking or strolling about – becomes an actually experienced, gestural simulation of the formerly urban activity of encounter, of movement amongst concrete existences.

So what escape can there be from a space thus shattered into images, into signs, into connected-yet-disconnected data directed at a 'subject' itself doomed to abstraction? For space offers itself like a mirror to the thinking 'subject', but, after the manner of Lewis Carroll, the 'subject' passes through the looking-glass and becomes a lived abstraction.

XIII

In this same abstract space, as it is being constituted, a substitution is effected that is no less significant than those mentioned above: the replacement of *residence* by *housing*, the latter being characterized by its functional abstraction. The ruling classes seize hold of abstract space as it comes into being (their political action occasions the establishment of abstract space, but it is not synonymous with it); and they then use that space as a tool of power, without for all that forgetting its other uses: the organization of production and of the means of production – in a word, the generation of profit.

The idea of *residing* has a poetic resonance – 'Man resides as a poet', says Hölderlin – yet this cannot obscure the fact that for many centuries this idea had no meaning outside the aristocracy. It was solely in the service of 'the great' – nobles and priests – that architects built religious edifices, palaces or fortresses. The private mansion or *hôtel particulier*, as developed by an already decadent aristocracy, and quickly aped by the bourgeoisie (of the 'high' variety, of course), calls for formal rooms sumptuously appointed but at the same time well set back from public thoroughfares – from streets, squares or boulevards. These rooms give onto a main courtyard. The aristocrat is concerned neither with seeing nor with being seen – save on ceremonial occasions. He 'is' *per se*. The essence of a palace or mansion thus lies in its interior disposition. Its luxury retains something organic, something natural, whence its charm. The façade is strictly secondary and derivative. Often it is lacking altogether, its role usurped by the severity of a monumental porch or formal carriage entrance leading to the courtyard. Within, the household goes about its business: the lord is amidst his dependants – wife, children, relations at various removes; and these in turn are surrounded by their servants. There is no privacy here: the word has no meaning. Both privacy and the façade will come only with the advent of the bourgeoisie and the bourgeoisification of the nobility. Still, 'common' areas, stables or kitchens, are clearly distinct from the spaces occupied by the masters, whose pride, arrogance, needs and desires are deployed in places set aside for the purpose.

The bourgeois apartment is no doubt a parody of the aristocratic mansion, yet beyond this imitative aspect a quite different way of occupying space is to be discerned. The formal rooms – drawing-room, dining-room, smoking-room, billiard room – are lavish in their size, decoration and furnishings. Their disposition is quite different from that of the aristocratic residence, for doors, windows and balconies open these rooms to the street. The visible and the visual are already in command. The façade, designed both to be looked at and to provide a point of vantage, is organized, with its sculptures, balustrades and mouldings, around balconies. The street's continuity, meanwhile, is founded upon the alignment of juxtaposed façades. Though its function is now reduced to transit alone, the street retains a great importance. In designing a façade and its ornamentation, the architect helps animate the street and contributes to the creation of urban space. A perspectivist rationality still governs the ordering of streets and avenues, squares and parks. Though there is no longer much of the organic left, space has nevertheless preserved a certain unity. The bourgeois apartment building is not yet a mere box. As for the bodily 'functions' of eating and drinking, sleeping and making love, these are thrust out of sight. Adjudged strictly crude and vulgar, they are relegated to the rear of the house, to kitchens, bathrooms, water closets and bedrooms often to be found along or at the end of dark corridors or over small, ill-lit courtyards. In short, in the outside–inside relationship, it is the outside that predominates. Eros disappears, in paradoxical fashion, into this two-tiered interior of reception rooms and private rooms. A psychoanalysis of space would show that bourgeois space implies a filtering of the erotic, a repression of *libidines* that is at once caesura and censure. The servants or domestic staff, for their part, live under the eaves. In the inhabited space a moralizing solemnity is the order of the day (something unknown to the aristocracy), an atmosphere of family and conjugal life – in short, of genitality – all of which is nobly dubbed an *intimité*. If the outside dominates the inside–outside relationship, this is because the outside is the only thing that really matters: what one sees and what is seen. Nevertheless, the interior, where Eros dies, is also invested with value – albeit in a mystifying and mystified way. Heavy curtains allow inside to be isolated from outside, the balcony to be separated from the drawing-room, and hence for 'intimacy' to be preserved and signified. Occasionally a curtain is drawn, and light bathes the façade: festivity is thus announced. In another sphere (or perhaps better: for the benefit of another sphere), this picture is completed by the addition of things called *objets d'art*; sometimes these are painted or sculpted nudes which add the cachet of a touch of nature or of libertinage – in order, precisely, to keep all such ideas at arm's length.

The lived experience of space is not divorced from theory. Clearly it would be trite indeed to stress everyday lived experience only to elevate it immediately to the level of theory. Describing the ill effects wreaked by the advent of lifts, which allowed the well-to-do to monopolize the upper storeys of buildings while at the same time avoiding the encounters to which the use of stairways and landings had formerly obliged them, does not get us very far. Theory does not have to place lived experience in brackets in order to promote its concepts, however. On the contrary, lived experience partakes of the theoretical sphere, and this means that the division between conceptualization and life (though not the need to draw distinctions and exercise discernment) is artificial. The analysis of bourgeoisified space validates the theory of abstract space. What is more, inasmuch as this theory unifies the lived and the conceptualized it exposes the content of abstraction while at the same time reuniting the sensory and the theoretical realms. If the senses themselves become theoreticians, theory will indeed reveal the meaning of the sensory realm.

For the working class, as is well known, the primary product of capitalism in its 'ascendant' phase – the capitalism of the *belle époque*, with its competitiveness, its princely rate of profit, and its blind but rapid accumulation – was slums at the edge of the city. This trend quickly destroyed the space of traditional residential buildings, where bourgeois lived on the lower floors, and workers and servants in the garrets. The one-room slum dwelling that had once been found, typically, at the end of a dark passageway, in a back courtyard or perhaps even in a cellar, was thus banished to peripheral neighbourhoods or suburbs. If this was a *belle époque*, it belonged to the bourgeoisie.

It was at this juncture that the idea of *housing* began to take on definition, along with its corollaries: minimal living-space, as quantified in terms of modular units and speed of access; likewise minimal facilities and a programmed environment. What was actually being defined here, by dint of successive approximations, was the lowest possible *threshold of tolerability*. Later, in the present century, slums began to disappear. In suburban space, however, detached houses contrasted with 'housing estates' just as sharply as the earlier opulent apartments with the garrets of the poor above them. The idea of the 'bare minimum' was no less in evidence. Suburban houses and 'new towns' came close to the lowest possible *threshold of sociability* – the point beyond which survival would be impossible because all social life would have disappeared. Internal and invisible boundaries began to divide a space that nevertheless remained in thrall to a global strategy and a single power. These boundaries did not merely separate levels – local, regional, national and worldwide. They also separated zones where people were supposed to be reduced to their 'simplest expression', to their 'lowest common denominator', from zones where people could spread out in comfort and enjoy those essential luxuries, time and space, to the full. As a matter of fact 'boundaries' is too weak a word here, and it obscures the essential point; it would be more accurate to speak of fracture lines revealing the true – invisible yet highly irregular – contours of 'real' social space lying beneath its homogeneous surface.

This reality is concealed by the widely promoted image of a hierarchy of levels, a neat ordering of variables and dimensions. A logical impli-cation, a purely formal conjunction/disjunction, is thus substituted for the concrete relationship between homogeneous and broken up. Space is spoken of as though it were able, in a more or less harmonious fashion, to 'organize' its own component factors: modular units and plans, the composition and density of occupation, morphological (or formal) *versus* functional elements, urbanistic and architectural features, and so on. The dominant discourse on space – describing what is seen by eyes affected by far more serious congenital defects than myopia or astigmatism – robs reality of meaning by dressing it in an ideological garb that does not appear as such, but instead gives the impression of being non-ideological (or else 'beyond ideology'). These vestments, to be more specific, are those of aesthetics and aestheticism, of rationality and rationalism.

A classical (Cartesian) rationality thus appears to underpin various spatial distinctions and divisions. Zoning, for example, which is respon-sible – precisely – for fragmentation, break-up and separation under the umbrella of a bureaucratically decreed unity, is conflated with the rational capacity to discriminate. The assignment of functions, and the way functions are actually distributed 'on the ground', becomes indistinguishable from the kind of analytical activity that discerns differ-ences. What is being covered up here is a moral and political order: the specific power that organizes these conditions, with its specific socio-economic allegiance, *seems* to flow directly from the Logos – that is, from a 'consensual' embrace of the rational. Classical reason has appar-ently undergone a convulsive degeneration into technological and tech-nocratic rationality; this is the moment of its transformation into its opposite – into the absurdity of a pulverized reality. It is 'on the ground' too that the state-bureaucratic order, itself a cloak for state capitalism (except when it is a cloak for state socialism), simultaneously achieves self-actualization and self-concealment, fuzzying its image in the crystal-clear air of functional and structural readability. The unity of reason (or of *raison d'état*) is thus draped effectively over the plethora of juxtaposed and superimposed administrative divisions, each of which corresponds to a particular 'operation'.

Abstract space is thus repressive in essence and *par excellence* – but thanks to its versatility it is repressive in a peculiarly artful way: its intrinsic repressiveness may be manifested alternately through reduction, through (functional) localization, through the imposition of hierarchy and segregation – or through art. The fact of viewing from afar, of *contemplating* what has been torn apart, of arranging 'viewpoints' and 'perspectives', can (in the most favourable cases) change the effects of a strategy into aesthetic objects. Such art objects, though generally abstract, which is to say non-figurative, nevertheless play a figurative role in that they are truly admirable representations of a 'surrounding' space that effectively kills the surroundings. All of this corresponds only too well to that urbanism of maquettes and overall plans which is the perfect complement to the planning of sewers and public works: the creator's gaze lights at will and to his heart's content on 'volumes'; but this is a fake lucidity, one which misapprehends both the social practice of the 'users' and the ideology that it itself enshrines. None of which prevents it in the slightest degree from presiding over the spectacle, and forging the unity into which all the programmed fragments must be integrated, no matter what the cost.

XIV

The breaking-up of space gives rise to conflict when two disconnected contents, each from its own angle of approach, tend towards a single form (organization). Take, for example, a *company* and its space. A company is often surrounded by an agglomeration that serves it, and to which it has given rise: a mining village or company town. In such cases the community comes under the absolute rule of the company, i.e. the rule of the company's (capitalist) owners. Employees tend to lose their status as free workers (or 'proletarians' in Marx's sense), retaining mastery over whatever time they do not give up in the form of labour time to the capitalist – who buys labour power, but not the worker as a physical being and a human individual. To the extent that capitalist enterprises create enclaves of complete dependence and subjection of workers, these remain isolated even within the space where the 'freedom' of the individual, and that of (commercial and industrial) capital itself, hold sway. But to the extent that these enclaves tend to link up, they constitute a fabric well suited to the emergence of a totalitarian capitalism (founded on the fusion of the economic and the political).

Big-city space is in no way analogous to the space of a company town – and it is for this reason that a city cannot be run on such a model, no matter how *big* a company one envisages. Workers in the city are as a rule 'free' workers (relatively speaking, of course, and always bearing in mind the abstractly philosophical meaning of 'freedom'). This is what makes it possible for urban workers to live side by side with other social classes. The social division of labour predominates over its technical division. Otherwise, the city would not allow the reproduction of labour power or the reproduction of production relations, nor would it allow the access of all to the various markets (and first and foremost to the market for consumer goods). And these are among the city's essential functions.

In other words, *liberty* engenders contradictions which are also spatial contradictions. Whereas businesses tend towards a totalitarian form of social organization, authoritarian and prone to fascism, urban conditions, either despite or by virtue of violence, tend to uphold at least a measure of democracy.

XV

The meanings conveyed by abstract space are more often prohibitions than solicitations or stimuli (except when it comes to consumption). Prohibition – the negative basis, so to speak, of the social order – is what dominates here. The symbol of this constitutive repression is an object offered up to the gaze yet barred from any possible use, whether this occurs in a museum or in a shop window. It is impossible to say how often one pauses uncomfortably for a moment on some threshold – the entrance of a church, office or 'public' building, or the point of access to a 'foreign' place – while passively, and usually 'unconsciously', accepting a prohibition of some kind. Most such prohibitions are invisible. Gates and railings, ditches and other material barriers are merely the most extreme instances of this kind of separation. Far more abstract signs and signifiers protect the spaces of elites – rich neighbourhoods or 'select' spots – from intruders. Prohibition is the reverse side and the carapace of property, of the negative appropriation of space under the reign of private property.

Space is divided up into designated (signified, specialized) areas and into areas that are prohibited (to one group or another). It is further subdivided into spaces for work and spaces for leisure, and into daytime and night-time spaces. The body, sex and pleasure are often accorded no existence, either mental or social, until after dark, when the prohibitions that obtain during the day, during 'normal' activity, are lifted. This secondary and derivative existence is bestowed on them, at night, in sections of the city (formerly, in Paris, around Pigalle and Montmartre, and more recently around Montparnasse and the Champs-Elysées) which are dedicated to that function, but which by the same token possess nothing aside from the accoutrements of entertainment, the infrastructure of this peculiarly sophisticated form of exploitation. In these neighbourhoods, and during these hours, sex seems to have been accorded every right; in actuality, the only right it has is to be deployed in exchange for cash. In accordance with this division of urban space, a stark contrast occurs at dusk as the lights come on in the areas given over to 'festivity', while the 'business' districts are left empty and dead. Then in a brightly illuminated night the day's prohibitions give way to profitable pseudo-transgressions.

XVI

How does this space, which we have described as at once homogeneous and broken up, maintain itself in view of the formal irreconcilability of these two characteristics? How can two such properties, 'incompatible' from a logical point of view, be said to enter into association with one another and constitute a 'whole' which not only does not disintegrate but even aids in the deployment of strategies?

We have already posed this question, though in a slightly different form, and also suggested an answer. We must come back to this issue, however. The solution is not to be found in space *as such* – as a thing or set of things, as facts or a sequence of facts, or as 'medium' or 'environment'. To pursue any such line of investigation is to return to the thesis of a space that is neutral, that is prior or external to social practice and hence on those grounds mental or fetishized (objectified). Only an *act* can hold – and hold together – such fragments in a homogeneous totality. Only action can prevent dispersion, like a fist clenched around sand.

Political power and the political action of that power's administrative apparatus cannot be conceived of either as 'substances' or as 'pure forms'. This power and this action do *make use of* realities and forms, however. The illusory clarity of space is in the last analysis the illusory clarity of a power that may be glimpsed in the reality that it governs, but which at the same time uses that reality as a veil. Such is the action of political power, which creates fragmentation and so controls it – which creates it, indeed, *in order to* control it. But fragmented reality (dispersion, segregation, separation, localization) may on occasion overwhelm political power, which for its part depends for sustenance on continual reinforcement. This vicious circle accounts for the ever more severe character of political authority, wherever exercised, for it gives rise to the sequence force–repression–oppression. This is the form under which state-political power becomes omnipresent: it is everywhere, but its presence varies in intensity; in some places it is diffuse, in others concentrated. In this respect it resembles divine power in religions and theologies. Space is what makes it possible for the economic to be integrated into the political. 'Focused' zones exert influences in all directions, and these influences may be 'cultural', ideological, or of some other kind. It is not political power *per se* that produces space; it does reproduce space, however, inasmuch as it is the locus and context of the reproduction of social relationships – relationships for which it is responsible.

XVII

The time has come to clarify the aims of the present discussion in terms of Marx and his thought – in terms, also, of political economy as science, and of the critique of political economy as ideology.

The best way to get Marx's thinking into perspective is to reconstitute it, to restore in its entirety, and to look upon it not as an end point or conclusion but rather as a point of departure. In other words, Marxism should be treated as one *moment* in the development of theory, and not, dogmatically, as a definitive theory. The fact is – and there is no reason not to repeat it here – that two errors or illusions have to be avoided in this connection. The first looks upon Marx's thought as a system, endeavours to integrate it into the body of established knowledge, and hence tries to apply epistemological criteria to it. The second seeks by contrast to demolish Marxist thought in the name of a radical critique, in the name of bringing criticism to bear on the very tools of criticism. Those who take the first approach are seduced by the idea of absolute knowledge, and accept the thesis, which is historically a Hegelian one, that such a knowledge exists and can be applied to a 'reality' itself already established. Partisans of the second view, meanwhile, fall under the spell of destruction and self-destruction, and become convinced that 'reality' can be destroyed by undermining the foundations of knowledge. Surely we should instead view Marxism today much as the theory of relativity views Newtonian physics – as a moment in the progress of thought, not only in the sense of a stage in that thought's historical genesis, to be recalled for pedagogical purposes, but also in the sense of a moment that is necessary because still immanent and essential, and indeed still evolving. In this way, the question of the *political* discontinuity or rift between the theory of the state (Hegel) and the radical critique of the state (Marx) is left open.

It is possible today to reconstruct the trajectory of political economy, its rise and fall, including the pinnacle it reached in the work of Marx. This brief and dramatic history cannot be detached from so-called economic 'reality' – that is, from the growth of the forces of production (the primitive accumulation of capital). The decline of economic thought began with the difficulties encountered by growth and by the ideology that justified and stimulated it – with the political empiricism and pragmatism of the solutions proposed to the problems associated with growth.

Before considering this history, it will be well to review a few concepts – that of *social labour*, for example, as first proposed by the great English political economists and later elaborated upon by others, notably Hegel and Marx. Social labour had an eventful career. Both reality and concept emerged along with the birth of modern industry, and both successfully imposed themselves, despite countervailing efforts and contingencies, to the point where they became crucial, in theory as in practice, for science and for society. Productive (industrial) labour, as reality, as concept and as ideology, gave rise to moral and artistic 'values', and hence production and productivity became not merely social motors but also the rational basis of a conception of the world linked to the philosophy of history and to the rising science of political economy. But soon obsolescence set in. Values and concepts derived from labour began to wear out. And as a theory of growth and a generator of models political economy disintegrated.

Something comparable had happened around the middle of the nineteenth century, but at that time Marx had given political economy a new lease on life in a way both unforeseen and incomprehensible to the economic pundits of the time. Simply stated, Marx supplied political economy with its own self-criticism as part of a global approach (to time, to history, and to social practice). This schema is well known today – even too well known, for its creative capacity (some would say its 'productive' capacity – and why not?) has been prejudiced in consequence. A creative capacity of this kind manifests itself between the time when a concept begins to perturb dominant tendencies and the time when it begins to promote these tendencies – when, in other words, it is incorporated into the established wisdom, into the public domain, into culture and pedagogy. Marx and Marxism have certainly not escaped a process of this kind, but the Marxist schema has retained much force. There is no knowledge, according to this schema, without a critique of knowledge – no knowledge aside from critical knowledge. Political economy as a science is not and cannot be 'positive' and 'positive' alone; political economy is also the critique of political economy – that is to say, the critique of the economic and of the political, and of their supposed unity or synthesis. An understanding of production implies its critical analysis, and this brings the concept of relations of production out of obscurity. These relations, once clearly identified, exert a retroactive influence upon the confused ensemble from which they have emerged – upon the concepts of productive social labour and of production. At this point a new concept is constituted, one which subsumes that of the relations of production but is not identical to it: the concept of *mode of production*. Between the relations of production and the mode of production is a connection that Marx never completely uncovered, never fully worked out. This created a lacuna in his thought that his successors have striven to fill. Whether they have succeeded in doing so is another matter.

What of the part played by the *land*, as concept and as reality, in this context? At the outset, for the physiocrats, the land was a determining factor, but subsequently it seemed fated quickly to lose all importance. Agriculture and agricultural labour were expected to fade away in face of industrial labour, as much from the quantitative point of view (wealth produced) as from the qualitative one (needs met by products of the land); agriculture itself, it was felt, could and should be industrialized. Furthermore, the land belonged to a class – aristocracy, landowners or feudal lords – which the bourgeoisie appeared certain either to abolish or else to subjugate into complete insignificance. Lastly, the town would surely come to dominate the country, and this would be the death knell (or the transcendence) of the whole antagonism.

The political economists wavered a good deal on the issues of land, of labour and agricultural products, of property and ground rent, and of nature, and their hesitations may easily be traced – including, naturally, those of Malthus as well as those of Ricardo and Marx.

Marx's initial intention in *Capital* was to analyse and lay bare the capitalist mode of production and bourgeois society in terms of a binary (and dialectical) model that opposed capital to labour, the bourgeoisie to the proletariat, and also, implicitly, profits to wages. This polarity may make it possible to grasp the conflictual development involved in a formal manner, and so to articulate it intelligibly, but it presupposes the disappearance from the picture of a *third* cluster of factors: namely the land, the landowning class, ground rent and agriculture as such. More generally speaking, this bringing to the fore of a binary opposition of a conflictual (dialectical) character implies the subordination of the historical to the economic, both in reality and in the conceptual realm, and hence too the dissolving or absorption, by the economic sphere proper, of a multiplicity of formations (the town, among others) inherited from history, and themselves of a precapitalist nature. In the context of this schema the space of social practice is imperceptible; time has but a very small part to play; and the schema itself is located in an abstract mental space. Time is reduced to the measure of social labour.

Marx quickly became aware – as he was bound to do – of resistance to this reductive schema (though many 'Marxists' – and all dogmatic Marxists without exception – have retained it, and indeed aggravated its problems instead of correcting for them).[11] Such resistance came from several sides, and in the first place from the very reality under consideration – namely, the Earth. On a world scale, landed property showed no signs of disappearing, nor did the political importance of landowners, nor did the characteristics peculiar to agricultural production. Nor, consequently, did ground rent suddenly abandon the field to profits and wages. What was more, questions of underground and above-ground resources – of the space of the entire planet – were continually growing in importance.

Such considerations account, no doubt, for the peculiarities of a 'plan' that is exceedingly hard to reconstruct – that of *Capital*. At the close of Marx's work, the issue of the land and its ownership re-emerges, and this in a most emphatic way, complete with consideration of ownership in the cases of underground resources, of mines, minerals, waters and forests, as well as in those of the breeding of livestock, of construction and of built-up land. Lastly, and most significantly, Marx now proposed

---

[11] The fate of Marxism has meant – and who by this time could still be unaware of it? – that all dispute, discussion or dialogue concerning the crucial areas of the theory has been prevented. For instance, any attempt to restore to its proper place the concept of ground rent has for decades been utterly squelched, whether in France, in Europe or in the world at large, in the name of a Marxism that has become mere ideology – nothing but a political tool in the hands of apparatchiks.

his 'trinity formula', according to which there were three, not two, elements in the capitalist mode of production and in bourgeois society. These three aspects or 'factors' were the Earth (Madame la Terre), capital (Monsieur le Capital), and labour (the Workers). In other words: rent, profit, wages – three factors whose interrelationships still needed to be identified and clearly set forth.[12] And *three*, I repeat, rather than *two*: the earlier binary opposition (wages *versus* capital, bourgeoisie *versus* working class), had been adandoned. In speaking of the earth, Marx did not simply mean agriculture. Underground resources were also part of the picture. So too was the nation state, confined within a specific territory. And hence ultimately, in the most absolute sense, politics and political strategy.

Capital, which was never completed, comes to a halt at this point. We are now beginning to understand the reasons why Marx failed to bring his work to a conclusion – a failure for which his ill health was only partly responsible.

What excuse could there be today for not going back to this exemplary if unfinished work – not with a view to consecrating it in any way but in order to put questions to it? This is especially needful at a time when capitalism, and more generally development, have demonstrated that their survival depends on their being able to extend their reach to space in its entirety: to the *land* (in the process absorbing the towns and agriculture, an outcome already foreseeable in the nineteenth century, but also, and less predictably, creating new sectors altogether – notably that of leisure); to the *underground* resources lying deep in the earth and beneath the sea-bed – energy, raw materials, and so on; and lastly to what might be called the *above-ground* sphere, i.e. to volumes or constructions considered in terms of their height, to the space of mountains and even of the planets. Space in the sense of the earth, the ground, has not disappeared, nor has it been incorporated into industrial production; on the contrary, once integrated into capitalism, it only gains in strength as a specific element or function in capitalism's expansion. This expansion has been an active one, a forward leap of the forces of production, of new modalities of production, but it has occurred without breaking out of the mode and the relations of the capitalist production system; as a consequence, this extension of production and of the productive forces has continued to be accompanied by a *reproduction of the relations of production* which cannot have failed to leave

[12] See also my *Espace et politique* (*Le droit à la ville, II*) (Paris: Anthropos, 1973), pp. 42ff. Marx's discussion is in *Capital*, vol. III, ch. 48.

its imprint upon the total occupation of all pre-existing space and upon the production of a new space. Not only has capitalism laid hold of pre-existing space, of the Earth, but it also tends to produce a space of its own. How can this be? The answer is: through and by means of urbanization, under the pressure of the world market; and, in accordance with the law of the reproducible and the repetitive, by abolishing spatial and temporal differences, by destroying nature and nature's time. Is there not a danger that the economic sphere, fetishized as the world market, along with the space that it determines, and the political sphere made absolute, might destroy their own foundation – namely land, space, town and country – and thus in effect self-destruct?

Some of the new contradictions generated by the extension of capitalism to space have given rise to quickly popularized *representations*. These divert and evade the problems involved (i.e. the problematic of space), and in fact serve to mask the contradictions that have brought them into being. The issue of pollution is a case in point. Pollution has always existed, in that human groups, settled in villages or towns, have always discharged wastes and refuse into their natural surroundings; but the symbiosis – in the sense of exchange of energies and materials – between nature and society has recently undergone modification, doubtless to the point of rupture. This is what a word such as 'pollution' at once acknowledges and conceals by metaphorizing such ordinary things as household rubbish and smoking chimneys. In the case of 'the environment', we are confronted by a typically metonymic manoeuvre, for the term takes us from the part – a fragment of space more or less fully occupied by objects and signs, functions and structures – to the whole, which is empty, and defined as a neutral and passive 'medium'. If we ask, '*whose* environment?' or 'the environment *of what?*', no pertinent answer is forthcoming.

Although these points have been made earlier, it seems important that they be reiterated. The reason is that in many quarters truly magical origins and powers continue to be attributed to ideologies. How, for example, could bourgeois ideology, if it were no more than a mirror-like reflection of reality, actually reproduce this reality and its production relations? By masking contradictions? It certainly does do that, but it also brings nations and nationalisms into being – hardly a *specular* effect. There is no need to evoke history (the genesis of the nation states), however: any close examination of what such pseudo-theory purports to explain will suffice to demonstrate its absurdity. In Marx's trinitarian scheme, by contrast, there is no rift between ideology and political practice: power holds earth, labour and capital together and reproduces them (whether in conjunction or disjunction) separately.

In Marx, the critique of political economy had an import and meaning that a latter-day productivist approach would overlook completely. It was the actual concept of political economy, as a form of knowledge, that Marx had in his sights. He showed that, in promoting and practising a science that claimed to understand production and productive forces, the economists mystified both their readers and themselves. What they were describing were the conditions of scarcity and palliatives to that scarcity. Directly or indirectly, cynically or hypocritically, they preached asceticism. Well before the sixteenth century, perhaps in the depths of the Middle Ages, perhaps even earlier, at the time of Rome's decline and of early Judaeo-Christianity, Western society chose to accumulate rather than to live, so opening a chasm, creating a contradiction between enjoying and economizing whose drama would thereafter hold society in an iron grip. Centuries after this basic choice had been taken somewhere back in the mists of time, political economy arose as a rationale for it. Its birth as a science coincided with the triumph of economics in the sphere of social practice – the triumph, in other words, of the concern with accumulation by means of and for the sake of profit, an accumulation that was forever expanding.

So just who were the economists, in Marx's view? They were the voices of (relative) want, of the transition from archaic scarcities to a now-conceivable abundance. They made a study of (relative) scarcities and contributed to an unjust distribution of 'goods'. Their pseudo-science, which was, as such, ideological in character, embodied and masked a practice. The economists were acquainted with scarcity *per se*; they were not so much the expression of that scarcity as the concrete consciousness – albeit poorly developed – of the insufficiencies of production. This was the sense that political economy had for Marx. Or perhaps better: it is in this sense that (political) economy was political. It enabled the henchmen of the state, political power, to organize the apportionment of want. The concrete relations of production could thus give rise to distribution and to consumption. The 'distribution' in question was carried out under the masks of liberty and equality, even under those of fraternity and justice. The law's function was to codify the rules. 'Summum jus, summa injuria.'

Law and justice presided over injustice, and the name of equality was applied to an inequality that became no less flagrant as a result – though it did become harder to combat.

Whether deliberately or not, consciously or not, the economists put the finishing touches to the results of the law of value (results produced in a blind and spontaneous manner), for they effected the allocation, within (national) spatial frameworks, and according to the needs of industry's various branches, of the labour power and productive capacity available to a given society (Britain or France, for instance) under the capitalist mode of production and under the state that had control over that production. To this end the economists constructed an abstract space or abstract spaces in which to place and promote their models of 'harmonious development'. The methodology of a Bastiat, in Marx's time, was not much cruder than this. The economists never succeeded in getting out of mental space, the space of their models, into social space. The management of society, to which for a long time they contributed not a little, thus proceeded along the road of development (expanded accumulation), but it did so under the control of the bourgeoisie; the bourgeois relations of production were retained in their essentials; and, most importantly, the negative aspects of the situation were made to appear positive and constructive.

During the period of the economists' ascendancy, the 'benefits of nature' and the 'elements' (water, air, light, space) received mention, if at all, only for the purpose of excluding them from the domain of political economy: on account of their abundance they had no exchange value; their 'use' embodied no value; they were the outcome of no social labour; and no one *produced* them.

What has occurred in this connection in more recent times? And what is the situation today? Certain goods that were once scarce have become (relatively) abundant, and vice versa. As a result use value, so long overshadowed by exchange value, has been relocated and, as it were, reinvested with value. Bread has lost the symbolic force it formerly had in Europe, where it once stood for food in general, for everything precious, and even for labour itself ('Give us this day our daily bread'; 'In the sweat of thy face shalt thou eat bread'). In the advanced countries, where agriculture has been industrialized, a permanent overproduction, whether overt or covert, has long been the order of the day, complete with the stocking of surplus grain and restraints, often subsidized, on the exploitation of productive land. Not that this has had the slightest impact on the suffering of millions – indeed, hundreds of millions – of human beings in the so-called underdeveloped nations, who are prey to malnutrition if not outright famine. Much the same sort of thing may be said of a host of objects of everyday utility in the major industrialized countries. Nobody today is unaware of the fact that the obsolescence of such products is planned, that waste has an economic function, or that fashion plays an enormous role, as does 'culture', in a functionalized consumption that is structured accordingly. These developments spelt the doom of political economy, and its place has been usurped by market research, sales techniques, advertising, the manipulation of needs, investment-planning guided by consulting-firms, and so forth. Manipulation of this kind, a practice only too compatible with political propaganda, has no more need of 'science' than it has of ideology; it calls for information rather than knowledge.

Thanks to a dialectical process, the (relative) abundance of industrial products in today's so-called consumer society is accompanied by an inverse phenomenon: new scarcities. This dialectic itself has been the subject of hardly any analysis or explanation, for its operation is concealed by the continual discussion of pollution, threats to the 'environment', ecosystems, the destruction of nature, the using-up of resources, and so forth. Such entities serve only as ideological shields. Meanwhile, the ever-increasing 'new scarcities' are liable to precipitate a crisis (or crises) of a type without precedent. Those commodities which were formerly abundant because they occurred 'naturally', which had no value because they were not products, have now become rare, and so acquired value. They have now to be produced, and consequently they come to have not only a use value but also an exchange value. Such commodities are 'elemental' – not least in the sense that they are indeed 'elements'. In the most modern urban planning, using the most highly perfected technological applications, *everything* is produced: air, light, water – even the land itself. Everything is factitious and 'sophisticated'; nature has disappeared altogether, save for a few signs and symbols – and even in them nature is merely 'reproduced'. Urban space is detached from natural space, but it re-creates its own space on the basis of productive capacity. Natural space, at least under certain socio-economic conditions, becomes a scarce commodity. Inversely, scarcity becomes spatial – and local. Everything thus affected by scarcity has a close relationship to the Earth: the resources of the land, those beneath the earth (petroleum) and those above it (air, light, volumes of space, etc.), along with things which depend on these resources, such as vegetable and animal products and energies of various kinds.

The 'elements' lose their natural determinations, including their siting and situation, as they are incorporated into the 'space envelopes' which are fast becoming the social building-blocks of space. They assume value – both use value and exchange value – because it is no longer possible to draw them directly from an everlasting source, namely nature. The demands made by current developments such as these are surely just as important as the potential exhaustion – still on the distant horizon – of industrial (e.g. mineral) resources. In the traditional industrial production process, the relationship to space had long been comprised of discrete points: the place of extraction or origin of raw materials, the place of production (factory), and the place of sale. Only the distribution networks of this system had a wider spatial dimension. Now that the 'elements' themselves are produced and reproduced, however, the relationship of productive activity to space is modified; it involved space now in another way, and this is as true for the initial stages of the process (for example, the management of water and water resources) as it is for the final stages (within urban space) and for all the steps in between.

The finiteness of nature and of the Earth thus has the power to challenge blind (ideological) belief in the infinite power of abstraction, of human thinking and technology, and of political power and the space which that power generates and decrees.

Once the 'elements' begin to circulate within systems of production, allocation or distribution, they necessarily become part of wealth in general, and so fall within the purview of political economy. But is this political economy still classical political economy? The new shortages are not comparable to the scarcities of earlier times, for the notable reason that the relationship to space has changed. They are located, more and more firmly, within space as a whole, within that space into which the old industrial production was inserted, with its discrete points, and which was subsequently completely occupied by expanding capitalism and by the reproduction of production relations. It is in this space, therefore, that a new demand now arises: the demand for the production or reproduction of 'elemental' materials (raw materials, energies). What will be the outcome? Will this new demand have a stimulating and integrating effect on capitalism, or will it rather be a force for disintegration over the shorter or longer term?

When it comes to space, can we legitimately speak of scarcity? The answer is no – because available or vacant spaces are still to be found in unlimited numbers, and even though a relative lack of space may have left its mark on some societies (particularly in Asia), there are others where just the opposite is true – where, as in North America, society bears the clear traces of the vastness of the space open to its demographic and technological expansion. Indeed the space of nature remains open on every side, and thanks to technology we can 'construct' whatever and wherever we wish, at the bottom of the ocean, in deserts or on mountaintops – even, if need be, in interplanetary space.

The fact is that the shortage of space is a distinctly socio-economic phenomenon, one which can only be observed, and which only occurs, in quite specific areas, namely in or near urban *centres*. These may have grown up from historically established centres, from the old cities, or they may have evolved out of new towns.

The question of centrality in general, and of urban centrality in particular, is not a very simple one. It inhabits every aspect of the problematic of space. Germane to mental space as much as to social space, it links the two in a manner that surmounts the old philosophical distinctions, rifts and disjunctions between subject and object and between intellectual and material (or comprehensible and sensible). This is not to say that new distinctions and differences are not introduced thereby. The notion of centrality has a mathematical origin, as witness its application in the analysis of abstract space. Any given 'point' is a point of accumulation: surrounding it is an infinite number of other points. Otherwise, we should have no certainty as to the continuity of space. At the same time, around each (isolated) point, a surface – preferably square – can be described and analysed, as can any variation following an infinitesimal change in its distance from the centre (ds^2). Thus each centre may be conceived of in two ways: as full or empty, as infinite or finite.

In order correctly to frame the question of centrality and attempt to resolve it, a dialectical approach is in order. The appropriate method, however, is no longer that of Hegel, nor is it that of Marx, which was based on an analysis of historical time, of temporality. If we find ourselves obliged to accept the idea of a dialectical centrality, or of a dialectic of centrality, this is because there is a connection between space and the dialectic; in other words, there are spatial contradictions which imply and explain contradictions in historical time, though without being reducible to them. Inversely, if the notion of contradiction (of actual conflict) is not restricted to temporality or historicity, if it does in fact extend to the spatial realm, this means that a dialectic of centrality exists. This dialectical process develops the logical characteristics of centres (hitherto understood solely as *points*).

In what does the dialectical movement of centrality consist? First of all, centrality, whether mental or social, is defined by the gathering-together and meeting of whatever coexists in a given space. What does coexist in this way? Everything that can be named and enumerated. Centrality is therefore a *form*, empty in itself but calling for contents – for objects, natural or artificial beings, things, products and works, signs and symbols, people, acts, situations, practical relationships. This means that centrality closely resembles a *logical* form – and hence that there is a logic of centrality. Centrality as a form implies simultaneity, and it is a result thereof: the simultaneity of 'everything' that is susceptible of coming together – and thus of accumulating – in an act of thinking or in a social act, at a point or around that point. The general concept of centrality connects the punctual to the global. According to the orientation of modern thought first adopted by Nietzsche and since taken up by a number of thinkers (among them Georges Bataille), the centre or focal point is the place of sacrifice, the place where accumulated energies, desirous of discharge, must eventually explode. Each period, each mode of production, each particular society has engendered (produced) its own centrality: religious, political, commercial, cultural, industrial, and so on. The relationship between mental and social centrality must be defined for each case. The same goes for the conditions under which a given centrality will come to an end – whether it ruptures, explodes, or is rent apart.

Centrality is movable. We have long known – and recent work, notably that of Jean-Pierre Vernant, has confirmed the fact, and elaborated upon it – that the centre of the Greek city was forever being moved: from the semicircular area where chiefs and warriors conferred about their expeditions and divided up their booty to the city temple, and from the temple to the agora, a place of political assembly (and later, thanks to annexed arches and galleries, of commerce). This means that in ancient Greece a complex relationship existed between urban space and the temporality (rhythms) of urban life. The same goes for modern cities, and it would not be difficult, for example, to inventory the various shifts in centrality that have taken place in Paris in the nineteenth and twentieth centuries: the boulevards, Montmartre, Montparnasse, the Champs-Elysées, and so on.

What makes present-day society different in this regard? Simply this: centrality now aspires to be *total*. It thus lays claim, implicitly or explicitly, to a superior political rationality (a state or 'urban' rationality). It falls to the agents of the technostructure – to the planners – to provide the justification for this claim. In so doing, they naturally spurn the dialectic; and indeed a centrality of this order expels all peripheral elements with a violence that is inherent in space itself. This centrality – or, perhaps better, this centralization – strives to fulfil its 'totalizing' mission with no philosophy to back it up aside from a strategic one (whether conscious or not). Despite countervailing forces, some subversive, some tolerable – and tolerated on various grounds (liberalization, flexibility, etc.) – the centre continues effectively to concentrate wealth, means of action, knowledge, information and 'culture'.

In short, everything. These capacities and powers are crowned by the supreme power, by the ability to concentrate all powers in the power of Decision. This decision-making system makes the (illegitimate) claim that it is rational.

Throughout history centralities have always eventually disappeared – some displaced, some exploded, some subverted. They have perished sometimes on account of their excesses – through 'saturation' – and sometimes on account of their shortcomings, the chief among which, the tendency to expel dissident elements, has a backlash effect. Not that these factors are mutually exclusive – as witness ancient Rome, which suffered both internal saturation and the assaults of peripheral forces.

The interplay between centre and periphery is thus highly complex. It mobilizes both logic and dialectics, and is hence doubly determined. If one takes logic, whether formal or applied, as one's frame of reference, the dialectic tends to be set aside. Contradictions, however, can never be eliminated. If, on the other hand, one applies the dialectic, which is the theory of contradictions, one ends up giving short shift to logic, to coherence and cohesiveness. The fact is that neither approach is dispensable here. Centrality may give birth to an applied logic (a strategy); it may also burst asunder and lose its identity utterly.

It is primarily in connection with the scarcity of space that the matter of centrality has arisen here. The tendency to establish 'centres of decision-making' which bring together, within a limited area, those elements that found society and are therefore usable by power for its own purposes promotes a scarcity of space in the area in question – that is, the area surrounding a central point. Shortage of space has original and new characteristics as compared with other kinds of shortages, whether ancient or modern. In so far as it results from a historical process, it occurs spontaneously, yet it is sustained, and often sought and organized, by centrally made decisions. It introduces a contradiction between past and possible future abundance on the one hand and actually reigning scarcity on the other. This contradiction is not extraneous to the production relations embodied in space as a whole, and even less so to the reproduction of those relations, which it is the express purpose of the centres of decision to maintain; at the same time, it is a contradiction *of* space itself – and not merely *in* space, after the fashion of the classical contradictions engendered by history and by historical time. This must emphatically not be taken as implying that contradictions and conflicts *in* space (deriving from time) have disappeared. They are still present, along with what they imply, along with the strategies and tactics to which they give rise, and along, in particular, with the class conflicts that flow from them. The contradictions *of* space, however, envelop historical contradictions, presuppose them, superimpose themselves upon them, carry them to a higher level, and amplify them in the process of reproducing them. Once this displacement has been effected, the new contradictions may tend to attract all the attention, diverting interest to themselves and seeming to crowd out or even absorb the old conflicts. This impression is false, however. Only by means of a dialectical analysis can the precise relationships between contradictions *in* space and contradictions *of* space be unravelled, and a determination made as to which are becoming attenuated, which accentuated. Similarly, the production of things *in* space has not disappeared – and neither have the questions it raises (ownership of the means of production, management and control of production) – in face of the production *of* space. But the production of space – including the production of the 'elements', as discussed above – does subsume and broaden the scope of the problems thrown up by the production of things. The process of condensation and the centralizing tendency may therefore be said also to affect pre-existing contradictions, which they duly concentrate, aggravating and modifying them in the process.

Space is marked out, explored, discovered and rediscovered on a colossal scale. Its potential for being occupied, filled, peopled and transformed from top to bottom is continually on the increase: the prospect, in short, is of a space being produced whose nature is nothing more than raw materials suffering gradual destruction by the techniques of production. What is more, we now have the means to gather all knowledge and information, no matter how close or how far away its source may be, at a single point where it can be processed; data collection and computer science abolish distance, and they can confidently ignore a materiality scattered across space (and time). The theory of centrality implies a completely new capacity for *concentration* such as was formerly possessed only by the brain – indeed only by the brains of geniuses. Mental centrality and social centrality are linked by a mediation which no doubt has this task as its chief function. That mediation is information – which in this context cannot become part of knowledge without effectively connecting the mental and the social. Paradoxically, it is precisely with the advent of this state of affairs that space shatters. Rendered artificially scarce anywhere near a centre so as to increase its 'value', whether wholesale or retail, it is literally pulverized and sold off in 'lots' or 'parcels'. This is the way in which space in practice becomes the medium of segregations, of the component elements of society as they are thrust out towards peripheral zones. And this is the space that is now sliced up by a science itself segmented by specialization into discrete disciplines, each of which – and first of all political economy in its current form – constitutes its own particular space, mental and abstract, to be laboriously confronted with social practice. The process of slicing-up, moreover, becomes a 'discipline' in its own right: the instrument of knowledge is taken for knowledge itself. The search for some unity here is confined to laboured interdisciplinary or multidisciplinary montages which never manage to fit any of the pieces back together. The analytic approach excels only in the handling of cutting-tools, and unification is beyond the reach of partial sciences which could only regain focus by transforming their methodology, their epistemology, their agenda and their ideologies.

Such is the context of an unfolding 'economic' process which no longer answers to classical political economy and which indeed defies all the computations of the economists. 'Real property' (along with 'construction') is no longer a secondary form of circulation, no longer the auxiliary and backward branch of industrial and financial capitalism that it once was. Instead it has a leading role, albeit in an *uneven* way, for its significance is liable to vary according to country, time or circumstance. The law of unevenness of growth and development, so far from becoming obsolete, is becoming worldwide in its application – or, more precisely, is presiding over the globalization of a world market.

In the history of capitalism real property has played but a minor role. For one thing, the relics of the former ruling class long owned not only the agricultural land but also the land suitable for building; and, secondly, the relevant branch of production was dominated by trades and crafts. The situation of this branch, and of the whole economic sector in question, has now changed almost everywhere, though most of all in the major industrialized countries. Capitalism has taken possession of the land, and *mobilized* it to the point where this sector is fast becoming *central*. Why? Because it is a new sector – and hence less beset by the obstacles, surfeits, and miscellaneous problems that slow down old industries. Capital has thus rushed into the production of space in preference to the classical forms of production – in preference to the production of the means of production (machinery) and that of consumer goods. This process accelerates whenever 'classical' sectors show the slightest sign of flagging. The flight of capital towards this favoured sector can threaten capitalism's delicate self-regulating mechanisms, in which case the state may have to intervene. But this does not mean the elimination of the production of space as a sector which presupposes the existence of other forms of circulation but which nevertheless tends to displace the central activities of corporate capitalism. For it is space, and space alone, that makes possible the deployment of the (limited but real) organizational capacity of this type of capitalism.

It sometimes happens, then, that the 'real property' sector is rather brusquely called to order. As a mix of production and speculation, often hard to separate out from 'development', the sector oscillates between a subordinate function as a booster, flywheel or back-up – in short as a regulator – and a leading role. It is therefore part of the *general unevenness* of development, and of the *segmentation* of the economy as a global reality. At the same time it retains an essential function in that it combats the falling rate of profit. Construction, whether private or public, generates higher-than-average profits in all but the most exceptional cases. Investment in 'real estate', i.e. in the production of space, continues to involve a higher proportion of variable as compared with constant capital. The organic composition of capital is weak in this sphere, despite the high level of investment called for and despite the rapidity of technological progress. Small and middle-sized businesses remain common, while excavation and framing call for a great deal of manpower (often immigrant labour). A mass of surplus value is thus generated, with most being added to the general mass but a significant portion returning to construction firms, and to the promoters and speculators. As to those problems which arise because obsolescence in this area tends to be slow, so putting a brake on the circulation of capital, they are tackled by a variety of means. The mobilization of space becomes frenetic, and produces an impetus towards the self-destruction of spaces old and new. Investment and speculation cannot be stopped, however, nor even slowed, and a vicious circle is thus set up.

A strategy based on space, even if we leave military and political projects out of the picture, must be considered a very dangerous one indeed, for it sacrifices the future to immediate interests while simultaneously destroying the present in the name of a future at once programmed and utterly uncertain.

The mobilization of space for the purposes of its production makes harsh demands. The process begins, as we have seen, with the land, which must first be wrenched away from the traditional form of property, from the stability of patrimonial inheritance. This cannot be done easily, or without concessions being made to the landowners (ground rents). The mobilization is next extended to space, including space beneath the ground and volumes above it. The entirety of space must

be endowed with *exchange value*. And exchange implies interchangeability: the exchangeability of a good makes that good into a commodity,
just like a quantity of sugar or coal; to be exchangeable, it must be
comparable with other goods, and indeed with all goods of the same
type. The 'commodity world' and its characteristics, which formerly
encompassed only goods and things produced in space, their circulation
and flow, now govern space as a whole, which thus attains the autonomous (or seemingly autonomous) reality of things, of money.

Exchange value – as Marx showed, in the wake of the 'classical
economists', apropos of products/things – is expressed in terms of
money. In the past one bought or rented land. Today what are bought
(and, less frequently, rented) are *volumes* of space: rooms, floors, flats,
apartments, balconies, various facilities (swimming-pools, tennis courts,
parking-spaces, etc.). Each exchangeable place enters the chain of commercial transactions – of supply and demand, and of prices. The connection of prices with 'production costs' – i.e. with the average social labour
time required by production – is an increasingly elastic one, moreover.
This relationship, like others, is disturbed and complicated by a variety
of factors, notably by speculation. The 'truth of prices' tends to lose its
validity: prices are more and more independent of value and of production costs, while the operation of economic laws – the law of value
and the law of supply and demand, or (if non-Marxist terminology is
preferred) the interactions between desirability and profit margins – is
compromised. Fraud itself now becomes a law, a rule of the game, an
accepted tactic.

The need for comparability has been met by the production of virtually
identical 'cells'. This is a well-known fact – one that no longer surprises
anyone. It seems 'natural', even though it has hardly ever been explained
– and then very poorly. Yet its apparent naturalness itself cries out for
explanation. This is the triumph of homogeneity. From the point of
view of the 'user', going from one 'cell' to another can mean 'going
home'. The theory of 'modules' and its practical application have made
it possible to reproduce such cells, taken as 'models', *ad infinitum*. Space
is thus produced and reproduced *as* reproducible. Verticality, and the
independence of volumes with respect to the original land and its
peculiarities, are, precisely, produced: Le Corbusier thrust built volumes
into abstraction, separating them from the earth by means of piles and
pillars, on the pretext that he was exposing them to open air and
sunshine. At the same time – literally – volumes are treated as surfaces,
as a heap of 'plans', without any account being taken of time. Not that
time disappears completely in the case of abstractions thus erected, made

vertical, and made visual. The fact is, however, that the 'needs' about which we hear so much are forced under the yoke – or, rather, through the filter – of space. The truth, in fact, is that these needs are results and not causes – that they are subproducts of space. Exchangeability and its constraints (which are presented as norms) apply not only to surfaces and volumes but also to the paths that lead to and from them. All of which is justified on plans and drawings by the 'graphic synthesis' of body and gesture that is allegedly achievable in architectural projects.[13] The graphic elements involved (in drawings, sections, elevations, visual tableaux with silhouettes or figures, etc.), which are familiar to architects, serve as *reducers* of the reality they claim to represent – a reality that is in any case no more than a modality of an accepted (i.e. imposed) 'lifestyle' in a particular type of housing (suburban villa, high-rise, etc.). A 'normal' lifestyle means a normalized lifestyle. Meanwhile, the reference to the body (the 'modulor'), along with the figures and the promotional patter, serve literally to 'naturalize' the space thus produced, as artificial as it may be.

For all that architectural projects have a seeming objectivity, for all that the producers of space may occasionally have the best intentions in the world, the fact is that volumes are invariably dealt with in a way that refers the space in question back to the land, to a land that is still privately (and privatively) owned; built-up space is thus emancipated from the land *in appearance only*. At the same time, it is treated as an empty abstraction, at once geometric and visual in character. This relationship – a real connection concealed beneath an apparent separation and constituting a veritable Gordian knot – is both a practice and an ideology: an ideology whose practitioners are unaware that their activity is of an ideological nature, even though their every gesture makes this fact concrete. The supposed solutions of the planners thus impose the constraints of exchangeability on everyday life, while presenting them as both natural (or normal) and technical requirements – and often also as moral necessities (requirements of public morality). Here – as ever – the economic sphere that Marx denounced as the organization of asceticism makes common cause with the moral order. 'Private' property entails private life – and hence privation. And this in turn implies a repressive ideology in social practice – and vice versa, so that each masks the other. Spatial interchangeability inevitably brings a powerful tendency towards quantification in its train, a tendency which naturally extends outwards into the surroundings of the housing itself

[13] See A. de Villanova, in *Espaces et Société*, no. 3, p. 238.

– into those areas variously represented as the environment, transitional spaces, means of access, facilities, and so on. Supposedly natural features are swallowed up by this homogenization – not only physical sites but also bodies – the bodies, specifically, of the inhabitants (or 'users'). Quantification in this context is technical in appearance, financial in reality, and moral in essence.

Should use value be expected to disappear? Could the homogenization of fragments scattered through space, along with their commercial interchangeability, lead to an absolute primacy of exchange and exchange value? And could exchange value come to be defined by the signs of prestige or 'status' – i.e. by differences internal to the system, regulated by the relationship of specific locations to centres – with the result that the exchange of signs would absorb use value and supersede practical considerations rooted in production and in production costs?

The answer to these questions must be negative: the acquirer of space is still buying a use value. What is that use value? First of all, he is buying an inhabitable space commensurate with other such spaces, and semiologically stamped by a promotional discourse and by the signs of a certain degree of 'distinction'. That is not all, however: also purchased is a particular *distance* – the distance from the purchaser's dwelling-place to other places, to centres of commerce, work, leisure, culture or decision. Here time once more has a role to play, even though a space that is both programmed and fragmented tends to eliminate it as such. Admittedly the architect, the promoter or even the occupier can compensate for the shortcomings of a given location by introducing signs: signs of status, signs of happiness, signs of 'lifestyle', and so on. Such signs are bought and sold despite their abstract nature, despite their concrete *insignificance*, and despite their *over-significance* (in that they proclaim their meaning – namely, compensation). Their price is simply added to the real exchange value. The fact remains that a home-buyer buys a daily schedule, and that this constitutes part of the use value of the space acquired. Any schedule has pros and cons, involves the losing or saving of time, and hence something other than signs – to wit, a *practice*. The consumption of space has very specific features. It differs, of course, from the consumption of things in space, but this difference concerns more than just signs and meanings. Space is the envelope of time. When space is split, time is distanced – but it resists reduction. Within and through space, a certain social time is produced and reproduced; but real social time is forever re-emerging complete with its own characteristics and determinants: repetitions, rhythms, cycles, activities. The attempt to conceive of a space isolated from time entails a further contradiction, as embodied in efforts to introduce time into space by force, to rule time from space – time in the process being confined to prescribed uses and subjected to a variety of prohibitions.

XVIII

If we are to clarify the categories and concepts relating to the production of space, we shall need to return to Marx's concepts – and not only to those of social labour and production. What is a commodity? A concrete abstraction. An abstraction, certainly – but not an abstraction *in spite of* its status as a thing; an abstraction, on the contrary, *on account of* its status as a social 'thing', divorced, during its existence, from its materiality, from the use to which it is put, from productive activity, and from the need that it satisfies. And concrete, just as certainly, by virtue of its practical power. The commodity is a social 'being-there', an 'object' irreducible to the philosophical concept of the Object. The commodity hides in stores, in warehouses – in inventory. Yet it has no mystery comparable to the mystery of nature. The enigma of the commodity is entirely social. It is the enigma of money and property, of specific needs and the demand–money–satisfaction cycle. The commodity asks for nothing better than to *appear*. And appear it does – visible/readable, in shop windows and on display racks. Self-exhibition is its forte. Once it is apparent, there is no call to decode it; it has no need of decipherment after the fashion of the 'beings' of nature and of the imagination. And yet, once it has appeared, its mystery only deepens. Who has produced it? Who will buy it? Who will profit from its sale? Who, or what purpose, will it serve? Where will the money go? The commodity does not answer these questions; it is simply *there*, exposed to the gaze of passers-by, in a setting more or less alluring, more or less exhibitionistic, be it in a nondescript small shop or in a glittering department store.

The chain of commodities parallels the circuits and networks of exchange. There is a language and a world of the commodity. Hence also a logic and a strategy of the commodity. The genesis and development of this world, this discourse and this logic were portrayed by Marx. The commodity assumed a role in society very early on, before history, but that role was limited, coexisting with those of barter and of the gift. Its status grew, however, in the cities of the ancient world, and above all in the medieval towns. It then gave rise to commercial capital, to the conquest of the oceans and of distant lands – and hence also to the first adumbration of the world market. Upon this historical basis industrial capitalism was founded – a great leap forward for the commodity, putting it on course for the conquest of the world – i.e. the conquest of space. Ever since, the world market has done nothing but expand (so to speak). The actualization of the worldwide dimension, as a concrete abstraction, is under way. 'Everything' – the totality – is bought and sold. 'Theological subtlety', wrote Marx apropos of the commodity and its characteristics. He was right to speak of subtlety, because the abstraction involved attains a most remarkable complexity. He was right, also, to use the word 'theological', because the commodity as concrete abstraction acts as the power of determinate 'beings' (human groups, fractions of classes). The commodity is a *thing*: it is *in space*, and occupies a location. Chains of commodities (networks of exchange) are constituted and articulated on a world scale: transportation networks, buying- and selling-networks (the circulation of money, transfers of capital). Linking commodities together in virtually infinite numbers, the commodity world brings in its wake certain attitudes towards space, certain actions upon space, even a certain concept of space. Indeed, all the commodity chains, circulatory systems and networks, connected on high by Gold, the god of exchange, do have a distinct homogeneity. Exchangeability, as we have seen, implies interchangeability. Yet each location, each link in a chain of commodities, is occupied by a *thing* whose particular traits become more marked once they become fixed, and the longer they remain fixed, at that site; a thing, moreover, composed of matter liable to spoil or soil, a thing having weight and depending upon the very forces that threaten it, a thing which can deteriorate if its owner (the merchant) does not protect it. The space of the commodity may thus be defined as a homogeneity made up of specificities. This is a paradox new to our present discussion: we are no longer concerned either with the representation of space or with a representational space, but rather with a practice. Exchange with its circulatory systems and networks may occupy space worldwide, but consumption occurs only in this or that particular place. A specific individual, with a specific daily schedule, seeks a particular satisfaction. Use value constitutes the only real wealth, and this fact helps to restore its ill-appreciated importance. The paradigmatic (or 'significant') opposition between exchange and use, between global networks and the determinate locations of production and consumption, is transformed here into a dialectical contradiction, and in the process it becomes spatial. Space thus understood is both *abstract* and *concrete* in character: abstract inasmuch as it has no existence save by virtue of the exchangeability of all its component parts, and concrete inasmuch as it is socially real and as such localized. This is a space, therefore, that is *homogeneous yet at the same time broken up into fragments.*

The commodity in its social expression and the commodity world must not be allowed to obfuscate a truth even more concrete than social existence. We know that there are many markets, existing on many levels (local, regional, national, worldwide): the market for (material) commodities, the labour market, the market for capital, the market in leases (whether for agricultural land or for building-land), the market for works, signs and symbols, and so forth. These different markets constitute a unity, namely the world market understood in the broadest possible sense. They are all connected, yet all retain their distinctness. They are superimposed upon one another without becoming confused with one another, spaces interpenetrating according to a law already evoked above – the law of composition of non-strategic spaces, which is analogous to the physical law of superimposition and composition of small movements. There are two markets whose conquest represents the ultimate triumph of the commodity and of money: the market in *land* (a precapitalist form of property) and the market in *works* (which, as 'non-products', long remained extra-capitalist).

The commodity, along with its implications – networks of exchange, currency, money – may be looked upon as a component of social (practical) existence, as a 'formant' of space. Considered in isolation, 'in itself', however, it does not have the capacity, even on a world scale, to exist socially (practically). And it is in this sense that it remains an *abstraction*, even though, *qua* 'thing', it is endowed with a terrible, almost deadly, power. The 'commodity world' cannot exist for itself. For it to exist, there must be *labour*. It is the result of a productive activity. Every commodity is a *product* (of a division of labour, of a technical means, of an expenditure of energy – in short, of a force of production). Under this aspect also the concept must be spatialized if it is to become concrete. The commodity needs its space too.

XIX

A curious fate has been reserved for Marxism, for Marxist thought, for the categories, concepts and theories referred to as 'Marxist'. No sooner is Marx pronounced dead than Marxism experiences a resurgence. On reinspection, the classical texts emerge as far richer than had been supposed: often confused, even contradictory, they yield new meanings.

CONTRADICTORY SPACE 343

Some, such as the *Grundrisse* not many years ago, or the *Manuscripts of 1844* around 1930, have successfully revived a seemingly exhausted line of thought.

Each period in the development of modern society – and perhaps even each country – has had 'its' Marxism. 'Mainstream' Marxism, meanwhile, has made many wrong turns, deviating variously into philosophism, historicism or economism. By contrast, a number of concepts whose 'theoretical status' originally occasioned much controversy – that of alienation, for example – have ultimately enjoyed brilliant careers as truly enlightening notions.

The scientific and technological changes of the modern world have now made a reconsideration of Marxist thought inevitable. The thesis presented here might be summarized as follows. Each of the concepts of Marxism may be taken up once more, and carried to a higher level, without any significant moment of the theory as a whole being lost. On the other hand, if they are considered in the setting of Marx's exposition, these concepts and their theoretical articulation no longer have an object. The renewal of Marx's concepts is best effected by taking full account of space.

XX

For Marx, *nature* belonged among the forces of production. Today a distinction is called for that Marx did not draw: namely, that between the *domination* and the *appropriation* of nature. Domination by technology tends towards non-appropriation – i.e. towards destruction. This is not to say that such destruction must inevitably occur, but merely that there is a conflict between domination and appropriation. This conflict takes place in space. There are dominated spaces and there are appropriated spaces.

That is not the whole story, however. Nature appears today as a source and as a resource: as the source of energies – indispensable, vast, but not unlimited. It appears, more clearly than in Marx's time, as a source of *use value*. The tendency toward the destruction of nature does not flow solely from a brutal technology: it is also precipitated by the economic wish to impose the traits and criteria of interchangeability upon places. The result is that places are deprived of their specificity – or even abolished. At an even more general level, it will be recalled that the products of labour become commodities in the process of exchange. This means that their material characteristics are placed in abeyance,

CONTRADICTORY SPACE

along with the needs to which they correspond. Only at the moment when the exchange cycle is completed, the moment just prior to consumption, do we observe the re-emergence of the product's materiality, and of the need it answers – the re-emergence, in other words, of whatever *natural* (material, immediate) aspects still attach to the products of industry and of social labour. As source and resource, nature *spatializes* the concepts associated with it, among them the concept of productive consumption, of which Marx made great use but which has since been abandoned. Productive consumption always eliminates a material or natural reality – an energy, a quantity of labour power, or an apparatus. It uses (up): it is a use and a use value. It also *produces*.

Let us for a moment consider the *machine*. Marx (along with Charles Babbage, whose work he drew upon) was one of the first to bring out the importance of the machine – a mechanism differing from a simple tool, as from a set of tools brought together in a workshop where both workers and tools are subject to a division of labour. A machine draws energy from a natural source (at first water, then steam, and later still electricity) and uses it to perform a sequence of productive tasks. The worker, instead of manipulating a tool, now serves a machine. The result is a radical but contradictory transformation of the productive process: whereas labour is ever more divided and segmented, the machine is organized into an ensemble that is ever vaster, ever more cohesive, ever more unified, and ever more productive.

Machines originated in the country, not in the towns. The windmill and the loom – prototypical machines – were rural inventions. The earliest machines were improved and perfected on the basis of the type of energy (hydraulic, for example) that they employed, and with respect to the kind of materials that they processed (wool, cotton, etc.). From their beginnings, however, machines had the potential to generate something completely new, namely the automation of the production process; hence also a new rationality, and – ultimately – an end to labour itself.

With the rise of industry, the extension of the market, the advent of the commodity world – in short, with the new importance taken on by the economic sphere, by capitalism – the old towns, finding themselves assailed from all sides, had to make room for something else. All their compartmentalizations – physical walls, guilds, local oligarchies, restricted markets and controlled territories – had to be dismantled. Meanwhile, the machine developed apace, in tandem with capital investment. The generally accepted periodization proposed for the decline of the towns (palaeo-technical, neo-technical, pre-modern and technological stages) does not, however, give us an exact or complete idea of what actually took place. Had the town of the precapitalist era been pre-machine, so to speak, in its essence, it would surely have disappeared completely, just like its various compartmentalizations; in actuality, so far from disappearing, it endured and – albeit transformed – expanded. The fact is that the town itself was already a vast machine, an automaton, capturing natural energies and consuming them productively. Over the centuries, the town's internal and external arrangements – the functions, forms and structures of its productive consumption – have metamor-phosed. History, in a rather simple sense of the word, has lent impetus to extensions and elaborations of these spatial arrangements as well as to the introduction of connections – of sewers, water supply, lighting, transportation, energy delivery (or flow), information channels, and so forth. Urban productivity has increased incessantly, thanks to the proximity of, and the links between, the needed elements: in this regard the town has come over time to resemble an industrial plant rather than a workshop, though it has never become identical to such a plant. It is clear, therefore, that the town at a very early stage displayed certain characteristics of the machine or automaton, and even that it was a pioneer in this respect. The town is indeed a machine, but it is also something more, and something better: a machine appropriated to a certain use – to the use of a social group. As a 'second nature', as a produced space, the town has also retained – and this even during its crisis – certain natural traits, notably the importance assigned to use.

In the context of the expansion of capitalism, there is a need to reconsider the concept of fixed (or constant) capital, for this concept can no longer be confined in its connotations to the equipment, premises and raw materials of a given enterprise. According to Marx, *fixed capital is the measure of social wealth*. Quite obviously, this category must cover investment in space, such as highways or airports, as well as all sorts of infrastructural elements. How could the radar networks used to designate airways not be classified as fixed capital? These are aids of a new kind which the roads, canals and railways of an earlier period prefigured in only the faintest way. Transportation grids exemplify productive consumption, in the first place because they serve to move people and things through the circuits of exchange, and secondly because they constitute a worldwide investment of knowledge in social reality.

Such an extension of the notion of fixed capital allows us similarly to extend the notion of variable capital. And this with surprising results, for, contrary to many predictions, the incorporation of knowledge and technology into production has mobilized a considerable labour force, including the mass of workers, not highly skilled, needed for such tasks as excavation, construction and maintenance. This development has indeed offered relief to a capitalism suffering from the fact that the high organic composition of capital in the most advanced industries tends to reduce necessary labour time (for the working class to reproduce itself as labour power), and therefore to threaten the minimal necessary level of available manpower; this situation further frees an enormous quantity of social time (whence the importance assumed by leisure and by 'cultural' and other parasitic forms), as well as making for colossal surplus production, excess (floating) capital, and so on. Not that the production of space is solely responsible for the survival of capitalism: it is in no sense independent of the extension of capitalism to pre-existing space. Rather, it is the overall situation – spatial practice in its entirety – that has saved capitalism from extinction.

In describing the *organic composition* of capitalism, Marx added another socio-economic average to those averages whose functions and structures he had already analysed: average social labour and average rate of profit. When it takes the average organic composition of capital into account, theory rejoins social space – i.e. it ceases to operate in an abstract space. This average is only meaningful in connection with a specific space: the space, say, that is occupied by a branch of industry or, even better, by an economic entity of great scale – a country or a continent. At the level of a single factory it has no utility at all, save as it permits comparison of the organic composition of capital in that particular concern with the average in society at large. This concept comes truly into its own when applied worldwide, for there is a global organic composition of capital which subsumes the averages obtaining in specific countries or nations. The notion becomes concrete by becoming spatial (and vice versa: it is spatialized as it achieves concreteness). Here we find ourselves at the junction of political economy and its critique, as defined by Marx, on the one hand, and, on the other, a political economy of space (including its critique, which is a critique of states and of state powers holding sway over national territories). A theory based on the idea of organic composition would allow us to grasp the relationship between entities unequal in this regard, and to identify the consequences of those inequalities. For these result in transfers of value and of surplus value, and therefore of capital, as well as in contradictions within the capital market which give rise to monetary problems.[14] In the so-called underdeveloped countries, plundered, exploited, 'protected' in a multitude of ways (economic, social, political, cultural, scientific),

[14] See my *Au-delà du structuralisme* (Paris: Anthropos, 1971), pp. 400ff.

the obstacles in the way of growth and development become increasingly daunting. Meanwhile, the advanced countries use the more backward as a source of labour and as a resource for use values (energies, raw materials, qualitatively superior spaces for leisure activities); the Spain of today exemplifies this perfectly.

Space in its entirety enters the modernized capitalist mode of production, there to be used for the generation of surplus value. The earth, underground resources, the air and light above the ground – all are part of the forces of production and part of the products of those forces. The urban fabric, with its multiple networks of communication and exchange, is likewise part of the means of production. The town and its multifarious establishments (its post offices and railway stations – as also its storehouses, transportation systems and varied services) are fixed capital. The division of labour affects the whole of space – not just the 'space of work', not just the factory floor. And the whole of space is an object of productive consumption, just like factory buildings and plant, machinery, raw materials, and labour power itself.

In parallel with these developments, the realization of surplus value has ceased to occur solely within an area close to the point of production, confined to a local banking-system. Instead, this process takes place through a worldwide banking-network as part of the abstract relations (the manipulation of the written word) between financial agencies and institutions. The realization of surplus value has, so to speak, been 'deterritorialized'. Urban space, though it has thus lost its former role in this process, nevertheless continues to ensure that links are properly maintained between the various *flows* involved: flows of energy and labour, of commodities and capital. The economy may be defined, practically speaking, as the linkage between flows and networks, a linkage guaranteed in a more or less rational way by institutions and programmed to work within the spatial framework where these institutions exercise operational influence. Each flow is of course defined by its origin, its endpoint, and its path. But, while it may thus be defined separately, a flow is only effective to the extent that it enters into relationship with others; the use of an energy flow, for instance, is meaningless without a corresponding flow of raw materials. The coordination of such flows occurs within a space. As for the distribution of surplus value, this too is achieved spatially – territorially – as a function of the forces in play (countries, economic sectors) and as a function of the strategies and know-how of managers.

XXI

According to Marx (and his thesis is not unpersuasive), tools, machinery, premises, raw materials – in short, constant capital or, in current (and hence capitalist) parlance, investments – represent dead labour. In this way past activity crystallizes, as it were, and becomes a precondition for new activity. Current work, including brain work, takes up the results of the past and revivifies them. Under capitalism, however, what is dead takes hold of what is alive. In other words, the means of production belong to the individual capitalist and to the bourgeoisie as a class, and are used by them to retain their hold over the working class, to make that class work. In this context as in any other, a new society can only be defined as a turning of the world upon its head. But how could what is alive lay hold of what is dead? The answer is: through the production of space, whereby living labour can produce something that is no longer a thing, nor simply a set of tools, nor simply a commodity. In space needs and desires can reappear as such, informing both the act of producing and its products. There still exist – and there may exist in the future – spaces for play, spaces for enjoyment, architectures of wisdom or pleasure. In and by means of space, the work may shine through the product, use value may gain the upper hand over exchange value: appropriation, turning the world upon its head, may (virtually) achieve dominion over domination, as the imaginary and the utopian incorporate (or are incorporated into) the real. What we have called a 'second nature' may replace the first, standing in for it or superimposing itself upon it without wreaking complete destruction. So long, however, as the dead retains its hold over the living, destruction and self-destruction will be imminent threats. Being equally dependent on this whole (which, in the sphere of knowledge, is called 'reduction'), capitalism and the bourgeoisie can achieve nothing but abstractions: money and commodities, capital itself, and hence *abstract labour* (labour in general, the production of exchange value in general) within abstract space – the location and source of abstractions.

XXII

In summary, then, and taking the categories one by one while bearing in mind their theoretical links, we may say of social space that it simultaneously

1 has a part to play among the *forces of production*, a role originally played by nature, which it has displaced and supplanted;

2 appears as a product of singular character, in that it is sometimes simply *consumed* (in such forms as travel, tourism, or leisure activities) as a vast commodity, and sometimes, in metropolitan areas, *productively consumed* (just as machines are, for example), as a productive apparatus of grand scale;

3 shows itself to be *politically instrumental* in that it facilitates the control of society, while at the same time being a *means of production* by virtue of the way it is developed (already towns and metropolitan areas are no longer just works and products but also means of production, supplying housing, maintaining the labour force, etc.);

4 underpins the reproduction of production relations and property relations (i.e. ownership of land, of space; hierarchical ordering of locations; organization of networks as a function of capitalism; class structures; practical requirements);

5 is equivalent, practically speaking, to a set of institutional and ideological superstructures that are not presented for what they are (and in this capacity social space comes complete with symbolisms and systems of meaning – sometimes an overload of meaning); alternatively, it assumes an outward appearance of neutrality, of insignificance, of semiological destitution, and of emptiness (or absence);

6 contains potentialities – of works and of reappropriation – existing to begin with in the artistic sphere but responding above all to the demands of a body 'transported' outside itself in space, a body which by putting up resistance inaugurates the project of a different space (either the space of a counter-culture, or a counter-space in the sense of an initially utopian alternative to actually existing 'real' space).

XXIII

Space is already being reorganized as a function of the search for increasingly scarce resources: energy, water, light, raw materials of plant and animal origin. This tends – at least potentially – to restore the importance of use as opposed to exchange, albeit in and through a vast struggle. The production of space goes hand in hand with a new emphasis on 'nature' as source of use values (the materiality of things). Long a consumer of part of the surpluses of the exchange system (of social surplus production), the production of space has thus risen to prominence in parallel with the restoration of use value, a restoration occurring on a vast scale and affecting politics through and through – without, however, resolving itself into political strategies. For Marx nature was *the only true wealth*, and he carefully distinguished such wealth from fortunes measurable in terms of exchange value, in terms of money or specie. This idea remains true and profound, provided always that secondary (produced) space is not arbitrarily divorced, as if it embodied some particular significance, from the primary space of nature, which is the raw material and the matrix of production. The supreme good is time–space; this is what ensures the survival of being, the energy that being contains and has at its disposal.

Capitalism does not consolidate itself solely by consolidating its hold on the land, or solely by incorporating history's precapitalist formations. It also makes use of all the available abstractions, all available forms, and even the juridical and legal fiction of ownership of things apparently inaccessible to privative appropriation (private property): nature, the earth, life energies, desires and needs. Spatial planning, which uses space as a multipurpose tool, has shown itself to be extremely effective. Such an instrumental use of space is surely implicit in the 'conservative modernization' that has been introduced with varying degrees of success in many countries.

The foregoing remarks on scarcity, on centrality, on the 'mobilization of immovables' have at most offered only the barest outline of a political economy of space. The reason why such a political economy will not be further elaborated upon here is that it is an offshoot of a more powerful theory: the theory of the production of space. Does our present inquiry, focused as it is on space and on the set of problems attending it, point in the direction of a form of knowledge susceptible of replacing 'classical' political economy and its abstract models of development? Undoubtedly, yes. But it has to be made clear at the outset that the 'positive' and 'negative' (i.e. critical) aspects of such a theory will converge. The 'commodity world', which is an abstraction, cannot be conceived of apart from the world market, which is defined territorially (in terms of flows and networks) and politically (in terms of centres and peripheries). The notion of flows – a strictly economic notion that has been mistakenly generalized by some philosophers – is still not clearly understood; along with their spatial interconnections, flows, by reasons of their complexity, still lie beyond the analytic and programming capacities of the computer. The fetishism of an abstract economics is being transformed into the fetishism of an abstract economic space. Space-become-commodity develops the traits of commodities in space to the maximum.

In order to elevate experience of this space to the level of theoretical knowledge it will be necessary to introduce new categories while simultaneously refining some old and familiar themes. The analysis of *space envelopes* may be expected to take markets (local, national, and hence also worldwide) as its starting-point, and eventually to link up with the theory of networks and flows. And the theory of *use value*, so badly obscured and misapprehended since Marx, will be restored and returned, complete with its complexities, to its former standing.

How and why is it that the advent of a world market, implying a degree of unity at the level of the planet, gives rise to a fractioning of space – to proliferating nation states, to regional differentiation and self-determination, as well as to multinational states and transnational corporations which, although they stem this strange tendency towards fission, also exploit it in order to reinforce their own autonomy? Towards what space and time will such interwoven contradictions lead us?

We know with some degree of accuracy where surplus value is formed under present conditions; we have but a scant notion, however, of where it is realized, or how it is divided up, because banking and financial networks scatter it far from the places (factories, countries) that generate it. Finally, space is also being recast: in response to the growth of air transport, particularly in its geopolitical dimensions; in response to various new industries (computers, leisure, the extraction of petroleum and other resources); and in response to the expanding role of the multinationals.

It is to be hoped that, at the conclusion of an analytical and critical study such as the one here envisioned, the relationship between time and space would no longer be one of abstract separation coupled with an equally abstract confusion between these two different yet closely connected terms.

6

From the Contradictions of Space to Differential Space

I

Let us now review the theory of contradictory space by considering the contradictions in abstract space one by one. Just as white light, though uniform in appearance, may be broken down into a spectrum, space likewise decomposes when subjected to analysis; in the case of space, however, the knowledge to be derived from analysis extends to the recognition of conflicts internal to what on the surface appears homogeneous and coherent – and presents itself and behaves as though it were.

The first contradiction on our list is that between *quantity* and *quality*. Abstract space is measurable. Not only is it quantifiable as geometrical space, but, as social space, it is subject to quantitative manipulations: statistics, programming, projections – all are operationally effective here. The dominant tendency, therefore, is towards the disappearance of the qualitative, towards its assimilation subsequent upon such brutal or seductive treatment.

And yet in the end the qualitative successfully resists resorption by the quantitative – just as use resists resorption by value. Instead, it re-emerges in space. A moment comes when people in general leave the *space of consumption*, which coincides with the historical locations of capital accumulation, with the space of production, and with the space that is produced; this is the space of the market, the space through which flows follow their paths, the space which the state controls – a space, therefore, that is strictly quantified. When people leave this space, they move towards the *consumption of space* (an unproductive form of consumption). This moment is the moment of departure – the moment of people's holidays, formerly a contingent but now a necessary moment. When this moment arrives, 'people' demand a qualitative space. The qualities they seek have names: sun, snow, sea. Whether these are natural or simulated matters little. Neither spectacle nor mere signs are acceptable. What is wanted is materiality and naturalness as such, rediscovered in their (apparent or real) immediacy. Ancient names, and eternal – and allegedly natural – qualities. Thus the quality and the use of space retrieve their ascendancy – but only up to a point. In empirical terms, what this means is that neocapitalism and neo-imperialism share hegemony over a subordinated space split into two kinds of regions: regions exploited for the purpose of and by means of *production* (of consumer goods), and regions exploited for the purpose of and by means of the *consumption of space*. Tourism and leisure become major areas of investment and profitability, adding their weight to the construction sector, to property speculation, to generalized urbanization (not to mention the integration into capitalism of agriculture, food production, etc.). No sooner does the Mediterranean coast become a space offering leisure activities to industrial Europe than industry arrives there; but nostalgia for towns dedicated to leisure, spread out in the sunshine, continues to haunt the urbanite of the super-industrialized regions. Thus the contradictions become more acute – and the urbanites continue to clamour for a certain 'quality of space'.

In the areas set aside for leisure, the body regains a certain right to use, a right which is half imaginary and half real, and which does not go beyond an illusory 'culture of the body', an imitation of natural life. Nevertheless, even a reinstatement of the body's rights that remains unfulfilled effectively calls for a corresponding restoration of desire and pleasure. The fact is that consumption satisfies needs, and that leisure and desire, even if they are united only in a representation of space (in which everyday life is put in brackets and temporarily replaced by a different, richer, simpler and more normal life), are indeed brought into conjunction; consequently, needs and desires come into opposition with each other. Specific needs have specific objects. Desire, on the other hand, has no particular object, except for a space where it has full play: a beach, a place of festivity, the space of the dream.

The dialectical link (meaning the contradiction within a unity) between need and desire thus generates fresh contradictions – notably that between liberation and repression. Even though it is true that these dialectical processes have the middle classes as their only foundation, their only vehicle, and that these middle classes offer models of consumption to the so-called lower classes, in this case such mimesis may, under the pressure of the contradiction in question, be an effective stimulus. A passionate struggle takes place in art, and within artists themselves, the essential character of which the protagonists fail to recognize (it is in fact class struggle!): the struggle between body and non-body, between signs of the body and signs of non-body.

Mental space – the space of reductions, of force and repression, of manipulation and co-optation, the destroyer of nature and of the body – is quite unable to neutralize the enemy within its gates. Far from it: it actually encourages that enemy, actually helps to revive it. Which takes us far further than the often-mentioned contradictions between aesthetics and rationalism.

II

The above-mentioned quantity–quality contradiction is not grounded in a (binary) opposition but rather in a three-point interaction, in a movement from the space of consumption to the consumption of space via leisure and within the space of leisure; in other words, from the quotidian to the non-quotidian through festival (whether feigned or not, simulated or 'authentic'), or again from labour to non-labour through a putting into brackets and into question (in a half-imaginary, half-real way) of toil.

Another (binary) opposition seems highly pertinent, even though it serves to freeze the dialectical process. This is the opposition between production and consumption, which, though transformed by ideology into a structure, cannot completely mask the dialectical conflict suggested by the term 'productive consumption'. The movement glimpsed here is that between consumption in the ordinary sense, consumption necessitating the reproduction of things, and the *space of production*, which is traversed, and hence used and consumed, by *flows*; it is also the movement between the space of production and the space of reproduction, controlled by state power and underpinned by the reproducibility of things in space, as of space itself, which is broken up in order to facilitate this. Under neocapitalism or corporate capitalism institutional space answers to the principles of repetition and reproducibility – principles effectively hidden by semblances of creativity. This bureaucratic space, however, is at loggerheads with its own determinants and its own effects: though occupied by, controlled by, and oriented towards the reproducible, it finds itself surrounded by the non-reproducible – by nature, by specific locations, by the local, the regional, the national, and even the worldwide.

III

Where then is the principal contradiction to be found? Between the capacity to conceive of and treat space on a global (or worldwide) scale on the one hand, and its fragmentation by a multiplicity of procedures or processes, all fragmentary themselves, on the other. Taking the broadest possible view, we find mathematics, logic and strategy, which make it possible to represent instrumental space, with its homogeneous – or better, homogenizing – character. This fetishized space, elevated to the rank of mental space by epistemology, implies and embodies an ideology – that of the primacy of abstract unity. Not that this makes fragmentation any less 'operational'. It is reinforced not only by administrative subdivision, not only by scientific and technical specialization, but also – indeed most of all – by the retail selling of space (in lots).

If one needed convincing of the existence of this contradiction, it would suffice to think, on the one hand, of the pulverizing tendency of fragmented space and, on the other, of a computer science that can dominate space in such a fashion that a computer – hooked up if need be to other image- and document-reproducing equipment – can assemble an indeterminate mass of information relating to a given physical or social space and process it at a single location, virtually at a single point.

To present the homogeneous/fractured character of space as a binary relationship (as a simple contrast or confrontation) is to betray its truly dual nature. It is impossible to overemphasize either the mutual inherence or the contradictoriness of these two aspects of space. Under its homogeneous aspect, space abolishes distinctions and differences, among them that between inside and outside, which tends to be reduced to the undifferentiated state of the visible–readable realm. Simultaneously, this same space is fragmented and fractured, in accordance with the demands of the division of labour and of the division of needs and functions, until a threshold of tolerability is reached or even passed (in terms of exiguity of volumes, absence of links, and so on). The ways in which space is thus carved up are reminiscent of the ways in which the body is cut into pieces in images (especially the female body, which is not only cut up but also deemed to be 'without organs'!).

It is not, therefore, as though one had global (or conceived) space to one side and fragmented (or directly experienced) space to the other –

rather as one might have an intact glass here and a broken glass or mirror over there. For space 'is' whole and broken, global and fractured, at one and the same time. Just as it is at once conceived, perceived, and directly lived.

The contradiction between the global and the subdivided subsumes the contradiction between centre and periphery; the second defines the internal *movement* of the first. Effective globalism implies an established centrality. The concentration of 'everything' that exists in space subordinates all spatial elements and moments to the power that controls the centre. Compactness and density are a 'property' of centres; radiating out from centres, each space, each spatial interval, is a vector of constraints and a bearer of norms and 'values'.

IV

The opposition between exchange value and use value, though it begins as a mere contrast or non-dialectical antithesis, eventually assumes a dialectical character. Attempts to show that exchange absorbs use are really just an incomplete way of replacing a static opposition by a dynamic one. The fact is that use re-emerges sharply at odds with exchange in space, for it implies not 'property' but 'appropriation'. Appropriation itself implies time (or times), rhythm (or rhythms), symbols, and a practice. The more space is functionalized – the more completely it falls under the sway of those 'agents' that have manipulated it so as to render it unifunctional – the less susceptible it becomes to appropriation. Why? Because in this way it is removed from the sphere of *lived* time, from the time of its 'users', which is a diverse and complex time. All the same, what is it that a buyer acquires when he purchases a space? The answer is time.

Thus everyday life cannot be understood without understanding the contradiction between use and exchange (use value and exchange value). It is the political use of space, however, that does the most to reinstate use value; it does this in terms of resources, spatial situations, and strategies.

Is a system of knowledge – a science – of the use of space likely to evolve out of such considerations? Perhaps – but it would have to involve an analysis of rhythms, and an effective critique of representative and normative spaces. Might such a knowledge legitimately be given a name – that of 'spatial analysis', for example? That would be reasonable enough – but one is loth indeed to add yet another specialization to what is already a very long list.

V

The principal contradiction identified above corresponds to the contradiction discerned by Marx, at the very beginning of his analysis of capitalism, between the forces of production and the social relations of production (and of property). Though blunted now at the level of the production of things (in space), at a higher level – that of the production of space – this contradiction is becoming ever more acute. Technically, scientifically, formerly undreamt-of possibilities have opened up. A 'society' other than ours could undoubtedly invent, create or 'produce' new forms of space on this basis. Existing property and production relations erase these prospects, however; in other words, they shatter conceptions of space that tend to form in dreams, in imaginings, in utopias or in science fiction. Practically speaking, the possibilities are always systematically reduced to the triteness of what already exists – to houses in the suburbs or high-rises (individual boxes sprinkled with a few illusions *versus* hundreds of boxes stacked one on top of another).

These are very fundamental points, but the fact that they are so fundamental cannot be too often reiterated, because Marx's thinking tends to be weakened and diverted by all kinds of political attitudes. There are those who want a 'socialism' in the industrialized countries that would simply continue along the path of growth and accumulation – the path, in other words, of the production of things in space. Others would smash every single mechanism of the current mode of production for the sake of an 'extremist' revolutionary activism or 'leftism'. The appeal of the first group is to 'objectivity', that of the second to 'voluntarism' (subjectivity).

By furthering the development of the forces of production, the bourgeoisie played a revolutionary role. It was Marx's view – and to overlook this point is to misunderstand his whole thinking – that the advent of large-scale industry, along with scientific and technological advance, had shaken the world to its foundations. The productive forces have since taken another great leap – from the production of things in space to the production of space. Revolutionary activity ought, among other things, to follow this *qualitative leap* – which also constitutes a *leap into the qualitative* – to its ultimate consequences. This means putting the process of purely quantitative growth into question – not so much in order to arrest it as to identify its potential. The *conscious* production of space has 'almost' been achieved. But the threshold cannot be crossed so long as that new mode of production is pre-empted by the selling of space parcel by parcel, by a mere travesty of a new space.

VI

The violence that is inherent in space enters into conflict with knowledge, which is equally inherent in that space. Power – which is to say violence – divides, then keeps what it has divided in a state of separation; inversely, it reunites – yet keeps whatever it wants in a state of confusion. Thus knowledge reposes on the effects of power and treats them as 'real'; in other words, it endorses them exactly as they are. Nowhere is the confrontation between knowledge and power, between understanding and violence, more direct than it is in connection with intact space and space broken up. In the dominated sphere, constraints and violence are encountered at every turn: they are everywhere. As for power, it too is omnipresent.

Dominated space realizes military and political (strategic) 'models' in the field. There is more to it than this, however, for thanks to the operation of power practical space is the bearer of norms and constraints. It does not merely express power – it proceeds to repress in the name of power (and sometimes even in the name of nothing). As a body of constraints, stipulations and rules to be followed, social space acquires a normative and repressive efficacy – linked instrumentally to its objectality – that makes the efficacy of mere ideologies and representations pale in comparison. It is an essentially deceptive space, readily occupiable by pretences such as those of civic peace, consensus, or the reign of non-violence. Not that this space – dominating as well as dominated – is not inhabited as well by the agencies of the Law, of the Father, or of Genitality. Logic and logistics conceal its latent violence, which to be effective does not even have to show its hand.

Spatial practice regulates life – it does not create it. Space has no power 'in itself', nor does space as such determine spatial contradictions. These are contradictions of society – contradictions between one thing and another within society, as for example between the forces and relations of production – that simply emerge in space, at the level of space, and so engender the contradictions of space.

VII

The contradictions identified in the foregoing discussion have been presented in a conceptual and theoretical manner which may suggest that they are abstractions unrelated to plain facts, to the empirical realm. Nothing could be further from the truth. These formulations do correspond to facts – indeed, they are the distillation of an indeterminate number of experiences. The contradictions in question are readily verifiable: even the most fanatical of positivists could detect them with the naked eye. It is just that empiricism refuses to call them 'contradictions', preferring to speak only of inconsistencies or of 'dysfunctions'; the empiricist jibs at giving theoretical form to his observations, and confines himself to arranging his data into sets of logically connected facts.

Owners of private cars have a space at their disposition that costs them very little personally, although society collectively pays a very high price for its maintenance. This arrangement causes the number of cars (and car-owners) to increase, which suits the car-manufacturers just fine, and strengthens their hand in their constant efforts to have this space expanded. The productive consumption of space – which is productive, above all, of surplus value – receives much subsidization and enormous loans from government. This is just another way of barring all escape from a cruel spiral which optimists like to refer to as a 'regulatory system'; such 'systems' unquestionably play a 'self-regulating' role for society – provided that society is prepared to accept the side-effects. Enough said. As for 'green areas' – trees, squares that are anything more than intersections, town parks – these obviously give pleasure to the community as a whole, but who pays for this pleasure ? How and from whom can fees be collected ? Since such spaces serve no one in particular (though they do bring enjoyment to people in general), there is a tendency for them to die out. Non-productive consumption attracts no investment because all it produces is pleasure. Colossal sums, meanwhile, are invested in the most unproductive consumption imaginable: namely, the consumption of arms of all kinds, including rockets and missiles.

There are two ways in which urban space tends to be sliced up, degraded and eventually destroyed by this contradictory process: the proliferation of fast roads and of places to park and garage cars, and their corollary, a reduction of tree-lined streets, green spaces, and parks and gardens. The contradiction lies, then, in the clash between a consumption of space which produces surplus value and one which produces only enjoyment – and is therefore 'unproductive'. It is a clash, in other words, between capitalist 'utilizers' and community 'users'. (This account owes much to Alfred Sauvy – one of those who appears to see no contradictions here.[1])

VIII

Cases are legion where the empirical approach to a given process refuses to carry its description to a conceptual level where a dialectical (conflictual) dynamic is likely to emerge. For example, countries in the throes of rapid development blithely destroy historic spaces – houses, palaces, military or civil structures. If advantage or profit is to be found in it, then the old is swept away. Later, however, perhaps towards the end of the period of accelerated growth, these same countries are liable to discover how such spaces may be pressed into the service of cultural consumption, of 'culture itself', and of the tourism and the leisure industries with their almost limitless prospects. When this happens, everything that they had so merrily demolished during the *belle époque* is reconstituted at great expense. Where destruction has not been complete, 'renovation' becomes the order of the day, or imitation, or replication, or neo-this or neo-that. In any case, what had been annihilated in the earlier frenzy of growth now becomes an object of adoration. And former objects of utility now pass for rare and precious works of art.

Let us for a moment consider the space of architecture and of architects, without attaching undue importance to what is said about this space. It is easy to imagine that the architect has before him a slice or piece of space cut from larger wholes, that he takes this portion of space as a 'given' and works on it according to his tastes, technical skills, ideas and preferences. In short, he receives his assignment and deals with it in complete freedom.

That is not what actually happens, however. The section of space assigned to the architect – perhaps by 'developers', perhaps by government agencies – is affected by calculations that he may have some intimation of but with which he is certainly not well acquainted. This space has nothing innocent about it: it answers to particular tactics and strategies; it is, quite simply, the space of the dominant mode of production, and hence the space of capitalism, governed by the bourgeoisie. It consists of 'lots' and is organized in a repressive manner as a function of the important features of the locality.

[1] Alfred Sauvy, *Croissance zéro* (Paris: Calmann-Lévy, 1973).

As for the eye of the architect, it is no more innocent than the lot he is given to build on or the blank sheet of paper on which he makes his first sketch. His 'subjective' space is freighted with all-too-objective meanings. It is a visual space, a space reduced to blueprints, to mere images – to that 'world of the image' which is the enemy of the imagination. These reductions are accentuated and justified by the rule of linear perspective. Such sterilizing tendencies were denounced long ago by Gromort, who demonstrated how they served to fetishize the façade – a volume made up of planes and lent spurious depth by means of decorative motifs.[2] The tendency to make reductions of this kind – reductions to parcels, to images, to façades that are made to be seen and to be seen from (thus reinforcing 'pure' visual space) – is a tendency that degrades space. The façade (to see and to be seen) was always a measure of social standing and prestige. A prison with a façade – which was also the prison of the family – became the epitome and modular form of bourgeoisified space.

It may thus be said of architectural discourse that it too often imitates or caricatures the discourse of power, and that it suffers from the delusion that 'objective' knowledge of 'reality' can be attained by means of graphic representations. This discourse no longer has any frame of reference or horizon. It only too easily becomes – as in the case of Le Corbusier – a moral discourse on straight lines, on right angles and straightness in general, combining a figurative appeal to nature (water, air, sunshine) with the worst kind of abstraction (plane geometry, modules, etc.).

Within the spatial practice of modern society, the architect ensconces himself in his own space. He has a *representation of this space*, one which is bound to graphic elements – to sheets of paper, plans, elevations, sections, perspective views of façades, modules, and so on. This *conceived* space is thought by those who make use of it to be *true*, despite the fact – or perhaps because of the fact – that it is geometrical: because it is a medium for objects, an object itself, and a locus of the objectification of plans. Its distant ancestor is the linear perspective developed as early as the Renaissance: a fixed observer, an immobile perceptual field, a stable visual world. The chief criterion of the architectural plan, which is 'unconsciously' determined by this perceptual field, is whether or not it is realizable: the plan is projected onto the field of architectural thought, there to be accepted or rejected. A vast number

<hr>

[2] Cf. Georges Gromort, *Architecture et sculpture en France*, a volume in his *Histoire générale de l'art française de la Révolution à nos jours* (Paris: Librairie de France, 1923–5).

of representations (some would call them 'ideological' representations, but why bother with a term now so devalued by misuse?) take this route; any plan, to merit consideration, must be quantifiable, profitable, communicable and 'realistic'. Set aside or downplayed from the outset are all questions relating to what is too close or too distant, relating to the surroundings or 'environment', and relating to the relationship between private and public. On the other hand, subdivisions (lots) and specializations (functional localizations) are quite admissible to this practically defined sphere. Much more than this, in fact: though the sphere in question seems passive with respect to operations of this kind, its very passive acceptance of them ensures their operational impact. The division of labour, the division of needs and the division of objects (things), all localized, all pushed to the point of maximum separation of functions, people and things, are perfectly at home in this spatial field, no matter that it appears to be neutral and objective, no matter that it is apparently the repository of knowledge, *sans peur et sans reproche*.

Let us now turn our attention to the space of those who are referred to by means of such clumsy and pejorative labels as 'users' and 'inhabitants'. No well-defined terms with clear connotations have been found to designate these groups. Their marginalization by spatial practice thus extends even to language. The word 'user' (*usager*), for example, has something vague – and vaguely suspect – about it. 'User of what?' one tends to wonder. Clothes and cars are used (and wear out), just as houses are. But what is use value when set alongside exchange and its corollaries? As for 'inhabitants', the word designates everyone – and no one. The fact is that the most basic demands of 'users' (suggesting 'underprivileged') and 'inhabitants' (suggesting 'marginal') find *expression* only with great difficulty, whereas the *signs* of their situation are constantly increasing and often stare us in the face.

The user's space is *lived* – not represented (or conceived). When compared with the abstract space of the experts (architects, urbanists, planners), the space of the everyday activities of users is a concrete one, which is to say, subjective. As a space of 'subjects' rather than of calculations, as a representational space, it has an origin, and that origin is childhood, with its hardships, its achievements, and its lacks. Lived space bears the stamp of the conflict between an inevitable, if long and difficult, maturation process and a failure to mature that leaves particular original resources and reserves untouched. It is in this space that the 'private' realm asserts itself, albeit more or less vigorously, and always in a conflictual way, against the public one.

It is possible, nevertheless, if only in a mediational or transitional way, to form a mental picture of a primacy of concrete spaces – of semi-public, semi-private spaces, of meeting-places, pathways and passageways. This would mean the diversification of space, while the (relative) importance attached to functional distinctions would disappear. Appropriated places would be *fixed*, *semi-fixed*, *movable* or *vacant*. We should not forget that among the contradictions here a not unimportant part is played by the contradiction between the ephemeral and the stable (or, to use Heidegger's philosophical terminology, between Dwelling and Wandering). Although work – including a portion of household production (food preparation, etc.) – demands a fixed location, this is not true of sleep, nor of play, and in this respect the West might do well to take lessons from the East, with its great open spaces, and its low and easily movable furniture.

In the West the reign of the façade over space is certainly not over. The furniture, which is almost as heavy as the buildings themselves, continues to have façades; mirrored wardrobes, sideboards and chests still face out onto the sphere of private life, and so help dominate it. Any mobilization of 'private' life would be accompanied by a restoration of the body, and the contradictions of space would have to be brought out into the open. Inasmuch as the resulting space would be inhabited by *subjects*, it might legitimately be deemed 'situational' or 'relational' – but these definitions or determinants would refer to sociological content rather than to any intrinsic properties of space as such.

The restoration of the body means, first and foremost, the restoration of the sensory–sensual – of speech, of the voice, of smell, of hearing. In short, of the non-visual. And of the sexual – though not in the sense of sex considered in isolation, but rather in the sense of a sexual energy directed towards a specific discharge and flowing according to specific rhythms.

But these are no more than suggestions, or pointers.

IX

One of the most glaring paradoxes about abstract space is the fact that it can *be* at once the whole set of locations where contradictions are generated, the medium in which those contradictions evolve and which they tear apart, and, lastly, the means whereby they are smothered and replaced by an appearance of consistency. This gives space a function, practically speaking (i.e. within spatial practice), which was formerly filled by ideology, and which is still to some extent felt to require an ideology.

As long ago as 1961, Jane Jacobs examined the failures of 'city-planning and rebuilding' in the United States. In particular, she showed how the destruction of streets and neighbourhoods led to the disappearance of many acquired characteristics of city life – or, rather, characteristics assumed to have been permanently acquired: security, social contact, facility of child-rearing, diversity of relationships, and so on.[3] Jacobs did not go so far as flatly to incriminate neocapitalism, or as to isolate the contradictions immanent to the space produced by capitalism (abstract space). But she did very forcefully demonstrate how destructive this space can be, and specifically how urban space, using the very means apparently intended to create or re-create it, effects its own self-destruction.

Faced with the city's complexity and unintelligibility (whether real or merely apparent is of no consequence here), some in the United States were inspired to take the practical and theoretical initiative of creating specialists responsible for disentangling the web of problems and explaining them, though without necessarily proposing solutions. Such was the initial agenda of so-called 'advocacy planning', as opposed to the 'city-planning' of the authorities. The notion was that in this way 'users' and 'inhabitants', as a group, would secure the services of someone competent, capable of speaking and communicating – in short, an advocate – who would negotiate for them with political or financial entities.

The failure of this approach, as documented by Goodman, is rich in meaning.[4] When the interested parties – the 'users' – do not speak up, who can speak in their name or in their place? Certainly not some expert, some specialist of space or of spokesmanship; there is no such specialization, because no one has a right to speak for those directly concerned here. The entitlement to do so, the concepts to do so, the language to do so are simply lacking. How would the discourse of such an expert differ from that of the architects, 'developers' or politicians?

---

[3] Jane Jacobs, *The Death and Life of the Great American Cities* (New York: Random House, 1961).

[4] See Robert Goodman, *After the Planners* (Harmondsworth, Middx: Penguin, 1972), pp. 57 ff. Incidentally, it is worth noting Goodman's pertinent criticisms of Robert Venturi's theses, as set forth in *Complexity and Contradiction in Architecture* (New York: Museum of Modern Art/Doubleday, 1966): as Goodman effectively demonstrates (pp. 164ff.), Venturi's pseudo-dialecticalization of architectural space confuses the mildest of formal contrasts with true spatial contradictions.

The fact is that to accept such a role or function is to espouse the fetishization of communication – the replacement of use by exchange. The silence of the 'users' is indeed a problem – and it is the *entire* problem. The expert either works for himself alone or else he serves the interests of bureaucratic, financial or political forces. If ever he were truly to confront these forces in the name of the interested parties, his fate would be sealed.

One of the deepest conflicts immanent to space is that space as actually 'experienced' prohibits the expression of conflicts. For conflicts to be voiced, they must first be perceived, and this without subscribing to representations of space as generally conceived. A *theory* is therefore called for, one which would transcend representational space on the one hand and representations of space on the other, and which would be able properly to articulate contradictions (and in the first place the contradiction between these two aspects of representation). Socio-political contradictions are realized spatially. The contradictions of space thus make the contradictions of social relations operative. In other words, spatial contradictions 'express' conflicts between socio-political interests and forces; it is only *in* space that such conflicts come effectively into play, and in so doing they become contradictions *of* space.

X

The aforementioned contradiction between the *global* (the capacity to conceive of and deal with space on a wide scale, even on a world scale, as in the cases of computer science and the geopolitics of air transport) and the *fragmentary* (the subdivision of space for purposes of buying and selling) intensifies at the strategic level. In strategic spaces resources are always localized. Estimates are made in terms of units, whether units of production (firms) or units of consumption (households). Objectives and 'targets', by contrast, are always globalizing in tendency, and effectively worldwide in the case of the chief states and chief transnational corporations. Dispersion and subdivision, often carried to the point of complete segregation, are controlled and dominated by strategic aims, by wills-to-power of the highest order in terms both of the quantity of means employed and of the quality of goals pursued. Everything that is dispersed and fragmented retains its unity, however, within the homogeneity of power's space; this is a space which naturally takes account of the connections and links between those elements that it keeps, paradoxically, united yet disunited, joined yet detached from one another, at once torn apart and squeezed together.

It would be mistaken in this connection to picture a hierarchical scale stretching between two poles, with the unified will of political power at one extreme and the actual dispersion of differentiated elements at the other. For everything (the 'whole') weighs down on the lower or 'micro' level, on the local and the localizable – in short, on the sphere of everyday life. Everything (the 'whole') also *depends* on this level: exploitation and domination, protection and – inseparably – repression. The basis and foundation of the 'whole' is dissociation and separation, maintained as such by the will above; such dissociation and separation are inevitable in that they are the outcome of a history, of the history of accumulation, but they are fatal as soon as they are maintained in this way, because they keep the moments and elements of social practice away from one another. A spatial practice destroys social practice; social practice destroys itself by means of spatial practice.

At the strategic level, forces in contention occupy space and generate pressures, actions, events. The law of interpenetration of small movements does not obtain at this level.

This does not mean that the 'micro' level is any less significant. Though it may not supply the theatre of conflict or the sphere in which contending forces are deployed, it does contain both the resources needed and the stakes at issue. The goal of any strategy is still, as it always has been, the occupation of a space by the varied means of politics and of war.

A variety of conceptual grids may be developed to help decipher complex spaces. The broadest of these distinguishes between types of oppositions and contrasts in space: *isotopias*, or analogous spaces; *heterotopias*, or mutually repellent spaces; and *utopias*, or spaces occupied by the symbolic and the imaginary – by 'idealities' such as nature, absolute knowledge or absolute power. Though this classification is still rather crude, it does bring out a paradox – a contradiction not hitherto noticed: namely, the fact that the most effectively appropriated spaces are those occupied by symbols. Gardens and parks, which symbolize an absolute nature, are an example; or religious buildings, which symbolize power and wisdom – and hence the Absolute pure and simple.

A suppler and more concrete grid classifies places according to their attributions – private, public or mediational (passageways or pathways) – or, in other words, according to their use and their users.

A third type of grid would operate at the strategic level, and reveal the measure of order that exists beneath the chaotic surface of space:

the articulations between the market in space and the spaces of the market, between spatial planning and development and the productive forces occupying space, and between political projects and the obstacles they run into – that is to say, those forces that run counter to a given strategy and occasionally succeed in establishing a 'counter-space' within a particular space.

Why, then, should we not simply pursue this line of enquiry further, in the hope of arriving at a completely satisfactory grid? Two points are worth making by way of response to this question. First, there is no good reason for limiting the number of possible grids, or to deem one preferable in some absolute way to another. Secondly, the concept of the grid, like the concepts of the model and the code, is itself not above reproach. As tools of formal knowledge, all such concepts have a precise aim, which is to eliminate contradictions, to demonstrate a coherence, and to reduce the dialectical to the logical. Such an intent is immanent to a knowledge that aspires to be 'pure' and 'absolute' while remaining ignorant of its own *raison d'être* – which is to reduce reality in the interests of power.

XI

It is possible, on the basis of a particular knowledge – that of the production of space – to entertain the idea of a science of social space (a space both urban and rural, but predominantly rural).

What term would be most appropriate here? *Connaissance*? 'Science'? Or *savoir*?[5] I have used the term *savoir* above with an unfavourable connotation. This was not to suggest, however, that the term designates a knowledge now obsolete, relegated to history – gathering dust on the shelf alongside other outdated contributions. This use of the term is a little suspect, in any case, because there is an element of the arbitrary about it: anyone, after all, is free to decide what to file under outdated knowledge or received wisdom.

The negative connotation that I feel we *are* justified in attaching to *savoir* is the suggestion that such knowledge colludes to some degree with power, that it is bound up, whether crudely or more subtly, with political practice – and hence with the multifarious representations and rhetoric of ideology.

[5] [On the distinction between *connaissance* and *savoir*, see above, p. 10, note 16. – *Translator*.]

As for *connaissance*, knowledge in this sense at all times embodies both a self-criticism which relativizes it, and a critique of what exists, which naturally becomes more acute when political stakes (or politics at stake) and strategies are under scrutiny. *Connaissance* seeks to grasp the global. In this respect it is linked to philosophy, of which it is an extension, even though it makes common cause with social practice by virtue of its attachment to a specific, salient concept – the concept of production. We have now in effect defined *metaphilosophy*, which is grounded in philosophy but which opens philosophy up to the 'real' and the possible.

When the critical moment occurs, *connaissance* generates the *concrete universal*. The concepts necessary (among them that of *production*) are not sufficient unto themselves: they lead back to the practice that they hold up to view. When applied to such concepts, certain questions lose their validity: questions concerning either a specified *subject* (who is thinking? who is speaking? where is that person speaking from?) or an identifiable *object* (what space does it occupy? upon what site is it located?). It is not just by virtue of their content, but, just as importantly, by virtue of the theoretical *form* just described – that is, the link with lived experience, with practice, and with a radical critique – that these concepts are exempted from such questions.

The word 'science' continues to imply a detailed process of working-out and construction confined to a specified field and calling for strict adherence to predetermined methods. The result is scepticism towards all specialist dogmas, and notably towards the methods – the operational (or supposedly operational) concepts – used by particular specializations.

The science of space should therefore be viewed as a *science of use*, whereas the specialized sciences known as social sciences (including, for example, political economy, sociology, semiology and computer science) partake of exchange, and aspire to be sciences of exchange – that is, of communication and of the communicable. In this capacity, the science of space would concern itself with the material, sensory and natural realms, though with regard to nature its emphasis would be on what we have been calling a 'second nature': the city, urban life, and social energetics – considerations ignored by the simplistic nature-centred approaches with their ambiguous concepts such as the 'environment'. The tendency of such a science would run counter to the dominant (and dominating) tendency in another respect also: it would accord *appropriation* a special practical and theoretical status. *For* appropriation and for use, therefore – and *against* exchange and domination.

Co-optation, as already mentioned, should be looked upon as a practice intermediate between domination and appropriation, between exchange and use. To oppose it to production or to treat it as exclusive of production is to mistake its character. Properly understood, co-optation can lead to the production of a space. There are illustrious precedents for this. Consider, for example, Christianity's co-optation of the Roman basilica. Originally intended for a secular, civic and social function, as a place of encounter and of 'commerce' in the broadest sense of the word, this building was given a religious and political role; its transformation went hand in hand with its consecration, with its subordination to cryptal constraints and requirements. The adjoining areas of crypt and tombs slowly but surely gave it the form of the cross; the day would come when this form would give birth, in the light of the Word (the Logos resurrected), to the soaring upsurge of the Middle Ages. As for the structure itself, it underwent modifications that had no logical connection with those suffered by the function and the form. The invention of intersecting ribs was a turning-point, as everyone knows.

The form corresponds approximately to the moment of communication – hence to the realm of the *perceived*. The function is carried out, effectively or not, and corresponds to the *directly experienced* in a representational space. The structure is *conceived*, and implies a representation of space. The *whole* is located within a spatial practice. It would be inexact and reductionistic to define use solely in terms of function, as functionalism recommends. Form – the communicable, communication – is also an aspect of use, as is structure, which is always the structure of an object that we *make use of* and *use up*. Each time one of these categories is employed independently of the others, hence reductively, it serves some homogenizing strategy. Formalism puts all the emphasis on form, and thus on communicability and exchange. Functionalism stresses function to the point where, because each function has a specially assigned place within dominated space, the very possibility of multifunctionality is eliminated. And structuralism takes into account only structures, treating them as objects which are in the last analysis technological in character. The fact is, however, that *use* corresponds to a unity and collaboration between the very factors that such dogmatisms insist on dissociating.

Needless to say, no plan could conceivably maintain a perfect balance between these diverse moments or 'formants' of space. A given plan must of necessity highlight either function, or form, or structure. But the way that one or another of these moments or formants is brought into play to begin with does not imply the demise of the other two. On the contrary, considering that what appears first will later become mere appearance, the prospect is that the other moments will consequently become more 'real' in comparison. Herein, it would seem, lies the genius of art in the classical sense – an art which, though outdated as such, needs to be resumed and extended much as thought needs to resume and extend philosophy.

The initial analysis of a musical work has three moments or aspects: rhythm, melody, harmony. This tridimensionality ensures the possibility of endless production, even though the possibilities of each moment considered in isolation, or of each binary opposition, are finite. Works constructed around just one of these moments (for instance, around melody or percussion alone) are more readily communicable, but at the same time they are monotonous and unattractive. The great classical music maintained unity between the three moments: each player or work concentrates upon and accentuates one or another, only to bring the others into prominence sooner or later. This variation of effects is also to be found within a single composition, within a single sonata or symphony. The role of emphasis here, so far from being a homogenizing one, so far from serving to overwhelm all other possible aspects of the work, is simply to point up qualities and underscore differences. The result is movement instead of stagnation, as one moment always refers to the next, which it prepares for and informs. The simultaneous presence of materials (piano, strings, brass, etc.) and *matériel* (scales, modes, tones) opens up possibilities and amplifies differences, thus reversing the reductionist tendency, which is itself associated with the ideology of exchange and communication.

XII

Abstract space, which is the tool of domination, asphyxiates whatever is conceived within it and then strives to emerge. Though it is not a defining characteristic of abstract space, there is nevertheless nothing secondary or fortuitous about this proclivity. This space is a lethal one which destroys the historical conditions that gave rise to it, its own (internal) differences, and any such differences that show signs of developing, in order to impose an abstract homogeneity. The negativity that Hegelianism attributed to historical temporality alone is in fact characteristic of abstract space, and this in a double sense, or, rather, operating with redoubled force: it stands opposed to all difference whether actual or potential. Why has this lethal power been unleashed?

Is it related to the nuclear threat? To freewheeling technology? To rampant population growth? To the kind of development known to be undesirable yet desired by power? To ecological problems? Or, more obscurely, to the operation of abyssal forces or of self-destructive tendencies in the species or in the planet, to the operation of a death instinct?

But, then, how important is it that a cause or reason be found here? Granted, an answer would gratify the philosophers' age-old speculative instinct; the last remaining members of that species could focus their attention and interest on an ontologically privileged and illuminating area, and contemplate a supreme Cause or Reason – no longer for Being, but rather for Non-Being.

Would it not make more sense, however, instead of striving to discover the metaphysical *source* of the death sentence passed on itself by the 'world' – i.e. the Judaeo-Christian, Graeco-Roman world, 'overdetermined' by capitalism – to examine the *instrument* used? For neither the atomic bomb, nor the squandering of resources, nor demographic, economic or production-based growth – indeed, no single aspect of the threat – can define its instrument, which is space. All the above-mentioned causes or reasons converge in space. Space harbours them, receives and transforms them into efficacious (operational) agents. Space and space alone – instrumental space – with its specific effects and its strategic aims: the removal of every obstacle in the way of the total elimination of what is *different*.

At this level it becomes apparent just how necessary – and at the same time how inadequate – the theory of alienation is. The limitations of the concept of alienation lie in this: it is so true that it is completely uncontested. The state of affairs we have been describing and analysing validates the theory of alienation to the full – but it also makes it seem utterly trivial. Considering the weight of the threat and the level of terror hanging over us, pillorying either alienation in general or particular varieties of alienation appears pointless in the extreme. The 'status' of the concept, or of liberal (humanist) ideology, is simply not the real issue.

XIII

With regard to the difficult and still incomplete theory of difference, there is no need to do any more here than touch on a few points.

This theory covers the whole realm of knowledge (*connaissance*) and of thinking about knowledge. Its range extends from the *conceived* to the *directly lived*, which is to say from the concept without life to life without concepts. And from logic to the dialectic, linking the two and placing itself at their point of articulation. On the one hand it overlaps with the theory of coherence, and hence of identity (ultimately, tautological identity); on the other hand it overlaps with the theory of contradictions (ultimately, antagonistic contradictions).

Two inseparable distinctions have to be drawn in this connection: that between *minimal* and *maximal* differences, and that between *induced* and *produced* differences. The first of these distinctions belongs to logic, the second to the theory of dialectical movement. Within logico-mathematical sets, the difference between one and one (the first one and the second one) is strictly *minimal*: the second differs from the first only by virtue of the iteration that gives rise to it. By contrast, the difference between finite cardinal and ordinal numbers on the one hand and transfinite cardinal and ordinal numbers on the other is a *maximal* difference. An *induced* difference remains within a set or system generated according to a particular law. It is in fact constitutive of that set or system: for example, in numerical sets, the difference between the successive elements generated by iteration or recurrence. Similarly: the diversity between villas in a suburb filled with villas; or between different 'community facilities'; or, again, variations within a particular fashion in dress, as stipulated by that fashion itself. By contrast, a *produced* difference presupposes the shattering of a system; it is born of an explosion; it emerges from the chasm opened up when a closed universe ruptures. To a large extent, the theory of the production of differences is based on the theory of maximal differences: a given set gives rise, beyond its own boundaries, to another, completely different set. Thus the set of whole numbers generates first the set of fractions, then the sets of 'incommensurables' and 'transcendentals', and ultimately the set of transfinite numbers. As soon as logico-mathematical categories apply, production and induction in these senses come into play. Repetitions generate differences, but not all differences are equivalent. The qualitative arises from the quantitative – and vice versa.

Under the reign of historical time, differences induced within a given mode of production coexist at first with produced differences promoting the demise of that mode. A difference of the latter kind is not only produced – it is also productive. Thus those differences within medieval society that foreshadowed a new mode of production had themselves accumulated during the general process of accumulation; at last they precipitated a tumultuous transition and eventually shattered existing societies and their mode of production. The classical theory of dialectical development refers to this moment as a qualitative leap long prepared for by gradual (quantitative) changes.[6] This traditional view, however, has turned out to suffer from a number of shortcomings and lacunae, and if it is to be revived it must at the same time be given much more depth.

One more point: *particularities* are a function of primary nature, of sites, of resources. On the basis of their differences, unknown or misunderstood, they confront one another and clash with one another. Out of their struggles, which imply and complicate class struggles as well as conflicts between peoples and nations, there emerge differences properly so called. Drawing a clear distinction between particularities and differences makes it possible to dispense with such confused and dangerous metaphors as specificity, authenticity, and so on.

The formal theory of difference opens of itself onto the unknown and the ill-understood: onto rhythms, onto circulations of energy, onto the life of the body (where repetitions and differences give rise to one another, harmonizing and disharmonizing in turn).

XIV

Differences endure or arise on the margins of the homogenized realm, either in the form of resistances or in the form of externalities (lateral, heterotopical, heterological). What is different is, to begin with, what is *excluded*: the edges of the city, shanty towns, the spaces of forbidden games, of guerrilla war, of war. Sooner or later, however, the existing centre and the forces of homogenization must seek to absorb all such differences, and they will succeed if these retain a defensive posture and no counterattack is mounted from their side. In the latter event, centrality and normality will be tested as to the limits of their power to integrate, to recuperate, or to destroy whatever has transgressed.

The vast shanty towns of Latin America (*favelas, barrios, ranchos*) manifest a social life far more intense than the bourgeois districts of the cities. This social life is transposed onto the level of urban morphology, but it only survives inasmuch as it fights in self-defence and goes on the attack in the course of class struggle in its modern forms. Their poverty notwithstanding, these districts sometimes so effectively order their space

---

[6] For the theory of difference, see my *Logique formelle, logique dialectique*, 2nd edn (Paris: Anthropos, 1970), especially the 'Préface'. For 'induced' *versus* 'produced' differences, see my *Manifeste différentialiste* (Paris: Gallimard, 1971).

– houses, walls, public spaces – as to elicit a nervous admiration. *Appropriation* of a remarkably high order is to be found here. The spontaneous architecture and planning ('wild' forms, according to a would-be elegant terminology) prove greatly superior to the organization of space by specialists who effectively translate the social order into territorial reality with or without direct orders from economic and political authorities. The result – on the ground – is an extraordinary *spatial duality*. And the duality in space itself creates the strong impression that there exists a duality of political power: an equilibrium so threatened that an explosion is inevitable – and in short order. This impression is nonetheless mistaken – a measure, precisely, of the repressive and assimilative capacity of the dominant space. The duality will persist, certainly; and, failing any reversal of the situation, dominated space will simply be weakened. 'Duality' means contradiction and conflict; a conflict of this kind eventuates either in the emergence of unforeseen differences or in its own absorption, in which case only induced differences arise (i.e. differences internal to the dominant form of space). A conflictual duality, which is a transitional state between opposition (induced difference) and contradiction/transcendence (produced difference), cannot last forever; it can sustain itself, however, around an 'equilibrium' deemed optimal by a particular ideology.

XV

In the absence of any dialectical movement, a given logic (or, once again, a given strategy) may generate a space by generating a spiral or vicious circle (also deemed 'optimal' by ideology). A case in point is the spiral criticized by Goodman.[7] In the United States the federal government collects a certain percentage on petrol sales, so generating vast sums of money for urban and inter-urban highway construction. The building of highways benefits both the oil companies and the automobile manufacturers: every additional mile of highway translates into increased car sales, which in turn increase petrol consumption, hence also tax revenues, and so on. Goodman calls this 'asphalt's magic circle'. It is almost as though automobiles and motorways occupied the entirety of space.

Such are the workings of a 'logic' – i.e. a strategy. This sequence of operations implies a productive consumption: the consumption of a space, and one that is doubly productive in that it produces both surplus

---

[7] Goodman, *After the Planners*, part II, pp. 113ff.

value and another space. The production of space is carried out with the state's intervention, and the state naturally acts in accordance with the aims of capital, yet this production *seems* to answer solely to the rational requirements of communication between the various parts of society, as to those of a growth consistent with the interests of all 'users'. What actually happens is that a vicious circle is set in train which for all its circularity is an invasive force serving dominant economic interests.

XVI

Each spatial strategy has several aims: as many aims as abstract space – manipulated and manipulative – has 'properties'. Strategic space makes it possible simultaneously to force worrisome groups, the workers among others, out towards the periphery; to make available spaces near the centres scarcer, so increasing their value; to organize the centre as locus of decision, wealth, power and information; to find allies for the hegemonic class within the middle strata and within the 'elite'; to plan production and flows from the spatial point of view; and so on.

The space of this social practice becomes a space that *sorts* – a space that *classifies* in the service of a class. The strategy of classification distributes the various social strata and classes (other than the one that exercises hegemony) across the available territory, keeping them separate and prohibiting all contacts – these being replaced by the *signs* (or images) of contact. Two critical remarks are called for in this connection. The first concerns a kind of 'knowledge' that legitimates this strategy by treating it as an object of science. I refer to structuralism, which cites intellectual reasons of a high order for its interest in arrangements and classifications of the kind that we have been discussing; what it perceives here is intelligibility – the superior relationship of the (thinking) subject and the (constructed) object. In this respect (but not only in this respect) the ideology of structuralism, wearing the mantle of knowledge, serves power. The second point is that 'operational' notions of arrangement or classification govern the whole of space, and apply as much to private as to public space, as much to furnishings as to overall spatial planning. Such notions clearly serve power by contributing to a global homogenizing trend. After all, it is the state – 'public', and hence political, authority – that does the arranging and classifying. Operationalism of this kind actually conflates 'public' space with the 'private' space of the hegemonic class, or fraction of a class, that in the last analysis retains and maintains private ownership of the land and of the other means of production. It is therefore in appearance only that the 'private' sphere is organized according to the dictates of the 'public' one. The inverse situation (the world upside down – and waiting to be set on its feet) is the one that actually prevails. The whole of space is increasingly modelled after private enterprise, private property and the family – after a reproduction of production relations paralleling biological reproduction and genitality.

XVII

Mimesis has its role and function in this domination of space: imitation and its corollaries; analogy, and impressions to a greater or lesser degree informed by analogy; resemblances and dissimilarities; metaphor (substitution of one term for another) and metonymy (use of a part to refer to the whole). This role is a contradictory one, however: by assigning a model, which occupies a space, to an as-yet ill-defined desire, imitation ensures that violence (or rather counter-violence) will be done to that desire in its relationship with that space and its occupant. With its components and variants, mimesis makes it possible to establish an abstract 'spatiality' as a coherent system that is partly artificial and partly real. Nature is imitated, for example, but only *seemingly* reproduced: what are produced are the *signs* of nature or of the natural realm – a tree, perhaps, or a shrub, or merely the image of a tree, or a photograph of one. In this way nature is effectively replaced by powerful and destructive abstractions without any production of 'second nature', without any appropriation of nature; nature is left, as it were, in a no-man's-land. An actualized 'second nature', far removed from nature proper yet concrete at its own level, would be emancipated from artifice while at the same time retaining no suggestion of the 'natural'. Mimesis, on the other hand, pitches its tent in an artificial world, the world of the visual where what can be seen has absolute priority, and there simulates primary nature, immediacy, and the reality of the body.

As we saw earlier, social (spatial) practice in the first instance intuitively – i.e. in an initial *intuitus*, immediate and close to nature's immediacy – laid hold of a portion of nature which was already divided (and hence, too, of a portion of the body with its constitutive dualities): either the hole, the abyss, or else the mound, the shining hill; either the 'world' or the 'Cosmos'. And either the curve, the circle, the ring, or the straight line, ascending or descending. This able manoeuvre, which I sought to trace above, made it possible, beginning in the city of the ancient world, simultaneously to incorporate femaleness and to demote it, to establish dominion over it by assigning it a limited portion of space, and to reduce it to a 'femininity' subordinated to the principle of maleness, of masculinity or manliness. The *intuitus* whereby practice first produced a diversity of spaces was to be transformed into a *habitus* and then into an *intellectus*. These transformations were brought about on the basis of immediacy, of sensory impressions that already had a mental dimension (*intuitus*), that were already in some degree detached from 'pure' or 'natural' sensation, already amplified, broadened, elaborated – and hence already metamorphosed. Thus social space emerged from the earth and evolved, thanks to a stubbornly pursued process of 'intellectualization', until an abstract space was constructed, a geometric, visual and phallic space that went beyond spatiality by becoming the production of a homogeneous and pathogenic political 'medium' at once aberrational and norm-bound, coercive and rationalized: the 'medium' of the state, of power and its strategies. What is the destiny of this absolutely political 'medium', this space of absolute politics? At present, between metaphorizations and metonymizations, we are approaching tautology: we produce only the reproducible, and hence we produce only by reproducing or imitating past production. This is the ultimate contradiction: inasmuch as the capacity to produce space produces only reproductions, it can generate nothing but the repetitive, nothing but repetition. The production of space is thus transformed into its opposite: the reproduction of things in space. And mimesis (simulation, imitation) becomes merely a reproducibility grounded in received knowledge, technology and power, because reproducibility is what ensures the renewal (or reproduction) of existing social relations.

XVIII

What is commonly referred to as the 'political question' needs to be broken down, for like space itself it gives rise to a number of sub-questions, a number of different themes or problems: there is the question of the *political sphere* in a general sense, and of its function in social practice; there is the question of *politics* and its part in the capitalist mode of production; and there is the question of the *politicians* – statesmen or henchmen of the state – and of their qualifications and their selection (so to speak).

Questions concerning the state on the one hand and the political sphere (or particular political policies) on the other inevitably remain abstract – as do answers to them – so long as they are not couched in terms of the state's relationship to space.

That relationship, which has always been a real one, is becoming tighter: the spatial role of the state, whether in the past or in the present, is more patent. Administrative and political state apparatuses are no longer content (if they ever were) merely to intervene in an abstract manner in the investment of capital (in the properly economic sphere). So long as units of economic production and of social activity were scattered across the land, only the state was capable of binding them into a spatial unity – that of the nation. At the end of the Middle Ages in Western Europe, the towns and the urban systems substituted a secularized space for the absolute (religious) space of earlier centuries. It was in this political space, already unitary in character though still made up of scattered elements, that there arose the space of royal power, the space of the nation state in the making. This historical relationship between the state and space was considered earlier in our discussion.

Today the state and its bureaucratic and political apparatuses intervene continually in space, and make use of space in its instrumental aspect in order to intervene at all levels and through every agency of the economic realm. Consequently, (global) social practice and political practice tend to join forces in spatial practice, so achieving a certain cohesiveness if not a logical coherence. In France specific localized actions are linked up by the authorities (prefects) to global actions dictated by so-called planning-guidelines and national plans. Nothing that happens within the nation's borders remains outside the scope of the state and its 'services'. These cover space in its entirety.

Only those individuals who think and operate at the state level are familiar with all regional and local arrangements, with all the flows and networks (such as those which connect 'manpower deposits' to places where labour power is productively consumed).

The fact remains, however, that the proliferation of links and networks, by directly connecting up very diverse places, and by ending their isolation – though without destroying the peculiarities and differences to which that very isolation has given rise – tends to render the state redundant. Whence the clamour – sometimes high-pitched and superficial, sometimes stemming from the profoundest of motives – raised on all sides by those who want to loosen the grip of power, to decentralize, to manage (or self-manage) from the grass roots, whether at the level of production (the factory) or at the territorial level (town or city). The state's tendency to establish centres of decision armed with all the tools of power and subordinated to a single main centre, the capital, thus encounters stiff resistance. Local powers (municipalities, departments, regions) do not readily allow themselves to be absorbed. The state, moreover, can neither do everything, nor know everything, nor manage everything – indeed its maximum effectiveness consists in the destruction of whatever escapes its control: Hegel's absolute state cannot produce itself in this space because it is bound to destroy itself before it can bring the task to completion.

A certain 'pluralism' persists, therefore, but one which has no great significance so long as open conflict does not erupt among the forces in contention – that is to say, among the various groups, classes, or fractions of classes that have taken up defensive or offensive postures. This is why conflicts between local powers and central powers, wherever they may occur in the world, are of the greatest possible interest. Such conflicts – occasionally – allow something *other* to break the barriers of the forbidden. Not that hope should be placed, after the fashion of the American liberals, in pluralism *per se*, but it is not unreasonable to place some hope in things that pluralism lets by.

XIX

Innumerable groups, some ephemeral, some more durable, have sought to invent a 'new life' – usually a communal one. With their trials and errors, successes and failures, such communal experiments have so many denigrators and champions that we can get a fairly clear picture of them. Among the obstacles that they have run into and the reasons for their failure when it occurs must certainly be numbered the absence of an appropriated space, the inability to invent new forms. The communities of earlier times, monastic or otherwise, had contemplation, not enjoyment, as their *raison d'être* and goal. No doubt there is nothing more 'beautiful' than cloisters, but we need to remember that these structures were never built for the sake of beauty or art. Their significance and purpose was, rather, retreat from the world, ascesis, meditation. It is a curious and paradoxical fact that, while spaces dedicated to sensual delight have existed, they are few and far between: aside from the Alhambra with its gardens, and certain châteaux of the Loire, and perhaps a few villas of Palladio's, it is hard to think of real examples as opposed to literary and imaginary ones – the Abbey of Thélème, the palaces of the Arabian Nights, or the dreams of a Fourier. An architecture of pleasure and joy, of community in the use of the gifts of the earth, has yet to be invented. When one asks what agencies have informed social demands and commands, the answer is much more likely to be commerce and exchange, or power, or productive labour, or renunciation and death, than enjoyment and rest (in the sense of non-work).

Listening – even with half an ear – to the vengeful discourse of a Valerie Solanas in her *S.C.U.M. Manifesto*, powered as it may well be by deep resentments, it is hard to resist the conclusion that it is time for the sterile space of men, founded on violence and misery, to give way to a women's space. It would thus fall to women to achieve appropriation, a responsibility that they would successfully fulfil – in sharp contrast to the inability of male or manly designs to embrace anything but joyless domination, renunciation – and death.

Most if not all modern experiments in communal living have diverted an existing space to their own purposes and so lost their impetus on account of an inappropriate spatial morphology: bourgeois mansions, half-ruined castles, villages abandoned by the peasantry, suburban villas, and so forth.

In the end, the invention of a space of enjoyment necessarily implies going through a phase of *elitism*. The elites of today avoid or reject quantitative models of consumption and homogenizing trends. At the same time, though they cultivate the appearance of differences, these elites are in fact indistinguishable from one another. The 'masses', meanwhile, among whom genuine differences exist, and who at the deepest (unconscious) level seek difference, continue to espouse the quantitative and the homogeneous. The obvious reason for this is that the masses must *survive* before they can *live*.

Elites thus have a role, and first and foremost that role is to indicate to the masses how difficult – and indeed impossible – it is to live according to the strict constraints and criteria of quantity. It is true, of course, that the masses already experience this impossibility in their working lives; but this awareness has yet to be extended to the whole of life 'outside work'.

Whatever the outcome of the elitist quest for community, however, no matter how the relationship between elites and the labouring masses may turn out, the production of a new space commensurate with the capacities of the productive forces (technology and knowledge) can never be brought about by any particular social group; it must of necessity result from relationships between groups – between classes or fractions of classes – on a world scale.

There should therefore be no cause for surprise when a space-related issue spurs collaboration (often denounced on that basis by party politicians) between very different kinds of people, between those who 'react' – reactionaries, in a traditional political parlance – and 'liberals' or 'radicals', progressives, 'advanced' democrats, and even revolutionaries. Such coalitions around some particular counter-project or counter-plan, promoting a counter-space in opposition to the one embodied in the strategies of power, occur all over the world, as easily in Boston, New York or Toronto as in English or Japanese cities. Typically the first group – the 'reactors' – oppose a particular project in order to protect their own privileged space, their gardens and parks, their nature, their greenery, sometimes their comfortable old homes – or sometimes, just as likely, their familiar shacks. The second group – the 'liberals' or 'radicals' – will meanwhile oppose the same project on the grounds that it represents a seizure of the space concerned by capitalism in a general sense, or by specific financial interests, or by a particular developer. The ambiguity of such concepts as that of ecology, for example, which is a mixture of science and ideology, facilitates the formation of the most unlikely alliances.

Only a political party can impose standards for the recruitment of members and so achieve ideological unity. It is precisely the diversity of the coalitions just mentioned that explains the suspicious attitude of the traditional political parties towards the issues of space.

XX

A space in which each individual and/or collective 'subject', reconstituted on this new basis, would become acquainted with use and enjoyment is at present only in its infancy. Current notions of an 'alternative society' or 'counter-culture' are in no way free of confusion. What might a 'counter-culture' be, considering how much uncertainty surrounds the concept of 'culture' itself – just as much of a ragbag as the notion of the unconscious, because it is made the repository as easily of ideology as the results of history, of ways of life, or of the body's misconstrued demands? What might an 'alternative society' be, given the difficulty of defining 'society', and given that all such words lose any clear meaning if they do not designate either 'capitalism' or 'socialism' or 'communism' – terms which have themselves now become equivocal?

What runs counter to a society founded on exchange is a primacy of *use*. What counters quantity is quality. We know what counter-projects consist or what counter-space consists in – because practice demonstrates it. When a community fights the construction of urban motorways or housing-developments, when it demands 'amenities' or empty spaces for play and encounter, we can see how a counter-space can insert itself into spatial reality: against the Eye and the Gaze, against quantity and homogeneity, against power and the arrogance of power, against the endless expansion of the 'private' and of industrial profitability; and against specialized spaces and a narrow localization of function. Naturally, it happens that *induced* differences – differences internal to a whole and brought into being by that whole as a system aiming to establish itself and then to close (for example, the suburban 'world of villas') – are hard to distinguish either from *produced* differences, which escape the system's rule, or from *reduced* differences, forced back into the system by constraint and violence. Naturally, too, it happens that a counter-space and a counter-project simulate existing space, parodying it and demonstrating its limitations, without for all that escaping its clutches.

The only possibility of so altering the operation of the centralized state as to introduce (or reintroduce) a measure of pluralism lies in a challenge to central power from the 'local powers', in the capacity for action of municipal or regional forces linked directly to the territory in question. Inevitably such resistance or counter-action will tend to strengthen or create independent territorial entities capable to some degree of self-management. Just as inevitably, the central state will muster its own forces in order to reduce any such local autonomy by exploiting isolation and weakness. Hence a quite specific dialectical process is set in train: on the one hand, the state's reinforcement is followed by a weakening, even a breaking-up or withering-away; on the other hand, local powers assert themselves vigorously, then lose their nerve and fall back. And so on – in accordance with a cycle and with contradictions which must, sooner or later, achieve resolution. What form might that resolution take? Ultimately, perhaps, that of the replacement of the state's machinery by data-processing machines fed and managed from below. Putting the spatial problematic into terms of forces – the relative strength of socio-political forces – effectively gets us out of a number of ludicrous dilemmas: either the city is non-existent or else it is a system; either space is an inert underlay or else it is the 'medium' of a fully self-contained ecological reality; and either the urban sphere occupies a niche or else it is a subject. Just as economic pressure from the base – and such pressure alone, in the shape of unions, the making of demands, striking, and so forth – is able to modify the *production* of surplus value, so pressure grounded in spatial practice is alone capable of modifying the apportionment of that surplus value – i.e. the distribution of the portion of social surplus production allotted to society's collective 'interests', to so-called social services. Such grass-roots pressure, if it is to be effective in this regard, cannot be confined to attacking the state *qua* guardian of the 'general interest'. For this state, born of the hegemony of a class, has as one of its functions – and a more and more significant function – the organization of space, the regularization of its flows, and the control of its networks. It devotes to these purposes a considerable part of global surplus value, of the surplus production assigned to the running of society. Pressure from below must therefore also confront the state in its role as organizer of space, as the power that controls urbanization. the construction of buildings and spatial planning in general. This state defends class interests while simultaneously setting itself above society as a whole, and its ability to intervene in space can and must be turned back against it, by grass-roots opposition, in the form of counter-plans and counter-projects designed to thwart strategies, plans and programmes imposed from above.

XXI

The quest for a 'counter-space' overwhelms the supposedly ironclad distinction between 'reform' and 'revolution'. Any proposal along these lines, even the most seemingly insignificant, shakes existing space to its foundations, along with its strategies and aims – namely, the imposition of homogeneity and transparency everywhere within the purview of power and its established order. The silence of the 'users' mentioned earlier may be explained as follows: consumers sense that the slightest shift on their part can have boundless consequences, that the whole order (or mode of production) weighing down upon them will be seriously affected by the slighest movement on their part.

The situation has consequences that seem paradoxical at first. Certain deviant or diverted spaces, though initially subordinate, show distinct evidence of a true productive capacity. Among these are spaces devoted to leisure activity. Such spaces appear on first inspection to have escaped the control of the established order, and thus, inasmuch as they are spaces of play, to constitute a vast 'counter-space'. This is a complete illusion. The case against leisure is quite simply closed – and the verdict is irreversible: leisure is as alienated and alienating as labour; as much an agent of co-optation as it is itself co-opted; and both an assimilative and an assimilated part of the 'system' (mode of production). Once a conquest of the working class, in the shape of paid days' off, holidays, weekends, and so on, leisure has been transformed into an industry, into a victory of neocapitalism and an extension of bourgeois hegemony to the whole of space.

As an extension of dominated space, leisure spaces are arranged at once functionally and hierarchically. They serve the reproduction of production relations. Space thus controlled and managed constrains in specific ways, imposing its own rituals and gestures (such as tanning), discursive forms (what should be said or not said), and even models and modulations in space (hotels, chalets – the emphasis being on private life, on the genital order of the family). Hence this space too is made up of 'boxes for living in', of identical 'plans' piled one on top of another or jammed next to one another in rows. Yet, *at the same time*, the body takes its revenge – or at least calls for revenge. It seeks to make itself known – to gain recognition – as *generative*. (Of what? Of practice, of use, hence of space – and, by extension, of the human species.) A positivity, then, negated by its own consequences – and later restored. The beach is the only place of enjoyment that the human species has discovered in nature. Thanks to its sensory organs, from the sense of smell and from sexuality to sight (without any special emphasis being placed on the visual sphere), the body tends to behave as a *differential field*. It behaves, in other words, as a *total* body, breaking out of the temporal and spatial shell developed in response to labour, to the division of labour, to the localizing of work and the specialization of places. In its tendency, the body asserts itself more (and better) as 'subject' and as 'object' than as 'subjectivity' (in the classical philosophical sense) and as 'objectivity' (fragmented in every way, distorted by the visual, by images, etc.).

In and through the space of leisure, a pedagogy of space and time is beginning to take shape. As yet, admittedly, this is no more than a virtuality, and one which is denied and rejected, but it nevertheless indicates a trend (or rather a counter-trend). Time, meanwhile, retrieves its use value. And the critique of the space of labour, whether implicit or explicit, leads in turn to a critique of fractured (specialized) gestures, of silence, of discomfort and malaise.

Despite its anachronistic aspect, the return to immediacy, to the organic (and hence to nature), gives rise to startling differences. Through music – indecisively, clumsily, yet effectively – rhythms reclaim their rights. They can no longer be forgotten, even though simulation and mimesis have replaced any true *appropriation* of being and of natural space: and even though the appeal to the body is ever liable to turn into its opposite – total passivity on the beach, mere contemplation of the spectacle of sea and sun.

The space of leisure *tends* – but it is no more than a tendency, a tension, a transgression of 'users' in search of a way forward – to surmount divisions: the division between social and mental, the division between sensory and intellectual, and also the division between the everyday and the out-of-the-ordinary (festival).

This space further reveals where the vulnerable areas and potential breaking-points are: everyday life, the urban sphere, the body, and the differences that emerge within the body from repetitions (from gestures, rhythms or cycles). The space of leisure bridges the gap between traditional spaces with their monumentality and their localizations based on work and its demands, and potential spaces of enjoyment and joy; in consequence this space is the very epitome of contradictory space. This is where the existing mode of production produces both its worst and its best – parasitic outgrowths on the one hand and exuberant new branches on the other – as prodigal of monstrosities as of promises (that it cannot keep).

XXII

The degree to which a city can resist despoliation, the difficulty encountered by those who would lay it waste, is well illustrated by the case of Paris. As in any urban space, something is always going on – but not everything that is going on tends in the same direction. While neocapitalism and the centralizing state reorganize the city's supposedly historic section in accordance with their interests, neighbourhoods not far from the centre are in the process of becoming more working-class in character: around Belleville, for example, an area that is still very animated, immigrant workers and *colons* repatriated from North Africa rub shoulders – not without a measure of friction. Meanwhile, the Marais is experiencing the influx of an elite element, but this is an elite made up of intellectuals and of members of the (old and new) liberal professions, which does not look down its nose at the common people. In this respect, it differs from the old-style bourgeoisie, still solidly ensconced in the city's 'residential' *arrondissements* and suburbs. It is not inconceivable that the Marais and its vicinity will long retain some relationship with production – with craft industry, small or medium-size manufacturing – and a proletarian and even sub-proletarian population.

Paris has not completely lost the excitement that characterized it as a city of festival in earlier times. As 1968 showed, it is still a crucible, still a focal point. There is an acute contradiction here: it is not in the interests of the political establishment and the hegemonic class to extinguish this spark, for to do so would effectively destroy the city's worldwide reputation – based, precisely, on its daring, its willingness to expose the possible and the impossible, its so-called cultural development, and its panoply of actions and actors (working class, intelligentsia, students, artists, writers, and others). Yet at the same time the political powers and the bourgeoisie controlling the economy are afraid of all such ferment, and have a strong urge to crush it under suffocating central decision-making.

In Paris, as in any city worth the name, the allied effects of centralism and monumentality have not yet run their course. Each of these trends is based on simultaneous inclusion and exclusion precipitated by a specific spatial factor. The centre gathers things together only to the extent that it pushes them away and disperses them, while a monument exercises an attraction only to the degree that it creates distance. It is inevitable, therefore, that the reduction of old particularities, of ethnic groups, 'cultures' or nationalities, should produce new differences. It is impossible to bring urban reality to a complete stop. To do so would kill it – and in any case it puts up far too strong a resistance. Though dominated, ravaged, the urban realm successfully reconstitutes itself. Only in the most extreme circumstances could this reality be reduced to a state of inertia, flat on the ground (so to speak), utterly dispersed and deanimated. Furthermore this extreme state of affairs, so hard to arrive at, would present perils of its own. The contradiction between the passivity and the activity of people (of 'inhabitants' or 'users') is never completely resolved in favour of passivity.

There is nothing more contradictory than 'urbanness'. On the one hand, it makes it possible in some degree to deflect class struggles. The city and urban reality can serve to disperse dangerous 'elements', and they also facilitate the setting of relatively inoffensive 'objectives', such as the improvement of transportation or of other 'amenities'. On the other hand, the city and its periphery tend to become the arena of kinds of action that can no longer be confined to the traditional locations of the factory or office floor. The city and the urban sphere are thus the setting of struggle; they are also, however, the stakes of that struggle. How could one aim for power without reaching for the places where power resides, without planning to occupy that space and to create a new political morphology – something which implies a critique in acts of the old one, and hence too of the status of the political sphere itself (as of specific political orientations)? It is worth pointing out in passing that illegitimate hybrids of country and city in no way escape the domination of space, as some people – particularly those who inhabit such spaces – seem to believe. On the contrary, these bastard forms degrade both urban and rural space. So far from transcending the conflict between the two, they thrust both into a confusion which would be utterly without form were it not for the 'structure' imposed by the space of the state.

The appropriation of politically dominated space poses an enormous political problem, one that must remain insoluble so long as no critique of the political realm, of specific politics and of the state is forthcoming – so long, in fact, as no withering-away of the state occurs, no matter by what route or by virtue of what process. At this level the opposition between appropriation and domination becomes a dialectical contradiction, as the appropriation of space, the development of the urban sphere, the metamorphosis of everyday life and the transcendence of the conflictual split between city and country all clash head-on with the state and with politics.

Seen from this perspective, dominant/dominated space, as imposed by the state upon its 'subjects', be they faithful or not, is simply the space, seemingly devoid of violence, of a sort of *pax estatica* (or, in the case of the Western countries, a *pax capitalistica*) reminiscent of the Pax Romana. Though seemingly secured against any violence, abstract space is in fact inherently violent. The same goes for all spaces promising a similar security: residential suburbs, holiday homes, fake countrysides and imitations of nature. The Marxist theory of the withering-away of the state gets a new lease on life when placed in the context of the following central insight: state management of space implies a logic of stability that is both destructive and self-destructive.

XXIII

In this connection it is worth reconsidering the grid mentioned earlier (see pp. 155–8), according to which there are three interacting and interwoven levels of space: the public or global, the private, and the mixed (mediating or intermediary) levels. The fact is that this grid deciphers and apportions social space in a way quite different from political thinking. According to the perspective of politics, no part of space can or may be allowed to escape domination, except in so far as appearances are concerned. Power aspires to control space in its entirety, so it maintains it in a 'disjointed unity', as at once fragmentary and homogeneous: it divides and rules. The grid embodies a different perspective, if only because it does not keep the spatial elements separate from one another within an abstract space. It reintroduces immanent differences and envisions spaces at once 'compact' and highly elaborated, places of encounter and places of transition (passages), as well as places appropriated to meditation and solitude. And it is akin to another analysis of levels, one which discriminates – without sundering them – between a 'micro' level (architecture; residence *versus* housing; neighbourhood), a 'medium' level (the city; town-planning; the town–country dichotomy), and finally a 'macro' level (spatial strategies, town and country planning, land considered in national, global or worldwide terms). We should remember, nevertheless, that 'grids' of this kind are still confined to the classification of fragments in space, whereas authentic knowledge of space must address the question of its production.

XXIV

Political power as such harbours an immanent contradiction. It controls flows and it controls agglomerations. The mobility of the component parts and formants of social space is constantly on the increase, especially in the 'economic' realm proper: flows of energy, of raw materials, of labour, and so on. But such control, to be effective, calls for permanent establishments, for permanent centres of decision and action (whether violent or not). There are certain essential activities, moreover, some pedagogical in character, some even related to play, that also require durable facilities. (Note that the mobility of flows and agglomerations has little to do with the rhythms and cycles of nature.) A novel and quite specific contradiction thus arises between what is transient and what is durable. The diversity of spatial forms and the flexibility of practice can only become more marked, along with the variety of functions, with multifunctionality – and indeed with dysfunctionality. Can the body in its quest for vindication use the resulting interstices as its way back? And what of primary and 'second' nature?

XXV

It is signs and images – the world of signs and images – that tend to fill the interstices in question. Signs of happiness, of satisfaction. Signs and images of nature, of Eros. Images and signs of history, of authenticity, of style. Signs of the world: of the other world, and of another – a different – world. Neo-this and neo-that, consumed as novelties, and signs of the old, the venerated, the admirable. Images and signs of the future. Signs and images of the urban, of 'urbanness'.

This world of images and signs, this tombstone of the 'world' ('Mundus est immundus') is situated at the edges of what exists, between the shadows and the light, between the conceived (abstraction) and the perceived (the readable/visible). Between the real and the unreal. Always in the interstices, in the cracks. Between directly lived experience and thought. And (a familiar paradox) between life and death. It presents itself as a transparent (and hence pure) world, and as reassuring, on the grounds that it ensures concordance between mental and social, space and time, outside and inside, and needs and desire. On the grounds, too, that it is unitary: that it instates a (rediscovered) unity of discourse, of language as systematic, of thought as logical. The world of signs passes itself off as a true world, and perhaps after all it has the right to do so – which would involve further compromise of the True (the absolute). The rule of this world is founded, then, on transparency. It leads, however, into opacity and into naturalness (not that of 'nature', but that of the signs of nature). This is a fraudulent world, indeed the most deceptive of all worlds – the world-as-fraud. A world where that which *contains* is hidden in corners or lurks on the sidelines. When there is talk of art and culture, the real subject is money, the market, exchange, power. Talk of communication actually refers only to solitudes. Talk of beauty refers to brand images. Talk of city-planning refers to nothing at all.

The world of images and signs exercises a fascination, skirts or submerges problems, and diverts attention from the 'real' – i.e. from the possible. While occupying space, it also signifies space, substituting a mental and therefore abstract space for spatial practice – without, however, doing anything really to unify those spaces that it seems to combine in the abstraction of signs and images. Differences are replaced by differential signs, so that produced differences are supplanted in advance by differences which are induced – and reduced to signs.

The evanescent space of images and signs does not, however, manage to attain consistency. It is a world that flees, a world with a perpetual, indeed a dizzying, need for rejuvenation. It even seems at times that this world is about to disappear bag and baggage down a hole, into some cleft that, with just a little widening, would swallow it up. Unfortunately, to suppose that the right word or gesture could tumble everything down the rubbish chute amounts to an existential (or existentialist) illusion. Anyone tempted to subscribe to such an illusion would be well advised to recall that in the booby-trapped space of images illusions are among the booby traps. Dispelling the fictitious yet real world of images and signs is going to take more than a magic formula or a ritual gesture, more than the words of a philosopher or the arm-wavings of a prophet.

Factors or causes may be discerned within 'reality', however, that may be expected in the long run to interfere with the smooth running of the fascinating and ambiguous world of images. In tandem with the division of labour, though not identifiable with it, a diversification of products and of operations related to production may be observed. Activities ancillary to manufacture proper have become more and more important, with a corresponding decrease in the significance of manual labour and of those tasks carried out on the shopfloor itself. Some people have even spoken in this regard of a 'tertiarization' of industry. The product's conception has much to do with this, for it now has to take 'needs' into account – whether these are assumed to exist or deliberately created, genuinely present or simply manipulated – and hence must deal with a mass of information. The organization of productive labour gets increasingly complex in consequence, as conceptual considerations and considerations of profit have to be reconciled and as product cycles themselves diversify more and more. There is a proliferation, too, of business services, and much more widespread subcontracting of auxiliary tasks. Another outcome is that urban centres (formerly known as cities) tend to take over all the intellectual aspects of the productive process (formerly known as science's role in production – or knowledge as one of the forces of production). This leads in turn to struggles for influence, power and prestige among the scientific and business groups concerned.

It may be asserted with reasonable confidence that the process of producing *things in space* (the range of so-called consumer goods) tends to annul rather than reinforce homogenization. A number of differentiating traits are thus permitted to emerge which are not completely bound to a specific location or situation, to a geographically determinate space. The so-called economic process tends to generate diversity[8] – a fact which supports the hypothesis that homogenization today is a function of political rather than economic factors as such; abstract space is a tool of power. Spatial practice in general, and the process of urbanization in particular (the explosion of the old cities, the extension of the urban fabric, and the formation of centres) cannot be defined uniquely in terms of industrial growth seen from the standpoint either of its quantitative results or of its technological features. The 'city' can be conceived of neither as a productive enterprise and unit, as a kind of vast factory, nor as a consumption unit subordinated to production.

It will be clear from the foregoing analysis that social space (spatial practice) has by now achieved – potentially – a measure of freedom from the abstract space of quantifiable activities, and hence too from the agendas set by reproduction pure and simple.

XXVI

The more carefully one examines space, considering it not only with the eyes, not only with the intellect, but also with all the senses, with the total body, the more clearly one becomes aware of the conflicts at work within it, conflicts which foster the explosion of abstract space and the production of a space that is *other*.

Spatial practice is neither determined by an existing system, be it urban or ecological, nor adapted to a system, be it economic or political. On the contrary, thanks to the potential energies of a variety of groups capable of diverting homogenized space to their own purposes, a theatricalized or dramatized space is liable to arise. Space is liable to be eroticized and restored to ambiguity, to the common birthplace of needs and desires, by means of music, by means of differential systems and valorizations which overwhelm the strict localization of needs and desires in spaces specialized either physiologically (sexuality) or socially (places set aside, supposedly, for pleasure). An unequal struggle, sometimes furious, sometimes more low-key, takes place between the Logos and the Anti-Logos, these terms being taken in their broadest possible sense – the sense in which Nietzsche used them. The Logos makes

---

[8] These remarks are inspired by Radovan Richta, *La civilisation au carrefour* (Paris: Seuil, 1974), translated from the Czech: *Civilizácia na rázcestí* (Bratislava: Vydavatel'stvo literatury, 1966).

inventories, classifies, arranges: it cultivates knowledge and presses it into the service of power. Nietzsche's Grand Desire, by contrast, seeks to overcome divisions – divisions between work and product, between repetitive and differential, or between needs and desires. On the side of the Logos is rationality, constantly being refined and constantly asserting itself in the shape of organizational forms, structural aspects of industry, systems and efforts to systematize everything, and so forth. On this side of things are ranged the forces that aspire to dominate and control space: business and the state, institutions, the family, the 'establishment', the established order, corporate and constituted bodies of all kinds. In the opposite camp are the forces that seek to appropriate space: various forms of self-management or workers' control of territorial and industrial entities, communities and communes, elite groups striving to change life and to transcend political institutions and parties.

The psychoanalytical account of conflict between a pleasure principle and a reality principle gives only an abstract and feeble idea of this great struggle. The full-blown conception of the revolution has to compete with a variety of corruptions, among them economistic and productivistic interpretations, and versions founded on the work ethic. The maximal version derives directly from Marx and his project of a total revolution entailing the end of the state, of the nation, of the family, of politics, of history, and so on, and adds to the central idea of an ever-greater automation of the productive process the related notion of the production of a space that is different.

Implicit in the great Logos–Eros dialectic, as well as in the conflict between 'domination' and 'appropriation', is a contradiction between technology and technicity on the one hand, and poetry and music on the other. A dialectical contradiction, as it is surely needless to recall, presupposes unity as well as confrontation. There is thus no such thing as technology or technicity in a pure or absolute state, bearing no trace whatsoever of appropriation. The fact remains, though, that technology and technicity tend to acquire a distinct autonomy, and to reinforce domination far more than they do appropriation, the quantitative far more than they do the qualitative. Similarly, although all music or poetry or drama has a technical – even a technological – aspect, this tends to be incorporated, by means of appropriation, into the qualitative realm.

The effect in space is the development of multifarious distortions and discrepancies – which should not, however, be mistaken for *differences*. Possibilities are blocked; mobility declines into fixedness. Does space also secrete a false consciousness? An ideology – or ideologies? Abstract space, considered together with the forces that operate within it, some of which serve to sustain and some to modify it, may accurately be said to bring manifestations of false consciousness and ideology in its wake. As a space that is fetishized, that reduces possibilities, and cloaks conflicts and differences in illusory coherence and transparency, it clearly operates ideologically. Yet abstract space is the outcome not of an ideology or of false consciousness, but of a practice. Its falsification is self-generated. Conflicts nevertheless manifest themselves on the level, precisely, of knowledge, especially that between *space* and *time*. The oppressive and repressive powers of abstract space are clearly revealed in connection with time: this space relegates time to an abstraction of its own – except for labour time, which produces things and surplus value. Time might thus be expected to be quickly reduced to constraints placed on the employment of space: to distances, pathways, itineraries, or modes of transportation. In fact, however, time resists any such reduction, re-emerging instead as the supreme form of wealth, as locus and medium of use, and hence of enjoyment. Abstract space fails in the end to lure time into the realm of externality, of signs and images, of dispersion. Time comes back into its own as privacy, inner life, subjectivity. Also as cycles closely bound up with nature and with use (sleep, hunger, etc.). Within time, the investment of affect, of energy, of 'creativity' opposes a mere passive apprehension of signs and signifiers. Such an investment, the desire to 'do' something, and hence to 'create', can only occur in a space – and through the production of a space. The 'real' appropriation of space, which is incompatible with abstract *signs* of appropriation serving merely to mask domination, does have certain requirements.

XXVII

The dialectical relationship between 'need' and 'desire' is only partly germane to our present theoretical investigation and discussion. Already obscure in itself, and even further obscured by the pronouncements of the ecologists, this relationship deserves to be clarified on its own terms. The concept of need implies or assumes certain determinants. There exist *needs*, in the plural, distinct one from the next; and, although the notion of a 'system of needs' was introduced as early as Hegel, such a system can only be conceived of as having a momentary reality, as formed within a totality and in accordance with the requirements of that totality (culture, ideology, ethical system, division of labour, etc.). Each need finds satisfaction in its object, in the consumption of that object, yet such satisfaction eliminates the need only temporarily, for a need is repetitive in character and after being satisfied will arise again and again, stronger and more urgent, until at last it reaches a saturation point or is extinguished.

As for desire, the concept never sloughs off its ambiguity, even if rhetoric tends to present it as a fullness. As applied to a reality prior to the emergence of needs, 'desire' refers to the energies available to the living being, energies that tend to be discharged explosively, with no definite object, in violent and destructive or self-destructive ways. Theological and metaphysical dogma has ever and always denied desire's initial lack of differentiation. For the most consistent theologians, desire *is* already, from the very beginning, a desire for desire and for eternity. For the psychoanalysts, desire 'is' sexual desire – desire for the mother or father. The problem here, however, is that desire, though originally *undifferentiated* – i.e. objectless, seeking an object and finding it, generally as a result of stimulation, in the surrounding space – is also *determined* as available (explosive) energy. This energy takes on definition – is objectified – in the sphere of need, and in the context of the complex relationship 'productive labour – lack – satisfaction'. *Beyond* this sphere of defined needs bound to objects (products), 'desire' denotes the concentration of still-available energies for a particular purpose or goal. Instead of a paroxystic moment of destruction or self-destruction, the aim is now creative: a love, a being, or a work. According to this view of matters (whose Nietzschean antecedents should be and are intended to be obvious), the doorway of Grand Desire (Eros) thus stands open to desire.

From this perspective, which is more clearly defined poetically, and hence qualitatively, than conceptually, things and products *in space* correspond to specific needs, if not to all needs: each need looks here for satisfaction, and finds and produces its object. Particular places serve to define the coming-together of a given need and a given object, and they are in turn defined by that meeting. Space is thus populated by visible crowds of objects and invisible crowds of needs.

What Girard says of 'objects' and 'subjects' applies equally well to most spaces: consecrated by violence, they derive their prestige from sacrifice or murder, war or terror.[9]

Needs (all needs and each separately) tend to recur, and hence require that their objects too be recurrent (this is so whether these objects are artificial or real – the distinction being hard to draw); at the same time,

---

[9] See René Girard, *La violence et le sacré* (Paris: Grasset, 1972).

however, needs also increase in number; and they die from repetition – from the phenomenon of saturation. Desire, which precedes needs and goes beyond them, is the yeast that causes this rather lifeless dough to rise. The resulting movement prevents stagnation and cannot help but produce differences.

XXVIII

In mathematics and the exact sciences, repetition (iteration, recurrence) generates difference. Induced or reduced, such difference tends towards formal identity, with whatever is left over being immediately assessed and subjected to a new, more thorough analysis. This sequence of operations is performed as nearly as possible in the clear light of strict logic. This is how numerical series come into being, from the number one to the transfinite numbers. In the experimental sciences, only a permanent apparatus and precisely repeated conditions make it possible to study variations and variables (i.e. remainders).

In music or poetry, by contrast, difference is what engenders the repetitive aspect that will make that difference effective. Art in general and the artistic sensibility bank on maximum difference, at first merely virtual, sensed, anticipated, and then, finally, produced. Art puts its faith in difference: this is what is known as 'inspiration', or as a 'project'; this is the motive of a new work – the thing that makes it *new*; only subsequently does the poet, musician or painter seek out means, procedures, techniques – in short, the wherewithal to realize the project by dint of repetition. Often enough, the project comes to naught, the inspiration turns out to have been vain: the posited and supposed difference turns out to have been an illusion, an appearance incapable of appearing – incapable, in other words, of objective self-production through the use of appropriate means (materials and *matériel*). The infinity of the project, easily mistaken (subjectively) for the infinity of meaning, aborts. The originality of the outline was a superfluity, its novelty a mere impression or conceit.

The enigma of the body – its secret, at once banal and profound – is its ability, beyond 'subject' and 'object' (and beyond the philosophical distinction between them), to produce differences 'unconsciously' out of repetitions – out of gestures (linear) or out of rhythms (cyclical). In the misapprehended space of the body, a space that is both close by and distant, this paradoxical junction of repetitive and differential – this most basic form of 'production' – is forever occurring. The body's secret is a dramatic one, for the time thus brought into being, though a bearer of the new, as in the progression from immaturity to maturity, also brings forth a terrible and tragic repetition – indeed the ultimate repetition: old age and death. This is the supreme difference.

Abstract space (or those for whom it is a tool) makes the relationship between repetition and difference a more antagonistic one. As we have seen, this space relies on the repetitive – on exchange and interchangeability, on reproducibility, on homogeneity. It reduces differences to induced differences: that is, to differences internally acceptable to a set of 'systems' which are planned as such, prefabricated as such – and which as such are completely redundant. To this reductive end no means is spared – not corruption, not terrorism, not constraint, not violence. (Whence the great temptation of counter-violence, of counter-terror, as a way of restoring difference in and through use.) Destruction and self-destruction, once accidental, have been transformed into laws of life.

Just like the fleshly body of the living being, the spatial body of society and the social body of needs differ from an 'abstract corpus' or 'body' of signs (semantic or semiological – 'textual') in the following respect: they cannot live without generating, without producing, without creating *differences*. To deny them this is to kill them.

Not far above this lower limit of 'being' are to be found certain struggling producers, among them architects, 'urbanists' and planners. There are others, however, who are perfectly at home here, in dominated space, manipulating exchangeable and interchangeable, quantities and signs – sums of money, 'real property', boxes for living in, technologies and structures.

The architect occupies an especially uncomfortable position. As a scientist and technician, obliged to produce within a specified framework, he has to depend on repetition. In his search for inspiration as an artist, and as someone sensitive to use and to the 'user', however, he has a stake in difference. He is located willy-nilly within this painful contradiction, forever being shuttled from one of its poles to the other. His is the difficult task of bridging the gap between product and work, and he is fated to live out the conflicts that arise as he desperately seeks to close the ever-widening gulf between knowledge and creativity.

The 'right to difference' is a formal designation for something that may be achieved through practical action, through effective struggle – namely, concrete differences. The right to difference implies no entitlements that do not have to be bitterly fought for. This is a 'right' whose only justification lies in its content; it is thus diametrically opposed to the right of property, which is given validity by its logical and legal form as the basic code of relationship under the capitalist mode of production.

XXIX

Some theorists of art and architecture (Umberto Eco, for instance) insist heavily and at length upon the differential role of semiological elements, including the curve and the straight line, the square form and the circular (or 'radical–concentric') form. This emphasis has a certain justification, and the concept of a semantic or semiotic 'differential' is not without its utility. Once the distinction between minimal (induced) differences and maximal (produced) differences is brought to the fore, however, things appear in a somewhat changed light. To build a few blocks of flats that are spiral in form by adding a handful of curves to the usual concrete angularities is not an entirely negligible achievement – but neither does it amount to very much. To take inspiration from Andalusia, and demonstrate a sensual use of curvatures, spirals, arabesques and inflexions of all kinds, so achieving truly voluptuous spaces, would be a different matter altogether. Neither the plant world nor the mineral world has as yet delivered itself of all the lessons it holds regarding space and the pedagogy of space. Within a given genus or species of plant, 'nature' *induces* differences; no two trees, nor even two leaves of a single tree, are completely identical – a fact noted by Leibniz in his exploration of the paradoxical relationship between identity and repetition on the one hand and dissimilarity and differentiation on the other. Yet nature, at another level, also *produces* differences: different species; different vegetable or animal forms; trees with a different texture, a different stance, or a different type of leaf. And all these differences are produced *within* the realm of the tree form, which is of course circumscribed by its own limiting conditions.

Why should spaces created by virtue of human understanding be any less varied, as works or products, than those produced by nature, than landscapes or living beings?

XXX

We can now begin to see the full implications of difference, which ultimately generates the contradiction between *true space* and the *truth of space*.

True space, the space of philosophy and of its epistemological off-shoot, seamless in all but an abstract sense, wrapped in the mantle of science, takes form and is formulated in the head of a thinker before being projected onto social and even physical 'reality'. Every effort is made to legitimize it by appealing to knowledge and to the formal kernel of knowledge. It is thanks to true space that we witness the rise of 'theoretical man' – the rise of the human realm reduced to the realm of knowledge, conceptualization passed off as direct experience. A kernel of knowledge thus claims necessary and sufficient status; and the centre aspires to be definite and definitive – and hence also absolute. It is of little consequence whether such claims are buttressed by political economy, by history, or by linguistics – whether or not ecology is called upon to fill in gaps in the picture – for the strategic approach is identical in every case. And so is the goal sought.The results are a super-dogmatism, sometimes unaccompanied by any clear-cut dogma, and an arrogant attitude which carries the old system-building of the philosophers to a new extreme. The stage of destruction and of self-destruction is soon reached. True space is a mental space whose dual function is to reduce 'real' space to the abstract and to induce minimal differences. Dogmatism of this kind serves the most nefarious enterprises of economic and political power. Science in general and each scientific specialization separately are the immediate servants of both administration and production within the framework of the dominant mode of production. The official account makes no bones about the fact that society's administrators feel the need for assistance from science when they find themselves confronted by 'an increasingly complex environment' with which they would like to establish a 'new relationship'. This 'public service' role assumed by a philosophy and science now installed and constituted as an official knowledge is legitimated by conflating mental space and political space, so constructing a 'system' whose long-lived and solid prototype is Hegelianism. In consequence, not only the idea of the True, but also that of meaning, and those of lived experience and of 'living', are severely compromised. Representational space disappears into the representation of space – the latter swallows the former; and spatial practice, put into brackets along with social practice as a whole, endures only as the unthought aspect of the thought that has now pronounced itself sovereign ruler.

By contrast, running counter to this dominant and official tendency, the *truth of space* ties space on the one hand to social practice, and on the other hand to concepts which, though worked out and linked theoretically by philosophy, in fact transcend philosophy as such precisely by virtue of their connection with practice. Social space calls for a theory of production, and it is this theory that confirms its truth.

The truth of space reveals what mental space and social space have in common – and consequently also the differences between them. There is no rift between the two, but there is a distance. There is no confusion between them, but they do have a common moment or element. Knowledge, consciousness and social practice may thus all be seen to share the *centre*. There is no 'reality' without a concentration of energy, without a focus or core – nor, therefore, without the dialectic: centre-periphery, accretion–dissipation, condensation–radiation, glomeration –saturation, concentration–eruption, implosion–explosion. What is the 'subject'? A momentary centre. The 'object'? Likewise. The body? A focusing of active (productive) energies. The city? The urban sphere? Ditto.

The *form of centrality* which, as a form, is empty, calls for a content and attracts and concentrates particular objects. By becoming a locus of action, of a sequence of operations, this form acquires a *functional* reality. Around the centre a *structure* of (mental and/or social) space is now organized, a structure that is always of the moment, contributing, along with form and function, to a practice.

The notion of *centrality* replaces the notion of *totality*, repositioning it, relativizing it, and rendering it dialectical. Any centrality, once established, is destined to suffer dispersal, to dissolve or to explode from the effects of saturation, attrition, outside aggressions, and so on. This means that the 'real' can never become completely fixed, that it is constantly in a state of mobilization. It also means that a general *figure* (that of the centre and of 'decentring') is in play which leaves room for both repetition and difference, for both time and juxtaposition.

What we have been considering, then, is an extension, after a hiatus, of traditional philosophy and of Marxist thought, an extension which embraces the radical critique of philosophy without, however, abandoning Hegel's teaching on the concrete universal and the import of the concept. We are concerned, in other words, with theory beyond system-building.

The truth of space thus leads back (and is reinforced by) a powerful Nietzschean sentiment: 'But may the will to truth mean this to you: that everything shall be transformed into the humanly-conceivable, the humanly-evident, the humanly-palpable! You should follow your own senses to the end. [*Eure eignen Sinne sollt ihr zu Ende denken.*][10] Marx,

---

[10] Friedrich Nietzsche, 'On the Blissful Islands', in *Thus Spoke Zarathustra*, tr. R. J. Hollingdale (Harmondsworth, Middx: Penguin, 1961), p. 110.

for his part, called in the *Manuscripts of 1844* for the senses to become theoreticians in their own right. The revolutionary road of the human and the heroic road of the superhuman meet at the crossroads of space. Whether they then converge is another story.

7

Openings and Conclusions

I

There is a question implicit in the foregoing analyses and interpretations. It is this: what is the mode of existence of social relations?

No sooner had the social sciences established themselves than they gave up any interest in the description of 'substances' inherited from philosophy: 'subject' and 'object', society 'in itself', or the individual or group considered in isolation. Instead, like the other sciences, they took *relationships* as their object of study. The question is, though, where does a relationship reside when it is not being actualized in a highly determined situation? How does it await its moment? In what state does it exist until an action of some kind makes it effective? Referring vaguely to global praxis is a distinctly inadequate way of responding to these questions. In analysing the social relationship, it is impossible simply to dub it a *form*, for the form as such is empty, and must have a content in order to exist. Nor can it be treated as a *function*, which needs objects if it is to operate. Even a *structure*, whose task it is to organize elementary units within a whole, necessarily calls for both the whole and the component units in question. Thus analytic thought finds itself returning, by virtue of its own dynamic, to the very entities and 'substantialities' that it had originally banished: to 'subject' and 'object', to the unconscious, to global praxis, and so on.

Granted, then, that *a social relationship cannot exist without an underpinning*, we still have to ask how that underpinning 'functions'. The 'material substrate' that historians and sociologists are inclined to see in the population, or among everyday objects of utility, does not supply an answer. What, it may be asked, is the relationship of the 'underpinning' to the relationship that it supports and bears? Thus to complicate the question by rephrasing it at a meta-level, while it brings us no closer to an answer, does at least show up the difficulty. The theoreticians of the Logos and of language (Hegel and Marx themselves) saw the problem clearly: there can be no thought, no reflection, without language, and no language without a material underpinning – without the senses, without mouths and ears, without the disturbance of masses of air, without voices and the emission of articulated signs. There are two antithetical ways of interpreting this. For some, among them Hegel and presumably Marx, these 'conditions' are realized because they 'express' a pre-existing rationality. For others, by contrast, meanings and signs 'express' nothing – they are arbitrary, and linked solely by the requirements of differences of an induced kind within a set of conventions. So far has this argument from the arbitrariness of the sign been carried that language itself has been brought into question, and it has become necessary to introduce new underlying factors such as the body, drives, and so on.

The solution based on the intervention of a pre-existing Logos, at once substantial and eternal, does not effectively put the question to rest, because it simply re-emerges on a different level. Both Hegel and Marx were thus led by their analyses to identify 'things/not-things' – or concrete abstractions: in Hegel's case, the concept; in Marx's, the commodity. *Things* – which for Marx are the product of social labour, destined to be exchanged, and invested for this reason with value in a double sense, with use value and exchange value – both embody and conceal social relations. Things would thus seem to be the underpinning of those relations. And yet, on the Marxist analysis, it is clear that things *qua* commodities cease to be things. And inasmuch as they remain things they become 'ideological objects' overburdened with meanings. *Qua* commodities, things can be resolved into relations; their existence is then purely abstract – so much so, indeed, that one is tempted to see nothing in them apart from signs and signs of signs (money). The question of the underpinning is thus not entirely answered by the postulation of a permanent material world. In the context of our present discussion, this question arises, in the first place, apropos of social space. This space qualifies as a 'thing/not-thing', for it is neither a substantial reality nor a mental reality, it cannot be resolved into abstractions, and it consists neither in a collection of things in space nor in an aggregate of occupied places. Being neither space-as-sign nor an ensemble of signs related to space, it has an actuality other than that of the abstract signs and real things which it includes. The initial basis or foundation of social space is nature – natural or physical space. Upon this basis are superimposed – in ways that transform, supplant or even threaten to destroy it – successive stratified and tangled networks which, though always material in form, nevertheless have an existence beyond their materiality: paths, roads, railways, telephone links, and so on. Theory has shown that no space disappears completely, or is utterly abolished in the course of the process of social development – not even the natural place where that process began. 'Something' always survives or endures – 'something' that is not a *thing*. Each such material underpinning has a form, a function, a structure – properties that are necessary but not sufficient to define it. Indeed, each one institutes its own particular space and has no meaning or aim apart from that space. Each network or sequence of links – and thus each space – serves exchange and use in specific ways. Each is *produced* – and serves a purpose; and each wears out or is consumed, sometimes unproductively, sometimes productively. There is a space of speech whose prerequisites, as we have seen, are the lips, the ears, the ability to articulate, masses of air, sounds, and so on. This is a space, however, for which such material preconditions are not an adequate definition: a space of actions and of inter-actions, of calling and of calling back and forth, of expressiveness and power, and – already at this level – of latent violence and revolt; the space, then, of a discourse that does not coincide with any discourse on or in space. The space of speech envelops the space of bodies and develops by means of traces, of writings, of prescriptions and inscriptions.

As for the commodity in general, it is obvious that kilograms of sugar, sacks of coffee beans and metres of fabric cannot do duty as the material underpinning of its existence. The stores and warehouses where these things are kept, where they wait, the ships, trains and trucks that transport them – and hence the routes used – have also to be taken into account. Furthermore, having considered all these objects individually, one still has not properly apprehended the material underpinning of the world of commodities. Nor do such notions as 'channel', derived from information theory, or 'repertoire', help us define such an ensemble of objects. The same goes for the idea of 'flows'. It has to be remembered that these objects constitute relatively determinate networks or chains of exchange within a space. The world of commodities would have no 'reality' without such moorings or points of insertion, or without their existing as an ensemble. The same may be said of banks and banking-networks vis-à-vis the capital market and money transfers, and hence vis-à-vis the comparison and balancing of profits and the distribution of surplus value.

Ultimately all these processes debouch into the space of the planet as a whole, with its multiplicity of 'layers', networks and sets of links: the world market and the division of labour that it subsumes and develops, the space of computer science, of strategic perspectives, and so on. Among the levels falling under the aegis of this planetary space are those of architecture, of urbanism, and of spatial planning.

The 'world market' is in no sense a sovereign entity, nor must it be thought of as an instrumental reality manipulated by imperialisms in full and absolute control. Solid in some respects, fragile in others, it has a dual character as commodities market on the one hand and as capital market on the other – and because of this duality it is impossible unconditionally to attribute logic or coherence to it. We know that the *technical division of labour* introduces *complementarities* (rationally linked operations), whereas its *social division* generates disparities, distortions and conflicts in a supposedly 'irrational' manner. Social relations do not disappear in the 'worldwide' framework. On the contrary, they are reproduced at that level. Via all kinds of interactions, the world market creates configurations and inscribes changing spaces on the surface of the earth, spaces governed by conflicts and contradictions.

Social relations, which are concrete abstractions, have no real existence save in and through space. *Their underpinning is spatial.* In each particular case, the connection between this underpinning and the relations it supports calls for analysis. Such an analysis must imply and explain a genesis and constitute a critique of those institutions, substitutions, transpositions, metaphorizations, anaphorizations, and so forth, that have transformed the space under consideration.

II

Propositions of this kind themselves imply and explain a project – namely, the quest for a knowledge at once descriptive, analytic and global. If one had to label such an endeavour, it might be termed 'spatio-analysis' or 'spatiology'. This would be consistent with – and in a sense offer a response to – certain terms already in use, such as 'semio-analysis' or 'socio-analysis' (not to mention 'psychoanalysis'). There is thus a certain advantage to be obtained by using one of these names, but the drawbacks are many. In the first place, the basic idea could be obscured, for the knowledge sought here is not directed at space itself, nor does it construct models, typologies or prototypes of spaces; rather, it offers an exposition of the *production of space*. A science of space or 'spatio-analysis' would stress the *use* of space, its qualitative properties, whereas what is called for is a knowledge (*connaissance*) for which the critical moment – i.e. the critique of established knowledge (*savoir*) – is the essential thing. Knowledge of space so understood implies the critique of space.

Lastly, a 'spatio-analytic' approach could confuse and hence compromise the idea of an *analysis of rhythms* – an idea that may be expected to put the finishing touches to the exposition of the production of space.

The whole of (social) space proceeds from the body, even though it so metamorphoses the body that it may forget it altogether – even though it may separate itself so radically from the body as to kill it. The genesis of a far-away order can be accounted for only on the basis of the order that is nearest to us – namely, the order of the body. Within the body itself, spatially considered, the successive levels constituted by the senses (from the sense of smell to sight, treated as different within a differentiated field) prefigure the layers of social space and their interconnections. The passive body (the senses) and the active body (labour) converge in space. The analysis of rhythms must serve the necessary and inevitable restoration of the total body. This is what makes 'rhythm analysis' so important. It also explains why such an approach calls for more than a methodology or a string of theoretical concepts, more than a system all of whose requirements have been satisfied.

III

With respect to traditional philosophy, the type of inquiry and theoretical activity in which we are engaged here may be described as *metaphilosophy*. The task of metaphilosophy is to uncover the characteristics of the philosophy that used to be, its language and its goals, to demonstrate their limitations and to transcend them. Nothing of the old philosophical quest will be abolished in the process – neither its categories, nor its basic theme, nor the set of problems with which it concerned itself. The fact is, however, that philosophy proper came to a halt when faced with contradictions that it had called forth but could not resolve. Thus space, for the philosophers, was split into two: into intelligible space on the one hand (the essence and transparency of the spiritual absolute), and unintelligible space on the other (the degradation of the spirit, absolute naturalness outside the spiritual realm). Consequently, they opted now for one, now for the other – now for space-as-form, now for space-as-substance, sometimes for the luminous space of the Cosmos, sometimes for the shadow-filled space of the world.

Philosophy *per se* cannot surmount these splits and separations; they are part and parcel of the philosophical attitude *per se*, which is by definition speculative, contemplative and systematizing – cut off from social practice and active political criticism. Metaphilosophy, so far from pursuing the metaphors of traditional philosophy, rejects them. The philosopher, 'caught in the web of words', is left behind as soon as meditation begins to deal with time and space instead of being imprisoned by them.

The critique of philosophy as an ideology is fraught with difficulty, for the concept of truth and the truth of the concept have to be saved from the degeneration and destruction towards which philosophical systems on their own downward path tend to drag them. This is a task that must remain unfinished here, but it will be taken up elsewhere, notably in the context of a confrontation between the most powerful of 'syntheses' – that of Hegel – and its radical critique; this critique is rooted on the one hand in social practice (Marx), and on the other hand in art, poetry, music and drama (Nietzsche) – and rooted, too, in both cases, in the (material) body.

As noted, philosophy stopped dead when it came face to face with the 'subject' and the 'object' and their relationship.

As to the 'subject', philosophically privileged in the Western tradition in the shape of the *cogito* of the thinking 'I' (whether in its empirical or transcendental version), it simply dissolved – and it did so as much practically as theoretically. Yet the problem of the 'subject', as raised by philosophy, remains a fundamental one. But what 'subject'? This question is echoed by another – what 'object'? – for a true account is equally needful in the case of the relationship to the 'object'. The object, just as easily as the subject, may assume a burden of ideology (of signs and meanings). By conceiving of the subject without an object (the pure thinking 'I' or *res cogitans*), and of an object without a subject (the body-as-machine or *res extensa*), philosophy created an irrevocable rift in what it was trying to define. After Descartes, the Western Logos sought vainly to stick the pieces back together and make some kind of montage. But the unification of subject and object in such notions as 'man' or 'consciousness' succeeded only in adding another philosophical fiction to an already long list of such entities. Hegel came close to a solution, but after him the dividing-line between the *conceived* and the directly *lived* was restored as the outer frontier of the Logos and the limit of philosophy as such. The theory of the arbitrariness of the sign, which once laid claim to impeccable scientific credentials as a sort of distillate of knowledge, served to exacerbate the rift between expressive and meaningful, signifier and signified, mental and real, and so on.

Western philosophy has *betrayed* the body; it has actively participated in the great process of metaphorization that has *abandoned* the body; and it has *denied* the body. The living body, being at once 'subject' and 'object', cannot tolerate such conceptual division, and consequently philosophical concepts fall into the category of the 'signs of non-body'. Under the reign of King Logos, the reign of true space, the mental and the social were sundered, as were the directly lived and the conceived, and the subject and the object. New attempts were forever being made to reduce the external to the internal, or the social to the mental, by means of one ingenious topology or another. Net result? Complete failure! Abstract spatiality and practical spatiality contemplated one another from afar, in thrall to the visual realm. In contrast, under the rule of *raison d'état*, as elevated in Hegel's philosophy to ultimate supremacy, knowledge and power contracted a solid – and legalized – alliance. Both desire with its subjectivism and ideas with their objectivism respected this alliance and followed a hands-off policy with regard to the Logos.

Today the body is establishing itself firmly, as base and foundation, *beyond philosophy*, beyond discourse, and beyond the theory of discourse. Theoretical thought, carrying reflection on the subject and the object beyond the old concepts, has re-embraced the body along with space, in space, and as the generator (or producer) of space. To say that such theoretical thinking goes 'beyond discourse' means that it takes account, for the purposes of a pedagogy of the body, of the vast store of non-formal knowledge embedded in poetry, music, dance and theatre. This store of non-formal knowledge (*non-savoir*) constitutes a potential true knowledge (*connaissance*). What 'beyond philosophy' means is: beyond the locus of substitutions and separations, beyond the vehicle of the metaphysical and the anaphoric. The realm beyond philosophy indeed finds its essential voice in the negation of anaphora – of that process by means of which philosophers have furthered the body's metamorphosis into abstractions, into signs of non-body. As for 'meta-philosophy', the term implies preserving philosophical concepts in their breadth while changing their connotations, while replacing their old 'objects' with new ones. We are speaking, therefore, of the abolition of Western metaphysics, of a tradition of thought running from Descartes to the present day via Hegel, a tradition that has been successfully incorporated into a society based on *raison d'état*, and at the same time into a particular conception of space and a particular spatial reality.

King Logos is guarded on the one hand by the Eye – the eye of God, of the Father, of the Master or Boss – which answers to the primacy of the visual realm with its images and its graphic dimension, and on the other hand by the phallic (military and heroic) principle, which belongs, as one of its chief properties, to abstract space.

The standing of time as it relates to this space is problematic, and has yet to be clearly defined. When religion and philosophy took duration under their aegis, time was in effect proclaimed a mental reality. But spatial practice – the practice of a repressive and oppressive space – tends to confine time to productive labour time, and simultaneously to diminish living rhythms by defining them in terms of the rationalized and localized gestures of divided labour.

Clearly time cannot achieve emancipation at one stroke, or *en bloc*. It is not so obvious, however, that such a liberation calls necessarily for morphological inventions or for a production of space. That could only be clearly established if it were possible to show that such an appropriation cannot be effected by diverting already existing spaces or morphologies.

IV

What many people look upon as the conclusion of a well-defined period, as the end of this or that (capitalism, poverty, history, art, etc.), or else as the institution of something new and definitive (an equilibrium, a system, etc.), should really be conceived of solely as a *transition*. Not exactly in Marx's sense, however. It is true that a theory of 'long-term' transition may also be found in Marx, for whom history as a whole – which he sometimes on this account refers to as 'prehistory' – serves as a transition between primitive and fully developed communism. This thesis is dependent upon Hegelian notions of the dialectic and of the negative. Our present approach is also based on an analysis of the overall process and its negative aspects, on an analysis that is tied to practice. The transition here considered is characterized first of all by its contradictions: contradictions between (economic) growth and (social) development, between the social and the political, between power and knowledge (*connaissance*), and between abstract and differential space. This short list includes only some of the contradictions concerned and is not intended as a ranking in any sense; its purpose is merely to give some idea of the poisonous flowers that adorn the present period. To define this period properly, we must also show whence it has come and whither it is bound – its *terminus a quo* and its *terminus ad quem*.

Its origins lie very far away from us, in an initial non-labour, in a nature that creates effortlessly, that gives instead of selling, a nature in which cruelty is hard to distinguish from largesse, in which pleasure and pain are not obviously separate. In this sense it is true to say, no matter how worn out and restricted in meaning the phrase may now be, that 'art imitates nature' – except for the fact that art seeks to separate sensual delight from suffering, and indeed to come down on the side of joy.

The period through which modernity struggles to make its way is headed towards another non-labour – that non-labour which is the goal of labour and the ultimate significance of the accumulation of means (technology, knowledge, machinery). A goal and a significance that are still far distant, however – and that will never be realized without risking catastrophe, or without bittersweet leave-takings of everything once valued, everything once triumphant. The bitter analytics of finiteness, as brought to the fore by post-Hegelian philosophy, and made fashionable by a host of 'moderns' since Valéry, is forever repeating the same message: the world is finite, time has run out, the reign of finitude is upon us.

The same dialectical process leads from primary and primordial nature to a 'second nature', from natural space to a space which is at once a product and a work, combining art and science within itself. The coming to maturity of this second nature is a slow and laborious process: its motor is automation, which is constantly pushing forward into the vast realm of necessity – the realm, that is to say, of the production of things in space. The process cannot be completed until the seemingly interminable period taken up by (infinitely divided) labour, by accumulation (of wealth, of materials and *matériel*) and by reductions (i.e. obstacles to development generated by established knowledge and power) has come to an end. This is a process of gigantic proportions, beset by risks and perils of all kinds, and liable to abort at the very moment when the door to new possibilities is opened.

The vast transition which we have thus characterized in terms of a few major rifts may be defined in many different but convergent ways. Space bears clear traces of the process – indeed more than traces: its very form stems from the dominance of the male principle, with its violence and love of warfare; and this principle has in turn been reinforced by the supposedly manly virtues, as promoted by the norms inherent to a dominated and dominating space. Whence the use and overuse of straight lines, right angles, and strict (rectilinear) perspective. The masculine virtues which gave rise to domination by this space can only lead, as we are only too well aware, to a generalized state of deprivation: from 'private' property to the Great Castration. It is inevitable in these circumstances that feminine revolts should occur, that the female principle should seek revenge. Were such a movement to take the form of a feminine 'racism' which merely inverted the masculine version, it would be a pity. Is a final metamorphosis called for that will reverse all earlier ones, destroying phallic space and replacing it with a 'uterine' space? We can be sure, at any rate, that this in itself will not ensure the invention of a truly appropriated space, or that of an architecture of joy and enjoyment. The contradiction *may* therefore be resolved in this way, and the split bridged. But not necessarily.

We may therefore justifiably speak of a transitional period between the mode of production of things in space and the mode of production of space. The production of things was fostered by capitalism and controlled by the bourgeoisie and its political creation, the state. The production of space brings other things in its train, among them the withering-away of the private ownership of space, and, simultaneously, of the political state that dominates spaces. This implies a shift from domination to appropriation, and the primacy of use over exchange (the withering-away of exchange value). If these events do not occur, the worst surely will – as suggested by a number of 'scenarios of the unacceptable' scripted by the futurologists. Meanwhile, it is thanks only to the notion of a conflict-laden transition from one mode of production (that of things) to another (that of space) that it is possible to preserve the Marxist thesis of the fundamental role of the forces of production while at the same time liberating this thesis from the ideology of productivity and from the dogma of (quantitative) growth.

V

Space is becoming the principal stake of goal-directed actions and struggles. It has of course always been the reservoir of resources, and the medium in which strategies are applied, but it has now become something more than the theatre, the disinterested stage or setting, of action. Space does not eliminate the other materials or resources that play a part in the socio-political arena, be they raw materials or the most finished of products, be they businesses or 'culture'. Rather, it brings them all together and then in a sense substitutes itself for each factor separately by enveloping it. The outcome is a vast movement in terms of which space can no longer be looked upon as an 'essence', as an object distinct from the point of view of (or as compared with) 'subjects', as answering to a logic of its own. Nor can it be treated as a result or resultant, as an empirically verifiable effect of a past, a history or a society. Is space indeed a medium? A milieu? An intermediary? It is doubtless all of these, but its role is less and less neutral, more and more active, both as instrument and as goal, as means and as end. Confining it to so narrow a category as that of 'medium' is consequently woefully inadequate.

Differential analysis has continually stressed the *constitutive dualities* of social space, dualities which underpin more complex – and, most importantly, triadic – determinations. These initial dualities (symmetries/asymmetries; straight lines/curves; and so on), have repeatedly re-emerged, embedded in each successive recasting of social space, acquiring new meanings in the process and invariably subordinated to the overall movement. As the underpinning of production and reproduction, abstract space generates illusions, and hence a tendency towards false consciousness, i.e. consciousness of a space at once imaginary and real. Yet this space itself, and the practice that corresponds to it, give rise, by virtue of a critical moment, to a clearer consciousness. No science has as yet offered an account of this generative process, and this is as true of ecology as it is of history. Differential analysis brings out the variations, pluralities and multiplicities which introduce themselves into genetically senior dualities, as well as the disparities, disjunctions, imbalances, conflicts and contradictions that emerge from them. Because of the diversity of the processes involved, the above exposition may have left the impression that abstract space has no clearly defined status. But that it is absolutely not so: theory has in fact pinpointed the truth of this space – namely, its contradictory character within the framework of the dominant tendency towards homogeneity (i.e. towards the establishment of a dominated space).

Where should we look for *logic* in this context? At what level is it located? At that of a praxeology of space? Within some particular system – spatial, planning-related, or urban? Or within the empirical sphere, as part of the employment of space as a tool? The answer, in all cases, is no. Rather, logic characterizes a double imposition of force: first in order to maintain a coherence and, later, in the shape of reductionism, in the shape of the strategy of homogenization and the fetishization of cohesiveness in and through reductions of all kinds. It is logic that governs the capacity – bound up with violence – to separate what has hitherto been joined together, to fracture all existing unities. This initial hypothesis concerning the relationship between logical and dialectical has been successively validated and upheld by argument and proof.

VI

It is impossible, in fact, to avoid the conclusion that space is assuming an increasingly important role in supposedly 'modern' societies, and that if this role is not already preponderant it very soon will be. Space's hegemony does not operate solely on the 'micro' level, effecting the arrangement of surfaces in a supermarket, for instance, or in a 'neighbourhood' of housing-units; nor does it apply only on the 'macro' level, as though it were responsible merely for the ordering of 'flows' within nations or continents. On the contrary, its effects may be observed on all planes and in all the interconnections between them. The theoretical error that consists in restricting the import of space to a single discipline – to anthropology, political economy, or sociology, for example – has been dealt with above. A number of theoretical conclusions still need to be drawn, however, from these observations.

Formerly each society to which history gave rise within the framework of a particular mode of production, and which bore the stamp of that mode of production's inherent characteristics, shaped its own space. We have seen by what means this was done: by violence (wars and revolutions), by political and diplomatic cunning, and, lastly, by labour. The space of any such society might justifiably be described as a 'work'. The ordinary meaning of this term, as applied to an object emerging from the hands of an artist, may very well be extended to the result of a practice on the plane of a whole society. As for a village – or a particular countryside – how could it *not* fall into this category? Already on this level, clearly, product and work are one and the same.

Today our concern must be with space on a world scale (and indeed – beyond the surface of the earth – on the scale of interplanetary space), as well as with all the spaces subsidiary to it, at every possible level. No single place has disappeared completely; and all places without exception have undergone metamorphoses. What agency shapes space worldwide? None – no force, no power. For forces and powers contend with one another within space, strategically, in such a way that history, historicity, and the determinisms associated with these temporal notions lose their meaning.

A number of causes and reasons emerge spontaneously from the obscurity of history in connection with this new situation, which is an increasingly important aspect of 'modernity'. They reveal themselves sufficiently, in fact, for reflective thought to get a sense of the multiplicity of their interactions. Among them are the world market (commodities, capital, manpower, etc.), technology and science, and demographic pressures – each striving for the status of an autonomous force. A paradox that we have already mentioned and emphasized is the fact that the political power which holds sway over 'men', though it dominates the space occupied by its 'subjects', does not control the causes and reasons that intersect within that space, each of which acts by and for itself.

Such more or less independent causes and reasons coexist in the space constituted by their effects, consequences and results; as enumerated by the experts, these include pollution of various kinds, the potential exhaustion of resources, and the destruction of nature. A good number of disciplines – ecology or demography, geography or sociology – describe these results, without going back to causes and reasons, as partial systems. What we have sought to do here is bring together causes and effects, consequences and reasons, in such a way as to transcend divisions between scientific domains and specializations, and to propose a unitary theory. 'Unitary' here must not be taken as implying that reasons or consequences, or causes and effects, are in any way confused or muddled on the basis of their spatial simultaneity or their more or less peaceful coexistence. Just the opposite, in fact. The theoretical conception we are trying to work out in no way aspires to the status of a completed 'totality', and even less to that of a 'system' or 'synthesis'. It implies discrimination between 'factors', elements or moments. To reiterate a fundamental theoretical and methodological principle, this approach aims both to reconnect elements that have been separated and to replace confusion by clear distinctions; to rejoin the severed and reanalyse the commingled.

A distinction has to be drawn between the problematic of space and spatial practice. The former can only be formulated on a theoretical plane, whereas the latter is empirically observable. It is not hard, however, for an ill-informed approach, one that misunderstands the method and the concepts involved, to confuse the two. The 'problematic' – the term is borrowed from philosophy – of space is comprised of questions about mental and social space, about their interconnections, about their links with nature on the one hand and with 'pure' forms on the other. As for spatial practice, it is observed, described and analysed on a wide range of levels: in architecture, in city planning or 'urbanism' (a term borrowed from official pronouncements), in the actual design of routes and localities ('town and country planning'), in the organization of everyday life, and, naturally, in urban reality.

Knowledge has been built up on the basis of (global) schemata. Once such schemata were *atemporal*, as in the case of classical metaphysics. After Hegel, however, they became *temporal* in character, which is to say that they proclaimed the priority of historical becoming, of mental duration, or of socio-economic time, over space. This theoretical posture cried out to be overturned – something that has indeed been attempted, though on indefensible grounds, by those eager to assert a priority of geographical, or demographic, or ecological space over historical time. In point of fact all these sciences are already the battleground of an immense confrontation between the temporal and the spatial. This confrontation is not one which could precipitate a crisis of knowledge, or force a reconsideration of the relationship of knowledge to a political power which is so effective with respect to people, yet so impotent as regards those determinations (technological, demographic, etc.) that put their stamp on abstract space, so producing that space as such and reproducing social relations within it.

Languages, each in particular and all of them in general, all linguistic systems, including that of established knowledge, are spoken and written in a mental time and space to which that knowledge tends to assign a privileged metaphysical status. They are clumsy in the way they give utterance to social time, to spatial practice. How could it be otherwise, considering that ordinary languages, whether lexically or syntactically viewed, have peasant origins, while even the more highly elaborated linguistic systems have theological–philosophical antecedents? As for industry and its techniques, as for the 'modern sciences', they have only just begun to affect vocabulary and grammar. Urban reality has hardly any influence at all – as witness the fact that we simply lack the words for it: the word *usager* ('user'), for example, which has been called upon for the purposes of our present discussion, as yet means very little in French, and it has no established meaning in this context in English. The fact is that languages and linguistic systems need to be dismantled and reconstructed. This task will be carried out by and in (spatial) social practice.

The salvation of knowledge (*connaissance*) depends entirely upon a methodological re-examination of its established forms (*savoir*), which congeal it by means of epistemology and seek to institute a supposedly absolute knowledge which is in fact no more than a pale imitation of divine wisdom. The only road for such a re-examination to take is the unification of critical knowledge with the critique of knowledge. The critical dimension of understanding must be brought to the fore. Collusion between 'knowledge' and 'power' must be forcefully exposed, as must the purposes to which bureaucracy bends knowledge's specialization. When institutional (academic) knowledge sets itself up above lived experience, just as the state sets itself up above everyday life, catastrophe is in the offing. Catastrophe is indeed already upon us.

In the absence of a reconstruction of this kind, knowledge must inevitably collapse under the blows of non-knowledge and the onslaughts of anti-knowledge (or anti-theory) – in short, it must relapse into the European nihilism that Nietzsche believed he had overcome.

To maintain an unselfcritical knowledge can only promote the decline of knowledge. Consider questions about space, for example: taken out of the context of practice, projected onto the plane of a knowledge that considers itself to be 'pure' and imagines itself to be 'productive' (as indeed it is – but only of verbiage), such questions assume a philosophizing and degenerate character. What they degenerate into are mere general considerations on intellectual space – on 'writing' as the intellectual space of a people, as the mental space of a period, and so on.

It is certainly impossible unreservedly to objectify representations or schemata worked out within a mental space and referring to that space itself, even – or rather especially – if they have been developed theoretically by philosophers or rationalized by epistemologists. On the other hand, who can grasp 'reality' – i.e. social and spatial practice – without starting out from a mental space, without proceeding from the abstract to the concrete? No one.

VII

The distinction between *infra* and *supra*, between 'short of' and 'beyond', is just as important as that between 'micro' and 'macro' levels. Thus there are countries and peoples, in the grip of deprivation and need, which must be said to exist 'short of' the everyday realm, because they can only *aspire* to a firmly grounded everyday life; the critique of everyday life becomes meaningful only once this threshold has been passed. Much the same sort of thing is true in the political sphere. *Short of* this sphere, people, groups or nations live and think who are still only part-way along the road that leads via politics to revolutions – or, alternatively, via revolutions to political life. *Beyond* political existence, meanwhile – and hence beyond an established nation state – politics becomes more specific, and political activity more specialized. Politics becomes a profession, and political machines (state and party apparatuses) are institutionalized. This situation in due course gives rise to political criticism – that is, to a radical critique of everyday life and its apparatuses as such; and eventually the political realm will begin to fade away. Once it reaches a certain level of intensity, politicization self-destructs: constant political activity eventually enters into contradiction with its own foundations.

What, then, of the political status of space? No sooner has space assumed a political character than its depoliticization appears on the agenda. A politicized space destroys the political conditions that brought it about, because the management and appropriation of such a space run counter to the state as well as to political parties; they call for other forms of management – loosely speaking, for 'self-management' – of territorial units, towns, urban communities, regions, and so on. Space thus exacerbates the conflict inherent in the political arena and in the state *per se*. It lends great impetus to the introduction of the anti-political into the political, and promotes a political critique which lends its weight to the trend towards the self-destruction of the 'moment of politics'.

VIII

Today everything that derives from history and from historical time must undergo a test. Neither 'cultures' nor the 'consciousness' of peoples, groups or even individuals can escape the loss of identity that is now added to all other besetting terrors. Points and systems of reference inherited from the past are in dissolution. Values, whether or not they have been organized into more or less coherent 'systems', crumble and clash. Sooner or later, the cultivated elites find themselves in the same situation as peoples dispossessed (alienated) through conquest and colonization. These elites find that they have lost their bearings. Why? Because nothing and no one can avoid *trial by space* – an ordeal which is the modern world's answer to the judgement of God or the classical conception of fate. It is in space, on a worldwide scale, that each idea of 'value' acquires or loses its distinctiveness through confrontation with the other values and ideas that it encounters there. Moreover – and more importantly – groups, classes or fractions of classes cannot constitute themselves, or recognize one another, as 'subjects' unless they generate (or produce) a space. Ideas, representations or values which do not succeed in making their mark on space, and thus generating (or producing) an appropriate morphology, will lose all pith and become mere signs, resolve themselves into abstract descriptions, or mutate into fantasies. Can a social group be expected to recognize itself in space merely because that space is held up before it like a mirror? Certainly not. The notion of appropriation implies far more and is far more exigent than the (highly speculative) thesis of a 'mirror-consciousness'. Long-lived morphologies (religious buildings, historical–political monuments) support our antiquated ideologies and representations. New ideas (socialism, for instance), though not without force, have difficulty generating their own space, and often run the risk of aborting; in order to sustain themselves, they may appeal to an obsolete historicity, or assume folkloric or quaint aspects. Viewed from this vantage point, the 'world of signs' clearly emerges as so much debris left by a retreating tide: whatever is not invested in an appropriated space is stranded, and all that remain are useless signs and significations. Space's investment – the production of space – has nothing incidental about it: it is a matter of life and death.

Historical formations flow into worldwide space much like rivers debouching into the ocean: some spread out into a swampy delta, while others suggest the turbulence of a great estuary. Some, in democratic fashion, rely on the force of inertia to ensure their survival; others look to power and violence (of a strategic – and hence military and political – kind).

Trial by space invariably reaches a dramatic moment, that moment when whatever is being tried – philosophy or religion, ideology or established knowledge, capitalism or socialism, state or community – is put radically into question.

With its confrontations and clashes, trial by space does not unfold in the same way for all historical formations, for things are affected by each formation's degree of rootedness in nature and by each's natural peculiarities, as well as by the relative strength of its attachments to the historical realm. And, though nothing and nobody eludes the dramatic moment just mentioned, it does not occur in identical fashion every-where. In other words, trial by space varies in character according to whether it concerns the old European nations, North or Latin America, the peoples of Africa or Asia, and so on. Still, there is no escaping a fate that weighs equally on religion and churches, on philosophy with its great 'systems' – and, of course, on dialectical (and historical) materi-alism. Provided Marx's formulations are not followed slavishly, and provided the most immediate influences upon him are set aside, something new and essential is to be derived from the persistent traces in his thinking of classical rationalism, teleology, and implicit metaphysics. The hypothesis of an ultimate and preordained meaning of historical becoming collapses in face of an analysis of the strategies deployed across the surface of the planet. At the terminal point as at the origin of this process of becoming is the Earth, along with its resources and the objectives that it holds out. Formerly represented as Mother, the Earth appears today as the centre around which various (differentiated) spaces are arranged. Once stripped of its religious and naïvely sexual attributes, the world as planet – as planetary space – can retrieve its primordial place in practical thought and activity.

IX

Confrontations and challenges to the established order can always be attributed ultimately to the 'class struggle'. It is no longer possible, however, to describe the frontiers along which battles rage (both practically and theoretically speaking) as if they corresponded simply to the dividing-line between the territory of the ruling class on the one hand and that of the exploited and oppressed classes on the other. The fact is that such disputed frontiers cross all spheres, including the spheres of the sciences and of knowledge in general, and all sectors of society, extrapolitical as well as political. The great theoretical struggles have strategic objectives which I have sought to point up: reunification of what has been split apart, and effective discrimination of what has been purposefully confused. The separation of quantity from quality, and the attribution to space of a quantity devoid of quality, bespeak misdirection and confusion in connection with the 'nature' of qualities. And vice versa. Philosophy in its decline, stripped now of any dialectical dimension, serves as a bulwark as much for illegitimate separations as for illegitimate confusions.

The answer to separation and dispersion is unification, just as the answer to forced homogenization is the discernment of differences and their practical realization. Struggles directed towards these goals, whether implicitly or explicitly, are waged on many fronts – and along many frontiers; they need have no obvious links with each other; they may be violent or non-violent in character; and some combat the tendency to separate while others combat the tendency to confuse. A politics that separates (by dividing and dispersing space) and fosters confusion (by conflating peoples, regions and spaces with states) continues to be opposed by political means.

X

This book has been informed from beginning to end by a *project*, though this may at times have been discernible only by reading between the lines. I refer to the project of a different society, a different mode of production, where social practice would be governed by different conceptual determinations.

No doubt this project could be explicitly formulated; to do so would involve heightening the distinctions between 'project', 'plan' and 'programme', or between 'model' and 'way forward'. But it is far from certain that such an approach would allow us to make forecasts or to generate what are referred to as 'concrete' proposals. The project would still remain an abstract one. Though opposed to the abstraction of the dominant space, it would not transcend that space. Why? Because the road of the 'concrete' leads via active theoretical and practical negation, via counter-projects or counter-plans. And hence via an active and massive intervention on the part of the 'interested parties'.

In the course of our discussion, we have discerned a host of causes and reasons for the absence of any such intervention, none of them seemingly definitive. The progression of what might be called a 'revolution of space' (subsuming the 'urban revolution') cannot be conceived of other than by analogy with the great peasant (agrarian) and industrial revolutions: sudden uprisings followed by a hiatus, by a slow building of pressure, and finally by a renewed revolutionary outburst at a higher level of consciousness and action – an outburst accompanied, too, by great inventiveness and creativity.

The obstacles faced by counter-plans may be enumerated. The most serious is the fact that on one side, the side of power, there are ranged resources and strategies on a vast scale – the scale, ultimately, of the planet – while in opposition to these forces stand only the limited knowledge and limited interests of generally medium-sized or small territorial spheres (in France for example, regions such as Occitanie, the Landes coast and Brittany). All the same, the necessary inventiveness can only spring from interaction between plans and counter-plans, projects and counter-projects. (Not that such interaction should be seen as excluding ripostes *in kind* to the violence of established political powers.)

The possibility of working out counter-projects, discussing them with the 'authorities' and forcing those authorities to take them into account, is thus a gauge of 'real' democracy. As for the frequently heard suggestion that a choice must be made between 'reductionism' and 'globalism', between restricted and total action, this is the perfect example of a false problem.

XI

These thoughts offer a partial response to the first and last question: 'How does the theory of space relate to the revolutionary movement as it exists today?'

A complete grasp of the theory and its essential articulations is necessary in order to answer the question properly. It is thus worth recalling that the theory of space refuses to take the term 'space' in any trivial or unexamined sense, or to conflate the space of social practice with space as understood by geographers, economists, and others. To accept any such conception of space, whether in the original form or as redefined by a particular discipline, is inevitably to view space as a tool or passive receptacle for the planners, with their talk of 'harmonious development', 'balance' and 'optimum use'.

Space assumes a regulatory role when and to the extent that contradictions – including the contradictions of space itself – are resolved.

Theory contributes to the dismantling of existing society by exposing what gnaws at it from within, from the core of its 'prosperity'. As it expands, this society (neocapitalism or corporate capitalism) can generate only chaos in space. The bourgeoisie, though it has successfully learned how to resolve a number of contradictions inherited from history, managing to achieve a measure of control over markets (something that Marx had not foreseen), and hence a relatively rapid development of the productive forces, will certainly not be able to resolve the contradictions of space (that is, the contradictions of *its* space).

The political organizations of today misconstrue or are ignorant of space and of issues relating to space. Why? This question has profound implications, for it pinpoints and defines the essence of the political. Political organizations are bequeathed to us by history; they prolong history and maintain it ideologically with their continual commemorations and reminders. And further than that they cannot go.

But might what is misunderstood today not be perfectly well understood tomorrow? Might it not indeed be the potential centre of future thought and action?

XII

From the point of view of their respective approaches to space, the Soviet model and the 'Chinese road to socialism' represent an opposition that is tantamount to a contradiction.

The Soviet model has as its starting-point a revision of the capitalist process of accumulation, coupled with a good intention – the desire to improve this process by speeding it up. This reinforced and intensified version of the capitalist model seeks to achieve rapid growth by relying on deliberately privileged 'strong points' – on large-scale enterprises and cities. All other places remain passive and peripheral relative to centres – centres of production, of wealth and of decision. The result is the creation of points of concentration or vortices: the strong points grow ever stronger, the weak ever weaker. Such vortices are seen as having a regulatory role because, once established, they 'function' automatically. Peripheral areas, meanwhile, abandoned to stagnation and (relative) backwardness, are more and more oppressed, controlled and exploited.

The Leninist law of uneven growth and development is thus in no way dealt with, nor are its negative effects countered. Just the opposite, in fact.

The 'Chinese road' testifies to a real concern to draw the people and space in its entirety into the process of building a different society. This process is conceived of as a multidimensional one, involving not only the production of wealth and economic growth but also the development and enrichment of social relationships – implying the production *in* *space* of a variety of goods as well as the production *of space* as a whole, the production of a space ever more effectively appropriated. The rift between strong and weak points would have no place in such a process. Uneven development would disappear or at least tend to disappear. This strategy means that political action will not result in the elevation of either the state or a political formation or party above society. This is the meaning generally given to the 'cultural revolution'. A further implication is dependence on agricultural towns, small or medium-sized, and on the whole range of production units, both agricultural and industrial, from the smallest to the largest, but always with special attention being paid to the smaller, even at the cost, if need be, of a slowing of the pace of production. This spatial orientation and strategy is designed to ensure (barring accidents) that the dichotomy between town and country with its attendant conflicts will dissolve thanks to a transformation of both poles rather than as a result of their degeneration or mutual destruction.

This is not of course to suggest that an industrial country could purely and simply – and without any particular effort – opt for the path followed by a predominantly agrarian one. It does show, however, that the theory of space is capable of accounting for revolutionary experience worldwide.

Revolution was long defined either in terms of a political change at the level of the state or else in terms of the collective or state ownership of the means of production as such (plant, equipment, industrial or agricultural entities). Under either of these definitions, revolution was understood to imply the rational organization of production and the equally rationalized management of society as a whole. In fact, however, both the theory and the project involved here have degenerated into an ideology of growth which, if it is not actually aligned with bourgeois ideology, is closely akin to it.

Today such limited definitions of revolution no longer suffice. The transformation of society presupposes a collective ownership and management of space founded on the permanent participation of the 'interested parties', with their multiple, varied and even contradictory interests. It thus also presupposes confrontation – and indeed this has already emerged in the problems of the 'environment' (along with the attendant dangers of co-optation and diversion).

As for the orientation of the process whose beginnings are thus discernible, we have sought to describe it above. It is an orientation that tends to surpass separations and dissociations, notably those between the *work* (which is unique: an object bearing the stamp of a 'subject', of the creator or artist, and of a single, unrepeatable moment) and the *product* (which is repeatable: the result of repetitive gestures, hence reproducible, and capable ultimately of bringing about the automatic reproduction of social relationships).

On the horizon, then, at the furthest edge of the possible, it is a matter of producing the space of the human species – the collective (generic) work of the species – on the model of what used to be called 'art'; indeed, it is still so called, but art no longer has any meaning at the level of an 'object' isolated by and for the individual.

The creation (or production) of a planet-wide space as the social foundation of a transformed everyday life open to myriad possibilities – such is the dawn now beginning to break on the far horizon. This is the same dawn as glimpsed by the great utopians (who, inasmuch as they demonstrated real possibilities, are perhaps not properly so described): by Fourier, Marx and Engels, whose dreams and imaginings are as stimulating to theoretical thought as their concepts.

I speak of an *orientation* advisedly. We are concerned with nothing more and nothing less than that. We are concerned with what might be called a 'sense': an organ that perceives, a direction that may be conceived, and a directly lived movement progressing towards the horizon. And we are concerned with nothing that even remotely resembles a system.

Afterword

David Harvey

The publication of Henri Lefebvre's magisterial *La production de l'espace*, in an excellent English translation by Donald Nicholson-Smith, is cause for considerable celebration. Few of Lefebvre's voluminous works (see the appended list) have seen the light of day in English, as compared to other foreign languages, and the work and life of one of the great French intellectual activists of the twentieth century is consequently little known to Anglo-American audiences.

Lefebvre was born in 1901 in Hagetmau in the Pyrenees (a region to which he long remained attached and which was later to be the site of his sociological enquiries into rural and peasant societies). His mother was, according to his autobiography, passionately, even fanatically, Catholic, while his father was urbanely anticlerical – the sort of contradiction he was to relish for the rest of his life. His political as well as his intellectual consciousness was shaped by the experience of the First World War, the Russian Revolution and that maelstrom of intellectual change which he invokes in *The Production of Space* as follows:

> ... around 1910 a certain space was shattered. It was the space
> of common sense, of knowledge, of social practice, of political
> power, a space hitherto enshrined in everyday discourse, just as in
> abstract thought, as the environment of and channel for communi-
> cations ... Euclidean and perspectivist space have disappeared as
> systems of reference, along with other former 'commonplaces' such
> as the town, history, paternity, the tonal system in music, tra-
> ditional morality, and so forth. This was a truly crucial moment.

Lefebvre attended the Sorbonne during the 1920s in a period of considerable intellectual effervescence and political turmoil. There he became one of a small group of *jeune philosophes* who, revolting against what they saw as the anachronistic and politically irrelevant establishment philosophy of the time (personified by Bergson), sought, largely through the pages of a radical journal, *Philosophies*, to redefine philosophical endeavours by means of intellectual encounters not only with the thought of Spinoza, Hegel and Nietzsche but also with the philosophical work of Heidegger and Marx (whose collected works were then in course of translation into French for the first time). Lefebvre and his companions refused to see philosophy as an isolated or wholly specialized activity. They thought it important to grapple not only with the progress of science (the theory of relativity, for example), but also with the qualities of daily life – the *quotidien*, as Lefebvre called it both then and in many later works. Eighteen months of military service in the wake of his opposition to France's colonial war in Morocco, followed by two years earning a living as a taxi driver in Paris (an experience which deeply affected his thinking about the nature of space and urban life), kept him from any temptation to an ivory-tower conception of philosophical work.

The *jeune philosophe* was immediately attracted to the artistic and cultural avant-garde movements of the 1920s. One of his first articles in *Philosophies* (1924) was a portrait of Dada which, though not that complimentary, was appreciative enough to bring him lifetime friendship with one of the leading figures of that movement, Tristan Tzara. His contacts with surrealists like Breton and Aragon marked him for life and played a particularly important role after his break with the Communist Party in 1956. His belief in the animating power of spectacle, of poetry, and of artistic practices became crucial in informing Lefebvre's attitude towards and active participation in the revolutionary movements of the 1960s.

Along with the other *jeune philosophes*, as well as many of the surrealists, Lefebvre gradually moved towards the positions espoused by the Communist Party, eventually joining in 1928, the year before he took up a regular position as *professeur de lycée* first at Privas (far from Paris in the Ardèche) and later, in 1932, at Montargis, which had the virtue of being much closer to the capital. His own adherence was to a large degree predicated upon a careful study of Marx's early writings, a growing appreciation of the importance of dialectical and historical materialist method (as manifest in Marx's *Capital*), and a strong feeling, much reinforced by the rise of fascism in Germany and elsewhere (including France), that collective resistance and international organization were essential for any progressive movement of the left.

His early years in the Party were marked both by militant or organizing activity and by the investigation of daily life in various industrial sectors (such as the silk industry in the region of Privas). Lefebvre was also preoccupied with the search for some kind of philosophical foundation or position that could be related to political practice. This proved no easy task, given the requirements and imperatives of the fight against fascism, the emergence of Popular Front politics in France and the growing Stalinism of the French Communist Party. *La conscience mystifiée*, which he published with Herbert Guterman in 1936, examined Marx's conception of alienation and the consciousness and politics which flowed therefrom, but was so badly received by the Party that Lefebvre was dissuaded from continuing with further volumes. His evaluation of Hitlerism was better received, as was his highly influential and widely disseminated expositional work on dialectical materialism, which first appeared in 1939 and which was seized and burned during the Nazi occupation.

By the outbreak of World War II he was already established as a major intellectual figure in the French Communist movement. Fleeing Paris in the face of the Nazi occupation and then removed by the Vichy government from a teaching position he had procured at St Etienne, he joined the Resistance, first in Marseilles and subsequently in the Pyrenees, in the valley of Campan, where he mixed Resistance activities with detailed studies of the life and history of peasant society. These were to make his reputation as a sociologist in the post-war period and ultimately resulted in *La Vallée de Campan*, published in 1963.

From 1945 to 1958, Lefebvre remained within the French Communist Party, but after a brief euphoric period in which he was widely regarded as the Party's leading philosopher (and used the weight of that position to attack, perhaps unwisely, what he saw as the unnecessary idealism of Sartre's existentialism), he found himself in a tense confrontation with the Party's resurgent Stalinism. For example, the French Party accepted Stalin's support of Lysenko's patently erroneous theories of plant breeding and criticised the use of the new high-yielding hybrid seeds then available from America as both bourgeois and counter-revolutionary – a position which Lefebvre thought nonsensical, in part for scientific reasons but also because he saw that by seeking to deny a source of extra productivity to the peasantry, the Party would in the end destroy its credibility with its peasant base (which duly happened in France, in contrast to Italy, where the Communists took a quite independent line).

This was not, however, an uncreative period for Lefebvre. Taking on the role of popularizer of Marxian ideas – a role he evidently relished – he used his position as an established researcher with the government-funded Centre National de Recherche Scientifique to publish a whole stream of critical but accessible evaluations of thinkers such as Descartes (1947), Diderot (1949), Pascal (two volumes in 1949 and 1954), the romantic poet and dramatist Alfred de Musset (1955), Rabelais (1955) and Pignon (1956). The point of these studies was not only to locate the thought and work of such creative writers in the material context of the day, but also to enquire into the creative potentiality of ideas and thought in history – a theme which was to become much more emphatic after he left the Party. These studies were complementary to major works on dialectical materialism (a multi-volume project, the first volume of which appeared in 1947 but which was then abandoned probably because of political pressures), Marxism, and 'the critique of everyday life', (in which he again came perilously close to touching upon the themes of alienation that had been so badly received in Communist circles with the publication of *La conscience mystifiée* in the 1930s).

The break with the Party came in the wake of the publication of the Khruschev Report of 1956, which revealed many of the horrors of Stalinism, but which the French Communist Party refused at first to acknowledge. Lefebvre, having access to the report via German colleagues, entered into an internal oppositional movement within the Party and was ultimately excluded in 1958. It is hard for most of us to understand what it might mean to be excluded from an organization to which one has belonged for some thirty years. The French Communist Party was not only a political party but the hub of its members' social and daily life (it has sometimes been likened to an extended and very close-knit family structure). Lefebvre effectively wrote his way out of the intense social and psychological difficulties generated by the break by writing *La somme et le reste* (1959), an autobiographical, autocritical, and evaluative summary of much of his own life's work in the context of the times.

Lefebvre did not leave the party by the right door but by the left. Liberated from Stalinist constraints, he could explore many of the ideas that had previously been latent by deepening his grasp and practice of Marx's dialectical method (a grasp which Jean-Paul Sartre described in his *Critique de la raison dialectique* as 'beyond reproach'), by exploring the history and sociology of daily life (the origins of modernity, the structure of peasant life, the significance of the 'urban revolution', and the origins of the Paris Commune as an exemplar of the manner in which popular movements could crystallize into an overwhelming revolutionary force), and continuing his enquiry into the role of romanticism, of aesthetic experience, of poetic and cultural endeavours, and of individual creative thought in revolutionary politics. As a professor of sociology first at the University of Strasbourg (1961–5) and then at Nanterre (1965–73), he argued against the structuralism of Althusser, the detachment from everyday life manifest in Foucault, the pessimistic undercurrent entering French philosophy through its engagement with Heidegger, and the historicism and scientism (positivism) becoming hegemonic in academic life. What marked him off from both the Marxist humanists (like Sartre and Merleau-Ponty) and the structuralist Althusserians was his refusal to see any division between the work of the so-called 'young Marx' (lauded by the humanists and denigrated by the Althusserians) and the 'mature Marx' (denigrated by the humanists and lauded by the Althusserians). Life is lived as a project, Lefebvre insisted, and Marx's life had to be seen as a totality of interests, flowing concurrently rather than as fragmented pieces. From that stance, he fought to rescue dialectical materialism from the Marxists, history from the historians, the capacity for revolutionary action from the structuralists and the social from the sociologists.

One of the key concepts he advanced in *La somme et le reste*, for example, was that of the 'moment' which he interpreted as fleeting but decisive sensations (of delight, surrender, disgust, surprise, horror, or outrage) which were somehow revelatory of the totality of possibilities contained in daily existence. Such movements were ephemeral and would pass instantaneously into oblivion, but during their passage all manner of possibilities – often decisive and sometimes revolutionary – stood to be both uncovered and achieved. 'Moments' were conceived of as points of rupture, of radical recognition of possibilities and intense euphoria. This idea was to be put to work to understand sublime moments of revolutionary fervour, such as the day the Paris Commune was declared. It was also to shape the consciousness of many students in the uprising of 1968. The doctrine foreshadowed and to some degree paralleled the ideas of the situationist movement which developed in Paris in the late 1950s. Lefebvre later fell out with the situationists. He provocatively though not altogether unfavourably depicted them as romantics while they accused him of plagiarising their ideas to interpret the Commune and of failing to appreciate the revolutionary potential of their own tactic of creating 'situations' as opposed to what they saw as Lefebvre's more passive stance of experiencing 'moments' when they happened to arise. The continued engagement with situationist ideas (as represented, for example, in Guy Debord's *La Société du spectacle*) seems to have had an important role. For example, Debord's critical observation that the 'moment', as Lefebvre initially conceived of it was purely temporal, as opposed to the spatio-temporality of the 'situation', is tacitly countered in Lefebvre's later works on urbanization and the production of space.

Much of this seemingly theoretical and abstract argument was lived out under the aegis of the student movement which culminated in the extraordinary 'moment' of May, 1968 – a moment which Lefebvre was to describe in intimate and reflective detail in his *L'irruption, de Nanterre au sommet* (1968). Lefebvre is sometimes depicted as 'father' of that movement, and certainly the spark that fired the thousand or so students who crammed into his lectures at Nanterre was important. The parallel between Marcuse's influence within the student movement in the United States and Lefebvre's role in the French context is probably a reasonable one. Both were thinkers and long-time activists who had something important to say to a restless and dissatisfied generation, but that is a far cry from crediting them with paternity of the entire event.

The years after 1968 were taken up with an intense enquiry into the nature of urbanization and the production of space. Seven books were written on these themes between 1968 and 1974, with *La production de l'espace* as the culminating work in the sequence. Lefebvre also co-founded the journal *Espace et Société* which brought together many distinguished young thinkers (the most well-known today being Manuel Castells) who were inspired by his interests. The two themes of urbanization and the production of space are interlinked in Lefebvre's thought. Increasingly during the 1960s, and particularly through the events of 1968, Lefebvre came to recognise the significance of urban conditions of daily life (as opposed to narrow concentration on work-place politics) as central in the evolution of revolutionary sentiments and politics. The significance of the outbreak in Nanterre – a suburban university close to the impoverished shanty-towns of the periphery – and the subsequent geography of street action in Paris itself, alerted him to the way in which these kinds of political struggle unfolded in a distinctively urban space. But consideration of the urban question quickly led him to deny that the city was any kind of meaningful entity in modern life. It had been superseded by a process of urbanization or, more generally, of the production of space, that was binding together the global and the local, the city and the country, the centre and the periphery, in new and quite unfamiliar ways. Daily life, the topic that had engaged his attention before 1968, as well as Marxist theory and revolutionary politics, had to be reinterpreted against this background of a changing production of space.

But it was characteristic of Lefebvre not to consider this purely from a technical, economic or even political standpoint, but to search for the ways in which to interpret revolutionary action, to generate new forms of representation of the possible, against a background of social processes that were redefining the very nature of human identity.

The Production of Space is a book that broaches many such questions and does so from multiple angles. Lefebvre here draws upon his intimate knowledge of philosophy, his reflections on Hegel, Marx, Nietzsche and Freud, his experiential encounters with poetry, art, song and carnival, his connections with the surrealists and situationists, his intense involvement in Marxism both as a current of thought and as a political movement, his sociological enquiries into urban and rural conditions of life, his particular conception of totality and dialectical method. The reader will find here not only innumerable lines of thought to be followed up, but tacit or implicit criticisms of structuralism, of critical theory and deconstruction, of semiotics, of Foucault's views on the body and power, and of Sartre's version of existentialism. Yet Lefebvre never rejects such formulations outright. He always engages with them in order to appropriate and transform the insights to be gained from them in new and creative ways. The book is, therefore, also an opening towards new possibilities of thought and action. Although the culmination of a lifetime of engagement, *The Production of Space* takes the form of a preliminary enquiry which contains much that is explosive, much that has the capacity to 'detonate' (a word he himself frequently choses) a situation that threatens to become fixed, frozen and ossified. It is, above all, an intensely political document.

Lefebvre insists that life should be lived as a project and that the only intellectual and political project that makes sense is a life. *The Production of Space* is by no means the end of that project, for he continues to write and work to this day. But it is a vital marker and one that deserves to be read widely and to be studied for the innumerable possibilities it contains.

A note on sources The two primary sources on Lefebvre's life and work are his own autobiographical study, written in 1959, entitled *La somme et le reste*, and the recent authorized biography, from which I have drawn extensively, by Rémi Hess: *Henri Lefebvre et l'aventure du siècle* (Paris: Editions A. M. Métailié, 1988.

BOOKS BY HENRI LEFEBVRE

1934 *Introduction aux morceaux choisis de Karl Marx* (with Norbert Guterman). Paris: NRF.

1936 *La conscience mystifiée* (with Norbert Guterman). Paris: Gallimard.

1937 *Le nationalisme contre les nations* (preface by Paul Nizan). Paris: Editions Sociales Internationales.

1938 *Hitler au pouvoir, bilan de cinq années de fascisme en Allemagne.* Paris: Bureau d'Editions.

1938 *Morceaux choisis de Hegel* (with Norbert Guterman). Paris: Gallimard.

1938 *Cahiers de Lénine sur la dialectique de Hegel* (with Norbert Guterman). Paris: Gallimard.

1939 *Nietzsche.* Paris: Editions Sociales Internationales.

1939 *Le matérialisme dialectique.* Paris: Alcan. Eng. tr.: *Dialectical Materialism.* London: Jonathan Cape, 1968.

1946 *L'existentialisme.* Paris: Editions du Sagittaire.

1947 *Logique formelle, logique dialectique.* Paris: Editions Sociales. 2nd edn, Paris: Anthropos, 1970.

1947 *Critique de la vie quotidienne, I: Introduction.* Paris: Grasset. 2nd edn, Paris: L'Arche, 1958. Eng. tr.: *Critique of Everyday Life, I: Introduction.* London: Verso, in press.

1947 *Marx et la liberté.* Geneva: Editions des Trois Collines.

1947 *Descartes.* Paris: Editions Hier et Aujourd'hui.

1948 *Pour connaître la pensée de Karl Marx.* Paris: Bordas.

1948 *Le marxisme.* Paris: Presses Universitaires de France.

1949 *Diderot.* Paris: Les Editeurs Français Réunis.

1949 *Pascal*, vol. I. Paris: Nagel.

1953 *Contribution a l'esthétique.* Paris: Editions Sociales.

1954 *Pascal*, vol. II. Paris: Nagel.

1955 *Musset.* Paris: L'Arche.

1955 *Rabelais.* Paris: Les Editeurs Français Réunis.

1956 *Pignon.* Paris: Editions Falaise.

1957 *Pour connaître la pensée de Lénine*. Paris: Bordas.
1958 *Problèmes actuels du marxisme*. Paris: Presses Universitaires de France.
1958 *Allemagne*. Paris/Zurich: Braun/Atlantis (with photographs by Martin Hurlimann).
1959 *La somme et le reste*, 2 vols. Paris: La Nef de Paris.
1962 *Critique de la vie quotidienne, II: Fondement d'une sociologie de la quotidienneté*. Paris: L'Arche.
1962 *Introduction a la modernité*. Paris: Editions de Minuit.
1963 *La Vallée de Campan, étude de sociologie rurale*. Paris: Presses Universitaires de France.
1963 *Karl Marx: Oeuvres choisies*, vol. I (with Norbert Guterman). Paris: Gallimard.
1964 *Karl Marx: Oeuvres choisies*, vol. II (with Norbert Guterman). Paris: Gallimard.
1964 *Marx*. Paris: Presses Universitaires de France.
1965 *Pyrénées*. Lausanne: Rencontre.
1965 *Métaphilosophie*. Paris: Editions de Minuit.
1965 *La proclamation de la Commune*. Paris: Gallimard.
1966 *Le langage et la société*. Paris: Gallimard.
1966 *La sociologie de Marx*. Paris: Presses Universitaires de France. Eng. tr.: *The Sociology of Marx* (Harmondsworth, Middx.: Penguin, 1968.
1967 *Position: contre les technocrates*. Paris: Gonthier.
1968 *Le droit à la ville*. Paris: Anthropos.
1968 *La vie quotidienne dans le monde moderne*. Paris: Gallimard. Eng. tr.: *Everyday Life in the Modern World*. Harmondsworth, Middx.: Penguin, 1971.
1968 *L'irruption, de Nanterre au sommet*. Paris: Anthropos. Eng. tr.: *The Explosion: Marxism and the French Revolution of May 1968*. New York: Monthly Review, 1969.
1970 *Du rural à l'urbain*. Paris: Anthropos.
1970 *La révolution urbaine*. Paris: Gallimard.
1970 *La fin de l'histoire*. Paris: Editions de Minuit.
1971 *Le manifeste differentialiste*. Paris: Gallimard.
1971 *Au-delà du structuralisme*. Paris: Anthropos.
1971 *Vers le cybernanthrope/Contre les technocrates*. Paris: Denoël-Gonthier.
1972 *La pensée marxiste et la ville*. Paris/Tournai: Casterman.
1972 *Trois textes pour le théâtre*. Paris: Anthropos.
1973 *Espace et politique (Le droit à la ville, II)*. Paris.

1973   *La survie du capitalisme, la reproduction des rapports de pro-duction.* Paris: Anthropos. Eng. tr.: *The Survival of Capitalism.* London: Allison and Busby, 1974.

1974   *La production de l'espace.* Paris: Anthropos.

1975   *Le temps des méprises.* Paris: Stock.

1975   *Hegel, Marx, Nietzsche, ou le royaume des ombres.* Paris/Tour-nai: Casterman.

1975   *L'idéologie structuraliste.* Paris: Seuil.

1976   *De l'Etat, I: L'Etat dans le monde moderne.* Paris: Union Génér-ale d'Editions.

1976   *De l'Etat, II: Théorie marxiste de l'Etat de Hegel à Mao.* Paris: Union Générale d'Editions.

1977   *De l'Etat, III: Le mode de production étatique.* Paris: Union Générale d'Editions.

1978   *De l'Etat, IV: Les contradictions de l'Etat moderne; La dia-lectique et/de l'Etat.* Paris: Union Générale d'Editions.

1978   *La révolution n'est plus ce qu'elle etait* (with Catherine Régulier). Paris: Editions Libres-Hallier.

1980   *La présence et l'absence.* Paris: Casterman.

1980   *Une pensée devenue monde.* Paris: Fayard.

1981   *Critique de la vie quotidienne, III: De la modernité au moder-nisme (Pour une métaphilosophie du quotidien).* Paris: L'Arche.

1985   *Qu'est-ce que penser?* Paris: Publisud.

1986   *Le retour de la dialectique, douze mots clefs pour le monde moderne.* Paris: Messidor/Editions Sociales.

1986   *Lukács 1955.* Paris: Aubier.

Index